NATIONAL ACADEMIES
Sciences
Engineering
Medicine

NATIONAL
ACADEMIES
PRESS
Washington, DC

The Chemistry of Fires at the Wildland–Urban Interface

Committee on the Chemistry of Urban Wildfires

Board on Chemical Sciences and Technology

Division on Earth and Life Studies

Consensus Study Report

NATIONAL ACADEMIES PRESS 500 Fifth Street, NW Washington, DC 20001

This activity was supported by contracts between the National Academy of Sciences and Centers for Disease Control and Prevention, the National Institute of Standards and Technology, and the National Institutes of Health, Department of Health and Human Services, under Contract No. HHSN263201800029I. Any opinions, findings, conclusions, or recommendations expressed in this publication do not necessarily reflect the views of any organization or agency that provided support for the project.

International Standard Book Number-13: 978-0-309-27705-1
International Standard Book Number-10: 0-309-27705-1
Digital Object Identifier: https://doi.org/10.17226/26460
Library of Congress Control Number: 2022949390

This publication is available from the National Academies Press, 500 Fifth Street, NW, Keck 360, Washington, DC 20001; (800) 624-6242 or (202) 334-3313; http://www.nap.edu.

Suggested citation: National Academies of Sciences, Engineering, and Medicine. 2022. *The Chemistry of Fires at the Wildland-Urban Interface*. Washington, DC: The National Academies Press. https://doi.org/10.17226/26460.

The **National Academy of Sciences** was established in 1863 by an Act of Congress, signed by President Lincoln, as a private, nongovernmental institution to advise the nation on issues related to science and technology. Members are elected by their peers for outstanding contributions to research. Dr. Marcia McNutt is president.

The **National Academy of Engineering** was established in 1964 under the charter of the National Academy of Sciences to bring the practices of engineering to advising the nation. Members are elected by their peers for extraordinary contributions to engineering. Dr. John L. Anderson is president.

The **National Academy of Medicine** (formerly the Institute of Medicine) was established in 1970 under the charter of the National Academy of Sciences to advise the nation on medical and health issues. Members are elected by their peers for distinguished contributions to medicine and health. Dr. Victor J. Dzau is president.

The three Academies work together as the **National Academies of Sciences, Engineering, and Medicine** to provide independent, objective analysis and advice to the nation and conduct other activities to solve complex problems and inform public policy decisions. The National Academies also encourage education and research, recognize outstanding contributions to knowledge, and increase public understanding in matters of science, engineering, and medicine.

Learn more about the National Academies of Sciences, Engineering, and Medicine at **www.nationalacademies.org**.

Consensus Study Reports published by the National Academies of Sciences, Engineering, and Medicine document the evidence-based consensus on the study's statement of task by an authoring committee of experts. Reports typically include findings, conclusions, and recommendations based on information gathered by the committee and the committee's deliberations. Each report has been subjected to a rigorous and independent peer-review process and it represents the position of the National Academies on the statement of task.

Proceedings published by the National Academies of Sciences, Engineering, and Medicine chronicle the presentations and discussions at a workshop, symposium, or other event convened by the National Academies. The statements and opinions contained in proceedings are those of the participants and are not endorsed by other participants, the planning committee, or the National Academies.

Rapid Expert Consultations published by the National Academies of Sciences, Engineering, and Medicine are authored by subject-matter experts on narrowly focused topics that can be supported by a body of evidence. The discussions contained in rapid expert consultations are considered those of the authors and do not contain policy recommendations. Rapid expert consultations are reviewed by the institution before release.

For information about other products and activities of the National Academies, please visit www.nationalacademies.org/about/whatwedo.

COMMITTEE ON THE CHEMISTRY OF URBAN WILDFIRES

Members

DAVID T. ALLEN (NAE), *Chair*, The University of Texas at Austin
OLORUNFEMI ADETONA, The Ohio State University
MICHELLE BELL (NAM), Yale University
MARILYN BLACK, Underwriters Laboratories Inc.
JEFFEREY L. BURGESS, University of Arizona
FREDERICK L. DRYER (NAE), University of South Carolina
AMARA HOLDER, US Environmental Protection Agency
ANA MASCAREÑAS, Independent Consultant
FERNANDO L. ROSARIO-ORTIZ, University of Colorado Boulder
ANNA A. STEC, University of Central Lancashire
BARBARA J. TURPIN, University of North Carolina at Chapel Hill
JUDITH T. ZELIKOFF, New York University

Staff

MEGAN E. HARRIES, Study Director (until August 2022)
LIANA VACCARI, Study Director
BRENNA ALBIN, Program Assistant
EMILY J. BUEHLER, Consultant
KESIAH CLEMENT, Research Associate (until July 2021)
CHARLES FERGUSON, Senior Board Director
MEGHAN HARRISON, Senior Program Officer (until July 2021)
ELLEN K. MANTUS, Scholar (until April 2021)
JEREMY T. MATHIS, Board Director (until March 2021)
MARILEE SHELTON-DAVENPORT, Senior Program Officer (until January 2021)
ABIGAIL ULMAN, Research Assistant (until May 2022)
BENJAMIN ULRICH, Senior Program Assistant (until March 2022)

Sponsors

CENTERS FOR DISEASE CONTROL AND PREVENTION
NATIONAL INSTITUTE OF ENVIRONMENTAL HEALTH SCIENCES
NATIONAL INSTITUTE OF STANDARDS AND TECHNOLOGY

v

Acknowledgments

This Consensus Study Report was reviewed in draft form by individuals chosen for their diverse perspectives and technical expertise. The purpose of this independent review is to provide candid and critical comments that will assist the National Academies of Sciences, Engineering, and Medicine in making each published report as sound as possible and to ensure that it meets the institutional standards for quality, objectivity, evidence, and responsiveness to the study charge. The review comments and draft manuscript remain confidential to protect the integrity of the deliberative process.

We thank the following individuals for their review of this report:

Tami Bond, Colorado State University
Shoba Iyer, California Environmental Protection Agency and San Francisco Department of the Environment
Samuel Manzello, REAX Engineering
Sarah McAllister, US Forest Service
Birgitte Messerschmidt, National Fire Protection Association
Jason Sacks, US Environmental Protection Agency
David Sedlak (NAE), University of California, Berkeley
Carsten Warneke, US National Oceanographic and Atmospheric Administration
Christine Wiedinmyer, University of Colorado Boulder

Although the reviewers listed above provided many constructive comments and suggestions, they were not asked to endorse the conclusions or recommendations of this report, nor did they see the final draft before its release. The review of this report was overseen by **Susan Brantley**, The Pennsylvania State University, and **Martin-Jose J. Sepulveda**, Florida International University. They were responsible for making certain that an independent examination of this report was carried out in accordance with the standards of the National Academies and that all review comments were carefully considered. Responsibility for the final content rests entirely with the authoring committee and the National Academies.

This study would not have been successful without the assistance of many. The committee is grateful to the people who helped provide research support to the report, including **Colette Schissel**, The University of Texas at Austin; **Yosuke Kimura**, The University of Texas at Austin; and the staff of the National Academies Research Center. We are especially grateful to the numerous expert individuals who spoke to the committee during an open information-gathering session or otherwise provided input (see Appendix D).

Acronyms and Abbreviations

AQI	Air Quality Index
ASTM	American Society for Testing and Materials
BBOP	Biomass Burn Observation Project
CAMS	continuous air quality monitoring station
CARB	California Air Resources Board
CDC	Centers for Disease Control and Prevention
CDD	chlorinated dibenzo-p-dioxin
CE	combustion efficiency
CF	combustion factor
CI	confidence interval
CMAQ	Community Multiscale Air Quality
COPD	chronic obstructive pulmonary disease
CWS	community water system
DBP	disinfection by-product
DOM	dissolved organic matter
ED	emergency department
EF	emission factor
EPA	US Environmental Protection Agency
ER	emission ratio
FASMEE	Fire and Smoke Model and Evaluation Experiment
FIREX-AQ	Fire Influence on Regional to Global Environments and Air Quality
GeoXO	Geostationary Extended Observations
GOES	Geostationary Operational Environmental Satellite
GREET	Greenhouse gases, Regulated Emissions, and Energy use in Technologies
HEPA	high-efficiency particulate air
HVAC	heating, ventilating, and air conditioning

IARC   International Agency for Research on Cancer
IMPROVE   Interagency Monitoring of Protected Visual Environments
ISO   International Organization for Standardization
IVOC   intermediate-volatility organic compound

MCE   modified combustion efficiency
MDA8   daily maximum eight-hour average
MERV   minimum efficiency reporting value
MODIS   Moderate Resolution Imaging Spectroradiometer

NAAQS   National Ambient Air Quality Standards
NIEHS   National Institute of Environmental Health Sciences
NIFC   National Interagency Fire Center
NIOSH   National Institute for Occupational Safety and Health
NIST   National Institute of Standards and Technology
NTU   nephelometric turbidity unit

OSB   oriented strand board
OSHA   Occupational Safety and Health Administration

PAH   polycyclic aromatic hydrocarbon
PBDE   polybrominated diphenyl ether
PCB   polychlorinated biphenyl
PCDD   polychlorinated dibenzo-p-dioxin
PCDF   polychlorinated dibenzofuran
PFAS   perfluoroalkyl and polyfluoroalkyl organic substance
PiG   plume-in-grid
PM   particulate matter
PM_{10} coarse particulate matter; particles with diameters of 10 micrometers or less
$PM_{2.5}$ fine particulate matter; particles with diameters of 2.5 micrometers or less
PTR-ToF-MS   proton-transfer-reaction time-of-flight mass spectrometry
PVC   polyvinyl chloride

RAP-Chem   Rapid Refresh with Chemistry
RISE   Research Institutes of Sweden
RR   relative rate
Rx-CADRE   Prescribed Fire Combustion-Atmospheric Dynamics Experiments

SOA   secondary organic aerosol
SP   SP Technical Research Institute of Sweden
SVOC   semi-volatile organic compound

TCEP   tris(2-chloroethyl) phosphate
TCPP   tris(1-chloro-2-propyl) phosphate
TEMPO   Tropospheric Emissions: Monitoring of Pollution
VIIRS   Visible Infrared Imager Radiometer Suite
VOC   volatile organic compound

WE-CAN   Western Wildfire Experiment for Cloud Chemistry, Aerosol Absorption, and Nitrogen
WUI   wildland-urban interface

ϕ equivalence ratio

CHEMICAL FORMULAS

BrO	hypobromite ion
C_2H_2	acetylene
C_2H_4	ethylene
C_3H_4O	acrolein
C_6H_6	benzene
C_7H_8	toluene
CF_3Br	bromotrifluoromethane
CH_2O or HCHO	formaldehyde
CH_3Br	bromomethane
CH_3Cl	chloromethane
CH_4	methane
CHOCHO	glyoxal
$ClNO_2$	chlorine nitrite
CO	carbon monoxide
CO_2	carbon dioxide
$COCl_2$	phosgene
COF_2	carbonyl fluoride
H_2O	water
H_2O_2	hydrogen peroxide
H_2S	hydrogen sulfide
H_2SO_4	sulfuric acid
H_3PO_4	phosphoric acid
HBr	hydrogen bromide
HCHO; see CH_2O	
HCl	hydrogen chloride
HCN	hydrogen cyanide
HF	hydrogen fluoride
HNO_2 or HONO	nitrous acid
HNO_3	nitric acid
HO_2	hydroperoxyl radical
HO_2NO_2	peroxynitric acid
HOCO	hydrocarboxyl radical
HONO, see HNO_2	
N_2O_5	dinitrogen pentoxide
NH_3	ammonia
NO	nitric oxide radical
NO_2	nitrogen dioxide
NO_3	nitrate radical
NO_x	nitrogen oxides
O_3	ozone
OH	hydroxyl radical
POF_3	phosphoryl fluoride
SO_2	sulfur dioxide
SO_3	sulfur trioxide
SO_x	sulfur oxides

Contents

xiii

Summary[1]

Wildland fire activity in the United States is increasing as climate change drives more frequent extreme weather events, including heat waves and droughts. Growth in wildland fire size and intensity has also been driven by historical land management practices that emphasized fire exclusion. Simultaneously, urban development has been expanding into wilderness areas, and fires at the interface between urban and wildland areas are also increasing.

The wildland-urban interface, or WUI, is conceptually defined as the area where structures and other human development meet undeveloped wildland or vegetation fuels. Many slight variations on this broad definition exist, and this report uses a specific definition based on the densities of structures and how those structures either intermix or interface with wildland. In these WUI areas, the combination of natural and human-made materials that burn in fires leads to additional atmospheric emissions, residues, and effluents not typically found in wildland fires. The reactive emissions from the combustion of urban structures will alter the chemistry of WUI fires from that of wildland fires. Wildland fires will also alter the combustion of urban structures, with structures at the WUI generally burning from the exterior inward, rather than the interior outward, leading to different emissions than those encountered in urban structure fires. WUI fires may also lead to higher human exposures than wildland fires because of their proximity to communities, and because individuals whose occupations bring them in proximity to wildland fires are increasingly being exposed to WUI-type fires. While WUI fires lead to emissions that are different than for wildland and urban fires, the precise nature and extent of those differences is not well understood.

As efforts are made to understand fire dynamics; fire progression; emissions, effluents, and residues from fires; and human exposure to toxicants at the WUI, the interdisciplinary quality of this problem has led to the involvement of multiple governmental agencies. To enhance their efforts, the National Institute of Standards and Technology, the Centers for Disease Control and Prevention, and the National Institute of Environmental Health Sciences asked the National Academies of Sciences, Engineering, and Medicine (the National Academies) to evaluate how information about chemistry can be used to inform the mitigation of acute and long-term health effects of WUI fires and to recommend chemistry research that could help to inform decision-makers charged with mitigating wildfire impacts. To address this request, the National Academies created the Committee on the Chemistry of Urban Wildfires, which prepared this report.

Three overarching themes of the committee's work were the growing importance of WUI fires, the complexity of the topic, and the lack of data specifically relevant to the chemistry of fires at the WUI. Tens of millions of homes

---

[1] This Summary does not include references. Citations for the findings presented in this Summary appear in subsequent chapters of the report.

are currently found in WUI communities; this number of structures is growing, and the risk to these communities from fires is growing. Health risks associated with toxicants from WUI fires can persist well beyond the active burning of the fire, due to contamination of the ecosystem (water and soil) and built environments. Health impacts can also extend well beyond the WUI communities, as smoke can be transported for hundreds to thousands of kilometers. While toxicants from WUI fires are a recognized problem, their variability and complexity present significant challenges. For example, the emissions from a WUI fire can vary depending on whether the fire burns homes, cars, or commercial areas; even a subset of these fuel types can vary, such as homes of different ages made of different materials. Emissions vary depending on fuel composition, fire characteristics, and the heating dynamics that the fuels experience. Human exposure can vary greatly, depending on weather patterns during and after the fire, personal activities, and the living and working circumstances of the people exposed. While few data are available on emissions and exposures from fires at the WUI, knowledge of the chemistry, emissions, exposures, and health impacts of wildland fires and urban fires can be combined to hypothesize characteristics of WUI fires. Throughout the report, the committee examines what is known of wildland fires and urban fires and uses that information to identify potential emissions, exposures, and health impacts of WUI fires. The committee also recommends that the collection of data is needed to more definitively characterize the variability and overall acute and chronic impacts of WUI fires to the environment and human health. This summary presents an overview of each topic and the proposed research agenda.

FUEL, COMBUSTION, EMISSIONS

Emissions from fires in the WUI can differ greatly from emissions from wildland fires. The fuels burned in WUI fires have different compositions, densities, and quantities of combustible materials than the vegetative biomass combusted in wildland fires. The urban materials and their characteristics impact the combustion conditions, the chemical reaction pathways that dominate during combustion, and the emissions to the air and the residues that remain.

Few data are available on the fuels, combustion characteristics, and emissions, effluents, and residues from WUI fires. Characteristics of these fires can currently be inferred only from the superposition of the characteristics of wildfires and urban fires. Numerous laboratory studies and field efforts have investigated wildland fire emissions, leading to estimates of emissions of multiple individual chemical species and of multiple types of particulate matter. A detailed understanding of the relationships between fuel loading, fire emissions, and fire conditions remains an area of active investigation. The research community's knowledge of emissions from the combustion of urban materials is derived largely from studies of the toxicity of emissions from enclosure fires (i.e., a fire within a room or compartment inside a building) or from laboratory test methods simulating enclosure fire conditions. These studies have demonstrated how both the material composition and the amount of oxygen available strongly impact the emissions of some toxicants. However, very little is known about urban fuel chemical composition and these fuels' effects on combustion processes, the types of species emitted, and species' interactions under different fire conditions.

Integrating the current knowledge of wildland and urban fires into approximations of what may occur in WUI fires can provide directional guidance about the chemistry leading to emissions from WUI fires, but the data needed to characterize that chemistry are presently very sparse. Although WUI fires are increasing in frequency and in magnitude, very little is known about the emissions from WUI fires. Much of the current research on WUI fires has centered on ignition mechanisms and the characteristics of the built environment that lead to destructive fires. Even when data are available for wildfires that affect structures, it is difficult to separate the emissions that are unique to the WUI from those that result from wildland or urban combustion.

Using what is known about the materials in the WUI environment, and combustion conditions, the committee examined how different fuels, the fuel loading and distributions, and a myriad of factors could affect predictions of combustion characteristics and emissions for WUI fires. The WUI fuels differ from the vegetative biomass found in wildlands, particularly in chemical elements and materials of concern, such as halogens, plastics, and metals, which exist in much higher concentrations in the WUI. Potential reaction pathways and resulting emissions, effluents, and residues from WUI fires are postulated. Following the approaches used in wildland and urban fires, emission factors are used to summarize this information, but emission factors for WUI fires are currently speculative. Major chemical classes emitted from urban fires and wildfires are also reviewed to lay the foundation for assessing the human health and environmental impacts.

ATMOSPHERIC TRANSPORT AND CHEMICAL TRANSFORMATIONS

Emissions from WUI fires can substantially, negatively impact human health and quality of life, not only in the vicinity of the fire but also hundreds to thousands of kilometers downwind. This can affect millions of people outside the fire zone, depending on the location of nearby cities. Smoke from major fires sometimes affects air quality on a continental scale.

National monitoring networks provide some data for assessing exposures of routinely monitored air pollutants, such as fine particulate matter, that are associated with emissions from WUI fires; however, data are very sparse on the gas- and particle-phase smoke composition specifically associated with WUI fires, and how it is transformed over short and long distances. WUI smoke composition is dynamic within the fire zone and differs from wildfire smoke because of the different materials combusted in the WUI. The resulting plume composition changes over time (from minutes to days) due to atmospheric chemistry and physical processing, leading to changes in pollutant composition, and in the resulting exposures, downwind of fires. The impact of the unique WUI emissions and chemistry on regional exposures is not well understood. While a dominant route of exposure is inhalation, it is important to also recognize that wet and dry atmospheric deposition can be a source of contaminated water and soil downwind of wildfires and thus also impact exposures through ingestion. Additional WUI emissions known to be water soluble or found in wet deposition may also contribute to water/soil contamination and exposure through WUI-associated deposition.

WATER, SOIL, AND BUILDING CONTAMINATION

Combustion reactions for materials at the WUI (e.g., household components such as siding, insulation, textiles, and plastic, as well as the combustion of biomass) result in significant emission of potential toxicants to the surrounding environment. Although a significant part of these emissions, effluents, and residues are in the gas phase, there are also pathways for the mobilization (or partitioning) of some of these toxicants into nearby buildings, soils, and water streams. There is also the possibility of toxicants in the plume to be deposited downwind of the fire area, as has been observed in long-term studies on the impact of wildland fires. Finally, the active process of firefighting could add contaminants to the immediate area, in the form of flame retardants and other compounds.

All of these processes can impact the immediate conditions after the fire, which are critical as first responders and community residents arrive back in the area to assess damage. There is a dearth of peer-reviewed data documenting the specific impacts of WUI fires on water, soil, and settled dust contamination in built environments, with only a few studies providing critical information on potential impacts.

HUMAN EXPOSURES, HEALTH IMPACTS, AND MITIGATION

The identification of chemicals of concern for human exposure, environmental justice and vulnerable populations, and health impacts from WUI fire emissions is affected by the data limitations described above. The committee assessed potential exposures at multiple scales, from the immediate fire zone, where emergency responders such as firefighters experience direct exposure to heat and high concentrations of fire emissions, to the regional and continental levels, where WUI fire smoke can have an extended impact. Adverse health effects associated with exposure to fire emissions have been measured hundreds of kilometers downwind, such as the 240-km-downwind impacts of the 2018 Camp Fire on California Bay Area residents, to well beyond 1,000 kilometers, such as the 2016 Horse River Fire, which started near Fort McMurray, Alberta, Canada, and resulted in smoke transport and impacts on air quality more than 4,000 kilometers away in New York City. Much of the information presented in the report draws on studies from wildland fires, with limited WUI-specific information. More data specific to WUI fires may provide additional insights to inform research, policies, mitigation practices, and practices that reduce exposure and negative health impacts.

The current literature largely focuses on exposures and health impacts related to smoke inhalation. These include myocardial infarction, ischemic heart disease, dysrhythmia, heart failure, pulmonary embolism, ischemic stroke, and transient ischemic attack. Asthma exacerbations are significantly associated with exposure to wildfire smoke. Exacerbations of chronic obstructive pulmonary disease are also significantly associated with greater contaminant amounts and exposure to wildfire smoke in most studies.

Within exposures to the general public, specific groups may experience greater impacts, such as children, pregnant people, older adults, and immunocompromised people, as well as communities who experience disproportionate environmental and social stressors, such as those of lower socioeconomic status, some non-white racial and ethnic groups, and tribal populations. Heightened vulnerability to wildfire emissions includes higher health responses to a given level of exposure. For example, children are especially vulnerable to air pollution effects due to their small body mass and rapid breathing rates, leading to higher exposure doses than adults. Adults may also experience multiple aspects of vulnerability such as differential exposures, ability to adapt, and health response, which can be related to the availability of language-appropriate and culturally appropriate information and medical care, access to quality health care, and other structural factors. Additionally, low-income communities often do not have adequate cooling in their homes to allow for keeping the windows closed during the summer heat when wildfires are typically present. They also may not be able to afford the purchase of home air cleaners for particle and chemical filtration.

Emergency responders, such as firefighters, are at risk for injury, death, and acute and chronic health impacts of wildfires, as well as mental health effects due to trauma. Outdoor workers, such as farmworkers and utility workers, are at elevated risk for respiratory effects of smoke and associated pollutants, and this may be coupled with preexisting medical conditions. Additional vulnerability factors related to occupational settings may include elevated outdoor air and dust exposures to chemicals and particles, amount of time spent outdoors, and available protective respiratory equipment and training. Post-fire clean-up can also lead to exposure to fire residues for a wide variety of populations.

As in other problem areas that seek to estimate the emissions, exposures, and health impacts associated with WUI fires, there is limited information on human exposures that is specific to WUI fires.

MEASUREMENT SCIENCE

Throughout the report, the committee outlined the diverse data needs associated with understanding the chemistry of WUI fires and their emissions, which range from collecting data on fuel characteristics of heterogeneous structures, to measuring the concentrations, exposures, and health impacts of large numbers of toxicants, many of which will be present at trace levels. Collecting much of these data is challenging, since materials burned in WUI fires are heterogeneous, and identifying their contributions to WUI fire plumes requires fine spatiotemporal resolution.

Data and measurement needs are also interconnected, as shown in Figure S-1. Information about at-risk communities and vulnerable populations can help define the types of structures and potential fuels at the WUI. Data on fuel compositions will determine combustion pathways. The chemical species formed during combustion will determine which atmospheric reaction pathways will be most important. The atmospheric chemistry and transport will determine the toxicants to which communities are exposed and the manner of the exposure. Exposures will determine health impacts. Because of this interconnection, a coordinated approach to data collection, from fuel and emission characterization to exposures, is desirable. Use of consistent measurement methods and collection of data on chemical species over their entire cycle from emission to exposure will enhance the value of all of the data that are collected.

Data and measurements are also needed on various timescales. Retrospective studies of fuels, fires, emissions, exposures, and health impacts, in the field, in laboratories, and via modeling, will be important to improve the understanding of the chemistry of WUI fires. These types of studies will benefit from coordinated databases and information repositories that allow efficient sharing of information and methods. Decision-makers charged with mitigating the impacts of WUI fires, and informing the public with information to mitigate exposure to toxicants, will also need access to well-organized information, but their information needs will be much more immediate, requiring rapid access and a near-real-time response.

RESEARCH AGENDA

WUI fires have the common characteristics of fuel mixes that are distinct from both wildland and urban fires, and a combustion chemistry that is spatially and temporally heterogeneous. These unique characteristics of WUI fires can lead to toxicants and exposures that are largely uncharacterized. The committee identified research that could (1) improve understanding of the fuels and emissions associated with WUI fires; (2) characterize the chemistry,

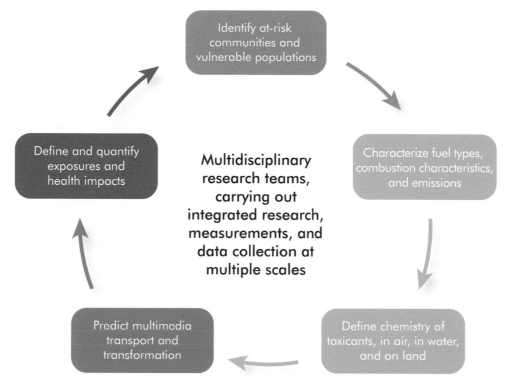

FIGURE S-1 The interdependence of multidisciplinary WUI fire emission research activities.

transport, and transformations of emissions; and (3) quantify exposures and health effects that result from WUI fire emissions. In addition, the committee identified improvements in measurement science and analytics that could make broad contributions to improve research capabilities. Priority research needs within this broad agenda are summarized in Table S-1. Priority research areas were identified based on the information needs of decision-makers, the timeliness and cost effectiveness of research initiatives, the magnitude of the effect being investigated, and the potential impact of broad scientific capabilities. The relative importance of these prioritization criteria were not explicitly weighted; individual chapters describe the criteria used in the choice of priority research areas and also define a broader set of research needs than those summarized in Table S-1. To be most effective, the research agenda should be carried out with multidisciplinary teams who integrate work that ranges from hazard identification and emission characterization to measurement of exposures and health outcomes.

> **Recommendation 1: To understand the chemistry, exposures, and health impacts of toxicants resulting from WUI fires, researchers and agencies that fund research should implement an integrated, multi-disciplinary research agenda. Agencies funding, and investigators performing, research on WUI fire emissions should coordinate their research plans, and they should create widely accessible repositories for data and information relevant to WUI fires.**

Implementing an effective research agenda will be challenging. Research has been conducted on emissions from wildland and urban fires for decades, and this research has led to significant advances in understanding of wildland fuels, wildland fire emissions, transformation and transport of emissions, and health impacts of toxicant exposures. There is still very little confidence, however, in making connections such as the impact of fire intensity to emission composition and downwind chemical evolution. Similar experiences are to be expected in research on

TABLE S-1 Research Priorities for Fires at the Wildland-Urban Interface (WUI)

	Fuels and Emissions	Chemistry, Transport, and Transformations	Exposure and Health	Measurement Science and Analytics
Collecting WUI-specific data	Assemble data on fuels, emissions, chemistry, transformations, exposure, and health impacts that are attributable to fires at the WUI, differentiated from wildland fires and urban fires, which will require novel measurements and analyses			
Fundamental measurements and data	- Map WUI communities, their material loadings, and the compositions - Identify combustion conditions and emissions typical of WUI fires - Examine interactions between human-made fuels and wildland fire fuels using mechanistic models and experiments at bench, and larger scales	- Identify primary toxicants emerging from WUI fires - Identify secondary species with toxic potential, formed from the atmospheric aging of WUI fire emissions - Gather existing data on air, water, and soil testing associated with WUI fires into an accessible database	- Improve understanding of indoor penetration and composition of WUI fire smoke - Evaluate health implications of smoke constituents, and exposure to constituents in water	- Develop new analytical capabilities for measuring chemical, particle, and biological indicators of WUI fire toxicants in studies on emission, exposure, and health outcomes
Field and population studies	- Assess fuels, consumption, and emissions of WUI fires - Perform coordinated, multi-platform, multimedia studies of WUI fire energetics and emissions	- Identify dominant daytime and nighttime atmospheric oxidants in WUI fire plumes - Identify the key precursors and formation pathways of secondary species with toxic potential, formed by the gaseous, aqueous, multiphase, and catalytic reactions in plumes - Identify key chemical species that can impact water and soils	- Characterize multi-route and multimedia exposures and health impacts - Improve understanding of acute and long-term health effects of WUI fire toxicant exposures - Improve exposure measurements for WUI fire emissions	- Optimize analytical methods for field deployment and increased accessibility - Develop biomarkers specific for WUI fires that can be used for exposure and toxicity assessment - Develop standard procedures for testing water and soil after WUI fires; establish databases of testing studies
Prediction, assessment, and exposure-mitigation capabilities	- Develop risk assessment procedures for WUI fires - Develop predictive models of WUI fire combustion and emissions - Identify strategies at a structure, neighborhood, and community level to mitigate WUI fire risk	- Develop condensed chemical mechanisms and sub-grid-scale processing needed for regional modeling of WUI fire emissions - Create improved retrospective and prospective models of WUI fire exposures - Evaluate risks to community water systems and response plans	- Measure the effectiveness of interventions for firefighters - Expand identification of vulnerable populations and culturally appropriate interventions - Develop health equity considerations for WUI fire exposures	- Deploy multi-scale sensing capabilities to assess chemical compositions of WUI fire plumes

WUI fire emissions. Research on emissions from WUI fires can build on the extensive knowledge base developed for wildland fires, but understanding the impacts of combustion of mixed wildland and urban fuels will require new information on fuels, emissions, transport and transformations, and health impacts of toxicant exposures. Making connections between research areas from emissions to exposure and health impacts will continue to be a challenge. The variability in the types of structures in WUI communities and the types of fire hazards that communities and vulnerable populations are exposed to also spans a wide range, and will add complexity. Building a comprehensive and fundamental body of knowledge on WUI fires will require a long-term effort.

The information needed to understand toxicants emerging from wildfires, structural fires, and WUI fires; their effects; and who they affect spans a broad range of scientific disciplines. Research findings from each of these disciplines need to be continuously communicated across disciplinary boundaries because each step in the chain—from hazard identification to quantification of exposures and health effects—depends on information emerging from other steps. This trans-disciplinary data and information sharing process could begin with information available on emissions from wildland and urban structure fires and be expanded to include information and data from WUI fires. This information sharing process could also include expansion of existing mechanisms for communication with decision-makers and the public, such as www.ready.gov/wildfires.

Recommendation 2: Those implementing research programs should design and implement a multidisciplinary WUI research program that includes the development of tools, resources, and messaging designed to inform a wide variety of decision-makers charged with mitigating wildland and WUI fire impacts. They should also create periodic summaries of policy-relevant research findings, and actionable messaging for decision-makers working with at-risk communities and vulnerable populations.

Policy-relevant research findings and actionable messaging could include recommendations for building materials to be used in WUI communities, strategies for reducing interior environment fire risks, public information regarding the effectiveness of measures to mitigate exposures, and community mappings of toxicant precursors, accessible to decision-makers.

More than 40 million homes in the United States are located at the WUI. Diverse and vulnerable populations will be increasingly exposed to hazards from WUI fires. The development of a multidisciplinary WUI research program, summarized in Table S-1, will have long-term benefits, but immediate action is also needed. The committee identified a number of areas where rapid action could have immediate benefits. These areas are listed in Box S-1. Commitment to both long-term progress and immediate action on improving the understanding of WUI fires would benefit communities throughout the United States and the world.

BOX S-1
Priorities for Near-Term Research

- Developing data systems to enable communities to predict the chemical composition of materials present in structures at risk from WUI fires. These data systems could include estimates of metal, halogen, and other chemical loadings in structures.
- Adding measurements of targeted WUI toxicants to air- and water-quality monitoring systems. These measurement systems could be rapidly deployed to areas impacted by WUI fires.
- Establishing information repositories on toxicant data, best practices for mitigation measures, and best practices for information dissemination. State agencies could lead in the coordination of data collection; data consistency, quality, and access could be addressed at a national level, and at all levels, communication and dissemination strategies for vulnerable and at-risk community populations could be developed.

1

Introduction

This chapter introduces the motivation for this consensus report. It begins with an overview of fires at the wildland-urban interface (WUI) and why they deserve attention. It discusses prior work, which this report builds upon, that helps frame the issue. It presents the statement of task for the National Academies of Sciences, Engineering, and Medicine (the National Academies) committee responsible for the findings and recommendations in this report and the committee's approach to this task, including its consideration of equity issues surrounding data availability and disparities in the impact of these fires. The chapter concludes with a description of the organization of the rest of the report.

WILDFIRES AT THE WUI

Wildfire activity in the United States is increasing as more frequent extreme weather events, including heat waves, droughts, and lightning activity, cause wildfires (Ahrens, 2013; Schwartz and Penney, 2020). Combined extreme heat and drought conditions have occurred more frequently across North America over the last century (Alizadeh et al., 2020), and fire seasons have lengthened by a global average of 18 percent from 1979 to 2013.[1] Growth in wildland fire size and intensity has also been driven by historical land management practices that emphasized fire exclusion and allowed the buildup of potential wildland fire fuels. Simultaneously, the WUI has continued to expand further into formerly wilderness areas (Radeloff et al., 2018). Since 1990, 41 percent of new housing units in the United States have been built in the WUI (Radeloff et al., 2018). Climate change is inextricably linked to these issues, as it contributes to extreme weather conditions that increase fire activity and is affected by land use changes like the expansion of the WUI (USGCRP, 2018). The combination of these factors has led to an increase in WUI fires.

The chemistry and ultimate health impacts of WUI fires are still poorly understood. WUI fires lead to higher human exposures than remote wildland fires because of their proximity to communities. They also have unique chemistry due to the combination of natural and human-made fuels that are burned, which may lead to the formation or release of toxic emissions not found in purely wildland fires.

---

[1] Fire season is defined by the US Forest Service as "1) Period(s) of the year during which wildland fires are likely to occur, spread, and affect resource values sufficient to warrant organized fire management activities. 2) A legally enacted time during which burning activities are regulated by state or local authority" (see https://www.fs.fed.us/nwacfire/home/terminology.html).

Economic impacts of WUI fires are also significant. A 2017 report by the National Institute of Standards and Technology (NIST) estimated the economic impacts associated with wildfires in the United States, including WUI fires, as between US$71 billion and US$347 billion per year. These economic impacts include both costs, such as preventive measures and disaster response, and losses, such as injuries, mental health impacts, disease or exacerbation of existing health conditions, and deaths. Economic valuation of loss of life accounted for US$28 billion to US$202 billion of the total economic impacts (Thomas et al., 2017). Chapter 2 expands on this discussion of the definition, causes, and impacts of WUI fires.

The expanding problem of WUI fires has led to research targeted at understanding fire dynamics and fire progression at the WUI. Post-incident analyses of operational responses, disaster management, and risk communication associated with WUI fires have also become more detailed as more real-time incident data are collected (Gaudet et al., 2020). The highly interdisciplinary character of this problem has led to the involvement of multiple agencies at the federal level for scientific research, emergency preparedness, and disaster response. The Wildland-Urban Interface Fire Group at NIST conducts research to understand and predict the spread of WUI fires, including the investigation and reconstruction of key case studies like California's Camp Fire in 2018 (Maranghides et al., 2021). The Centers for Disease Control and Prevention (CDC) runs multiple programs dedicated to wildfire disaster preparedness and response through their National Center for Environmental Health and Natural Disasters and Severe Weather division. The National Institute of Environmental Health Sciences (NIEHS) operates a Worker Training Program to support the health and safety of people involved in wildfire response operations. CDC and NIEHS recognize that those individuals whose occupations bring them in proximity to wildfires and their aftermath are increasingly being exposed to WUI-type fires.

To enhance their efforts, CDC, NIEHS, and NIST asked the National Academies to evaluate chemistry information that would improve the mitigation of acute and long-term health effects of residential burning during wildfires and to recommend chemistry research that could help to inform decision-makers charged with mitigating wildfire impacts on the general public. As a result of the request, the National Academies convened the Committee on the Chemistry of Urban Wildfires, which prepared this report.

THE COMMITTEE AND ITS TASK

The committee convened in response to the sponsors' request included experts in atmospheric chemistry, combustion chemistry, environmental chemistry, toxicology, and exposure science. Appendix B provides biographical information on the committee. The committee was asked to review the relevant state of the science and recommend chemistry research that could fill critical data gaps and help inform decision-makers charged with mitigating wildfire impacts on the general public. The Statement of Task is provided in Box 1-1.

BOX 1-1
Statement of Task

The National Academies of Sciences, Engineering, and Medicine will convene an ad hoc committee to describe chemistry information that would improve mitigation of acute and long-term health effects of residential burning during wildfires at the wildland-urban interface. That description will be based on decision-maker needs and the state of relevant science, such as (1) the chemical processes undergone by materials unique to urban structures and otherwise not present in wildland areas that undergo combustion during wildfires; (2) the identity of chemical species present in urban wildfire combustion products and debris; and (3) what is known about human exposure and relative importance of various human exposure pathways (in the air, water, and soil) for key chemical species, as well as challenges in collecting data to fill important data gaps. The committee's analysis will lead to a report with findings and recommendations that describe opportunities for chemistry research to fill decision-critical gaps and inform decision-makers charged with mitigating wildfire impacts on the general public.

THE COMMITTEE'S APPROACH TO ITS TASK

To accomplish its task, the committee convened a public workshop to gather information relevant to its task and held seven committee meetings to deliberate and reach its conclusions and recommendations. In addition to the workshop, committee members used their subject-matter expertise to independently compile available literature specific to WUI fires and, when in their judgment it was relevant, non-WUI wildfires. Early in its work, the committee identified the broadly interdisciplinary nature of its task as a barrier to applying consistency to the reporting of available evidence. For example, the types of studies available about combustion conditions and emission factors include laboratory-, bench-, and field-scale studies and often report dimensionless ratios, while studies examining health effects fall into different categories (e.g., epidemiological, observational, in vivo). The committee recognized that this problem exists in many interdisciplinary fields of research and is not unique to WUI fire chemistry; creating a new system for integration of the best available scientific information in each discipline was deemed beyond the study's scope. Therefore, the committee proceeded to use the highest-quality evidence in each scientific discipline while recognizing the eventual need to integrate information across multiple scientific disciplines. The report prioritizes information available in peer-reviewed sources.

The committee considered various issues and made several decisions in interpreting its charge. First, it considered the sequence of chemical processes that occur, from the initial burning of fuel sources in WUI fires to the chemical products generated by the fires. In considering fuel sources for fires at the WUI, the committee recognized that burned areas could include residential, commercial, and industrial sources but focused its attention on materials present in residences, because residences constitute a large fraction of structures burned at the WUI. The committee also recognized that combustion products depend on the intensity of the fire and the nature of heat transfer to fuels of both natural and human-made origin. The committee understood that fires at the WUI can be more intense than urban residential fires and that WUI fires often burn structures from the outside in, rather than the typical pattern of residential fires burning from the inside out. The committee recognized potential synergies in the combustion chemistry associated with materials involved in WUI fires: the chemical processes associated with the combustion of wildland fuels and fuels associated with the built environment will interact in WUI fires in potentially important ways.

Second, the committee considered human exposures during the active phase of the fire and post-fire activities. The populations it considered included the local communities in the burned region, first responders, communities that are exposed to regional plumes, and people who are involved in clean-up activities. The wide distribution of exposed communities made it important for the committee to consider chemical transformations of the initial combustion products over local and regional scales, and how those transformations are influenced by local and regional environmental conditions. Although a primary focus was on inhalation exposures, the committee also considered ingestion and dermal exposure routes that could be important in the exposed populations. The committee recognized the importance of considering affected communities through the lens of equity and environmental justice. Because the Statement of Task focuses on human exposure and health impacts, the committee decided that discussions of ecological impacts and long-term environmental fate were out of scope for this report.

Third, the committee recognized that WUI fires have the potential to generate a wide array of chemicals and that many chemical exposure pathways are relevant. To address its charge of identifying critical information gaps, the committee qualitatively considered categories of potential health effects associated with exposures but did not attempt to quantify effects where such analyses did not already exist in the literature. Exposures can be acute or chronic, depending on their magnitude and duration, and health effects can be acute or chronic, or have a delayed onset. The committee identified a number of potential effects on populations in the near-fire area, regionally, and on continental scales.

Finally, although the work of the committee can inform mitigation measures such as reducing the use of materials in structures at the WUI that might lead to combustion products that have adverse health effects, the committee did not directly evaluate mitigation policies or measures. Because the work of the committee assesses what is known and identifies critical information gaps associated with exposures to chemical releases from WUI fires, this report can help to inform communications with exposed populations during fire events.

THE COMMITTEE'S APPROACH TO EQUITY

The committee recognized that its discussions, findings, and recommendations must involve intentional considerations of equity and, during the course of the study, aimed to identify inequities and structural barriers and describe science-based opportunities to promote greater access and improved outcomes for vulnerable and marginalized groups. The committee acknowledges that health equity is a much larger issue that is a factor in types of exposures other than WUI fire exposure; however, it believes it is important to address in this report as it relates to the committee's task.

The disproportionate vulnerability of certain groups to effects from WUI fires can be seen through a few simple examples:

1. Smoke impacts community air quality at a regional and even continental scale, far outside the evacuation zone. Individuals' inhalation exposure to smoke is dependent in part on how airtight their dwelling is (how recently it was built; whether there is a heating, ventilating, and air-conditioning system) and whether they have the financial means to reduce indoor concentrations of smoke or to relocate to an area with better air quality.
2. Exposure to pollutants from WUI fires is increased for people who work outdoors. These jobs tend to be more physically demanding, increasing respiratory exchange with polluted air and therefore inhalation exposure.
3. Children, pregnant people, older adults, and people with preexisting conditions all have heightened susceptibility to health effects from exposure to toxicants from WUI fires, through inhalation or other routes of exposure.

The committee discussed inequity throughout the report; however, the majority of the report's related findings and recommendations can be found in Chapter 6, "Human Exposures, Health Impacts, and Mitigation."

PRIOR RELATED WORK BY THE NATIONAL ACADEMIES

This study focused specifically on fires at the WUI; however, the committee recognizes that a great deal can be learned from studying wildland fire research. The National Academies has held three workshops in recent years to gather experts together and share knowledge related to specific aspects of wildfire issues in the United States. A 2017 workshop and proceedings, titled *A Century of Wildland Fire Research: Contributions to Long-Term Approaches for Wildland Fire Management*, focused on fuel management and preventative measures that wildland fire managers, policy makers, and communities can implement to decrease wildfire incidence. A 2019 workshop and 2020 proceedings, titled *Implications of the California Wildfires for Health, Communities, and Preparedness*, focused on similar issues specific to California and expanded to include human health impacts of wildfires in the state. Most recently, a 2020 workshop and 2022 proceedings on *Wildland Fires: Toward Improved Understanding and Forecasting of Air Quality Impacts* focused on atmospheric chemistry and exposure science but specifically excluded the WUI. These three prior efforts, while touching on topics related to chemistry and the WUI, did not focus on WUI fires or their chemistry. This report builds on these workshops and focuses on the specific problem of the chemistry of WUI fires.

ORGANIZATION OF THIS REPORT

This report is organized into eight chapters. Chapter 2 describes the history and characteristics of the WUI and fires at the WUI, and provides working definitions for key terms that the committee uses throughout the report. It concludes with five context-setting examples of recent, major WUI fires.

Chapter 3 covers the materials, combustion, and emissions of fires at the WUI, emphasizing the differences between urban materials (often with higher concentrations of toxicants) and the vegetative biomass of wildlands. It summarizes what is known about the materials in the WUI environment, the combustion conditions, the dominant reaction pathways, and the resulting emissions from WUI fires. It highlights the complexity of the many intersecting factors that influence emissions: the material types and loading, and the arrangement of materials in the landscape; the many reaction pathways and the effects of conditions like temperature; and the combustion chemistry unique to WUI fires.

WUI fires can affect populations at great distances away from the fire. Chapter 4 discusses the atmospheric transport and chemical transformations of WUI fire emissions. The chapter explores the atmospheric chemistry and physical changes (such as dilution) that lead to dramatic changes in the pollutant composition downwind of fires, although little data exist on these processes for WUI fires. A section covers current practices in modeling fire plumes as they move away from the fire. Chapter 4 emphasizes the atmospheric chemistry that informs inhalation exposures and the resulting health effects associated with WUI fires.

While emissions from WUI fires can stay in the air, they also end up in the water and soil. Chapter 5 covers water and soil contamination, both near the fire and downwind. It discusses immediate impacts in the fire region and effects on community water systems, groundwater, and soil, as well as the deposition of contaminants downwind of a fire. It includes post-fire contamination associated with firefighting activities. Due to the lack of information on WUI fires, in some cases, information originally collected for wildland fires is extrapolated to WUI fires.

Chapter 6 explores pollutants of concern for human exposure, environmental justice, and vulnerable populations, as well as ways to reduce exposure and mitigate the impacts of WUI fires. It addresses distance scales from the immediate fire zone, where people including firefighters experience direct exposure to heat and fire emissions, to the regional and continental levels, where WUI fire smoke can have an extended impact. It explores routes of exposure and the acute and chronic health impacts for all near- and far-field populations, including firefighters. It discusses specific interventions for firefighters and affected communities. The chapter includes information from both wildland fires and, where available, WUI fires.

Chapter 7 describes the methods for collecting data on WUI fires and their emissions, focusing on the overall challenges in data collection and on the strengths and limitations of currently available techniques. Current methods include measuring the fire's emissions at near and far locations and measuring exposure to humans via sampling the nearby environment, personal exposure measurements, and biomonitoring. The chapter also describes the analytical instruments and modeling techniques used.

Chapter 8 summarizes the report and the information needs it identifies. It stresses the wide variety of types of information needed and the importance of multidisciplinary research teams. It presents the information needs as a research agenda. This agenda groups complementary tasks together and organizes priorities into four primary areas: (1) fuels and emissions; (2) chemistry, transport, and transformations; (3) exposure and health; and (4) measurement science and analytics.

The report concludes with five appendices. Appendix A provides a glossary of key terms used in this report. Appendix B presents biographical information about the committee. Appendix C contains a table summarizing the available data about the example fires described in Chapter 2 as a list of references. Appendix D contains the agenda for the public session held by the committee for information gathering purposes. Appendix E lays out the assumptions and engineering calculations that the committee used to create a table of combustible mass and energy content data in Chapter 3.

REFERENCES

Ahrens, M. 2013. *Lightning Fires and Lightning Strikes*. Quincy, MA: National Fire Protection Association. https://www.nfpa.org//-/media/Files/News-and-Research/Fire-statistics-and-reports/US-Fire-Problem/Fire-causes/oslightning.pdf.

Alizadeh, M. R., J. Adamowski, M. R. Nikoo, A. Aghakouchak, P. Dennison, and M. Sadegh. 2020. "A Century of Observations Reveals Increasing Likelihood of Continental-Scale Compound Dry-Hot Extremes." *Science Advances* 6 (39). https://doi.org/10.1126/sciadv.aaz4571.

Gaudet, B., A. Simeoni, S. Gwynne, E. Kuligowski, and N. Benichou. 2020. "A Review of Post-incident Studies for Wildland-Urban Interface Fires. *Journal of Safety Science and Resilience* 1 (1): 59–65. https://doi.org/10.1016/j.jnlssr.2020.06.010.

Maranghides, A., E. L. W. "Ruddy" Mell, S. Hawks, M. Wilson, W. Brewer, C. Brown, B. Vihnaneck, and W. D. Walton. 2021. *A Case Study of the Camp Fire – Fire Progression Timeline*. NIST Technical Note 2135. Gaithersburg, MD: National Institute of Standards and Technology. https://doi.org/10.6028/NIST.TN.2135.

NASEM (National Academies of Sciences, Engineering, and Medicine). 2017. *A Century of Wildland Fire Research: Contributions to Long-Term Approaches for Wildland Fire Management: Proceedings of a Workshop*. Washington, DC: The National Academies Press. https://doi.org/10.17226/24792.

NASEM. 2020. *Implications of the California Wildfires for Health, Communities, and Preparedness: Proceedings of a Workshop*. Washington, DC: The National Academies Press. https://doi.org/10.17226/25622.

NASEM. 2022. *Wildland Fires: Toward Improved Understanding and Forecasting of Air Quality Impacts: Proceedings of a Workshop*. Washington, DC: The National Academies Press. https://doi.org/10.17226/26465.

Radeloff, V. C., D. P. Helmers, H. A. Kramer, M. H. Mockrin, P. M. Alexandre, A. Bar-Massada, V. Butsic, T. J. Hawbaker, S. Martinuzzi, A. D. Syphard, and S. I. Stewart. 2018. "Rapid Growth of the US Wildland-Urban Interface Raises Wildfire Risk." *Proceedings of the National Academy of Sciences of the United States of America* 115 (13): 3314–3319. https://doi.org/10.1073/pnas.1718850115.

Schwartz, J., and V. Penney. October 23, 2020. "In the West, Lightning Grows as a Cause of Damaging Fires." *The New York Times*. https://www.nytimes.com/interactive/2020/10/23/climate/west-lightning-wildfires.html.

Thomas, D., D. Butry, S. Gilbert, D. Webb, and J. Fung. 2017. *The Costs and Losses of Wildfires: A Literature Survey*. NIST Special Publication 1215. Gaithersburg, MD: National Institute of Standards and Technology. https://doi.org/10.6028/NIST.SP.1215.

USGCRP (US Global Change Research Program). 2018. *Fourth National Climate Assessment. Volume II. Impacts, Risks, and Adaptation in the United States*. Edited by D. R. Reidmiller, C. W. Avery, D. R. Easterling, K. E. Kunkel, K. L. M. Lewis, T. K. Maycock, and B. C. Stewart. Washington, DC: US Global Change Research Program. https://doi.org/10.7930/NCA4.2018.

2

Defining and Contextualizing WUI Fires

This chapter describes how and why fires at the wildland-urban interface (WUI) have become more severe. Maps of WUI communities and fires in the contiguous United States illustrate the extent and distribution of WUI fires; however, quantifying the impacts of these fires, and even precisely defining the WUI interface, remains a challenge. An operational definition of the WUI as used in this report is provided, as are some descriptive examples of WUI fires. Key terms used throughout this chapter and the report are provided in Box 2-1.

DEFINING THE WUI

The WUI is conceptually defined as "the line, area, or zone where structures and other human development meet or intermingle with undeveloped wildland or vegetation fuels" (FMB, 2019). Many slight variations on this broad definition exist in government guidelines, academic research papers, and news coverage about the problem of WUI fires; however, this report required a more precise, technical definition. Therefore, this report uses

---

BOX 2-1
Key Terminology

Fire plume: Air mass downwind of combustion zone, containing elevated concentrations of combustion products

Interface WUI: Areas characterized by development (housing or other structures) located at the edge of a large area of wildland

Intermix WUI: Areas characterized by alternating development (housing or other structures) and wildlands

Wildland-urban interface (WUI): The community that exists where humans and their development meet or intermix with wildland fuel

---

a definition of the WUI adapted from the Federal Register (2001). The purpose of the Federal Register notice, from the US Department of Agriculture and Department of the Interior, was twofold: first, to precisely define the WUI, and second, to identify among WUI communities which areas are at high risk from wildfire based on a set of contributing factors. The definition of the WUI used in the Federal Register notice was modified from "A Report to the Council of Western State Foresters—Fire in the West—The Wildland/Urban Interface Fire Problem" dated September 18, 2000. Under this definition, "the urban wildland interface community exists where humans and their development meet or intermix with wildland fuel." There are three categories of communities that meet the Federal Register definition of the WUI: interface, intermix, and occluded. The Federal Register notice included a list (created in collaboration with states and tribal entities) of hundreds of communities in proximity to federally owned wildlands that were at high risk of wildfire based on this definition (Federal Register, 2001). The list included communities that later suffered extensive damage due to fire after the publication of the notice, as described later in this chapter.

The Federal Register also describes three factors that put WUI communities at higher risk of loss from wildfire: (1) the potential of the topology to intensify fires and the distribution of fuels; (2) the density of structures and the unique value of those structures; and (3) the lack of infrastructure to support firefighting response, primarily the accessibility of roads and sources of water. At Congress's request, this Federal Register notice identified high-risk WUI communities to help the Departments of Agriculture and the Interior plan and prioritize federal projects to reduce wildfire risk. Although these risk factors are important from a preventative perspective, this report addresses the chemistry taking place both during and after wildfire, not before; therefore, this report will not address these risk factors in detail.

Subsequent analyses by the US Forest Service operationalized the Federal Register listing of WUI communities. These analyses provided a precise definition of interface and intermix communities that could be applied to areas not covered by the original notice (Figure 2-1; Stewart et al., 2007). Intermix communities have alternating areas of housing and wildland, whereas interface communities are at the edge of a large area of wildland. This definition allowed for the development of mappings and the quantification of the extent of these communities over time.

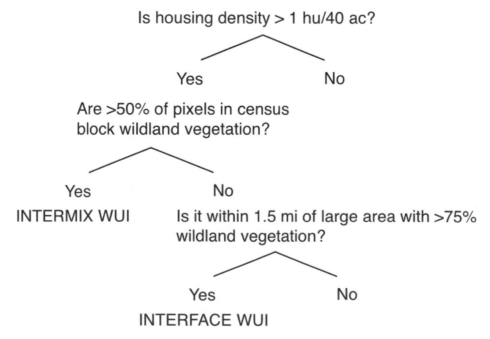

FIGURE 2-1 Definitions of WUI interface and intermix communities. SOURCE: Stewart et al. (2007).

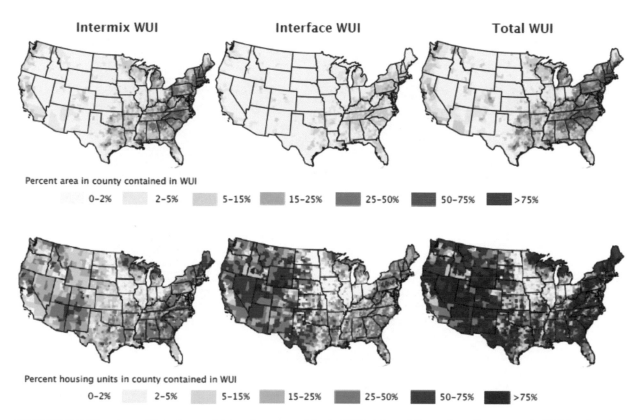

FIGURE 2-2 Mappings of interface and intermix communities in 2010. Adapted figure from Stewart et al. (2007) using data from Radeloff et al. (2018).

Figure 2-2 provides mappings of interface and intermix communities in the contiguous 48 states in 2010. Today, about a third of all housing units in the United States are in the WUI. This is true even though only 9 percent of the total land area within the lower 48 states is WUI (Marks, 2021). (In vast wildland areas, very little WUI exists, but every house in these wildlands is in the WUI, as seen in the upper and lower panels, respectively, of Figure 2-2.) While both intermix and interface WUI communities are found across the United States, housing units in the intermix WUI are more highly concentrated in eastern states, while interface WUI communities (houses at the edge of a large wildland area) appear to be slightly more common in the West. Because of limitations in data availability that currently preclude making a distinction between intermix and interface from a chemistry perspective, the remainder of this report is written about the WUI overall (rightmost panel of Figure 2-2).

In the two decades since the Federal Register definition was published, the WUI has evolved across the United States, primarily through expansion. Radeloff et al. (2018) used US Census data to estimate that from 1990 to 2010, the number of new houses in WUI communities in the United States grew from 30.8 to 43.4 million and land area grew from 581,000 to 770,000 km^2. This made WUI land area the largest growing land use type in the contiguous United States. Figure 2-3 shows WUI areas in 2010, highlighting two areas that experienced WUI fires that are described in the examples later in this chapter (Northern California and Gatlinburg, Tennessee). Although this report uses the definition of the WUI developed by the US Forest Service (Stewart et al., 2007), recent studies have suggested alternatives (Ager et al., 2019; Hanberry, 2020).

Finding: The number and spatial extent of WUI communities is growing rapidly.

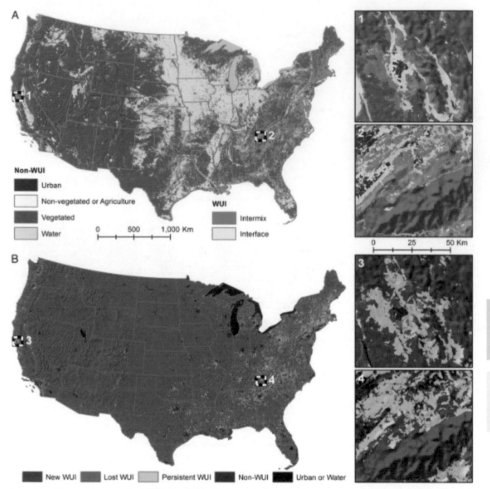

Fig. 1. The WUI in the United States was widespread in 2010 (*A*), as were changes in WUI area (*B*), for example, in and around Santa Rosa, California (1, 3), and Gatlinburg, Tennessee (2, 4), areas where wildfires destroyed many homes in 2017 and 2016, respectively.

FIGURE 2-3 WUI areas in 2010 (A) and changes in WUI areas from 1990 to 2010 (B); the numbered boxes are shown in detail at the right and highlight two areas that later experienced WUI fires (reproduced from Radeloff et al., 2018 [Figure 1]).

DEFINING SPATIAL AND TEMPORAL SCALES

The Federal Register notice and subsequent work provided the precise definition of the WUI used in this report; however, those WUI categories are primarily useful for assessing the risk of a wildfire before it occurs. For the purposes of this report, which focuses on the fire and post-fire period, the committee elected to describe the chemical processes and impacts in terms of order-of-magnitude length scales (Table 2-1).

The impacts of WUI fires occur on a variety of timescales. In this report, acute impacts will refer to phenomena during the time periods associated with the active fire. These time periods may extend from minutes and hours to weeks or longer. Chronic impacts will refer to phenomena that occur after the fire has been extinguished. These phenomena may persist for years.

FACTORS CONTRIBUTING TO INCREASING WUI FIRES

The US Forest Service estimates that in 2010, more than 70,000 communities and 43.4 million homes were at risk from WUI fires, a significant increase from the 30.8 million homes considered at risk in 1990 (Radeloff et al., 2018).

TABLE 2-1 Spatial Scales of Atmospheric Processes Impacting the Chemistry and Exposures Associated with WUI Fires

Length Scale (distance from flame front; km)	Descriptor	Atmospheric Processes Occurring at This Length Scale[a]
≤1	Immediate	Complex heat transfer from fire to fuels; rapid temperature evolution; effects of meteorologically induced and buoyancy-induced fluid dynamic interactions; chemical production of many of the emissions associated with health effects; chemical interaction with fire retardants; extremely high potential exposures
1–10	Near-field	Large-scale deposition of emissions; chemistry associated with mixing of fire plume with ambient air; some rapid initial aging of fire plume due to reactive intermediates; fire residues contaminating burned area
10–100	Local	Some aging may have occurred; concentrations and potential exposures are still high
100–1,000	Regional	Additional mixing of fire plume with ambient air; complex gas and particulate matter chemistry
>1,000	Continental	Long-range transport of plumes with complex gas and particulate matter chemistry

[a]This order-of-magnitude scale is used conceptually in this report to demarcate differences in chemical and physical processes and define how exposure may be impacted. These demarcations are not precise and are influenced by factors like meteorology and topography.

In the past ten years, wildfires have destroyed tens of thousands of structures (USFS, n.d.). Three main factors have driven the increase in WUI fires: (1) the expansion of areas at the WUI, (2) climate change, and (3) land management practices at the WUI.

The Expanding WUI

In 2000, the WUI made up 465,614 km² of the continental United States, a 52 percent expansion from 1970. Researchers project that its geographic size will grow by more than 10 percent by 2030 (Theobald and Romme, 2007). While population growth in general has increased the number of people living in the WUI, stronger effects have come from the deconcentration of populations (e.g., moving to the suburbs and exurbs), population shifts to the American West and Southeast, and an increased interest in "natural amenities" such as living in woodland areas. This is expected to continue (Hammer et al., 2009; Marks, 2021). Some research suggests that the main drivers of the migration to the WUI are demographic trends, including "socioeconomic objectives (e.g., rising housing prices in urban areas, access to resources and amenities, aesthetics, and increased access to nature) and structural societal limits (e.g., zoning laws, growth incentives, and land use)" (Peterson et al., 2021).

Increased development at the WUI has also increased human contact with wildland areas (Radeloff et al., 2018). Human-started wildfires accounted for 84 percent of all wildfires in two decades (1992–2012) of government records (Balch et al., 2017), so increased development at the WUI has also increased the risk of accidental fires at the WUI.

Simultaneously, as WUI areas expand, fires that start in wildland areas are more likely to burn wildland vegetation near WUI developments (Radeloff et al., 2005). These fires therefore have a greater chance of igniting homes or other structures at the WUI. Increased human contact with the WUI, as well as increased wildfire contact with human development and suburban/urban structures, both due to the rapid expansion of new construction at the WUI, resulted in 44.9 percent of the WUI and 17.5 million people between 2000 and 2010 being at risk of a wildfire occurring (Thomas and Butry, 2014). Furthermore, WUI communities are more vulnerable to fire events because geographical barriers make it harder to suppress fires in the WUI (Peterson et al., 2021).

Impacts of Climate Change

As the risk of wildfires coming into contact with communities at the WUI increases, so does the risk of wildfires igniting and spreading faster than humans can control them. Climate change has increased the frequency and length of droughts, particularly in the American West, which dry out soil and vegetation (USGCRP, 2018). This makes wildland vegetation more likely to ignite and to burn rapidly once ignited, reaching more communities at the expanded WUI (Peterson et al., 2021).

Greenhouse effects caused by increased atmospheric carbon dioxide concentrations have dried the air at the same time as droughts have become longer and more common (Liu et al., 2014). This makes it easier for fires to spread and reduces the likelihood of a precipitation event that would control or exterminate the fire. Based on climate projections, the risk of wildfires is projected to continue to increase in most areas of the world as climate change worsens (Bowman et al., 2020; Hurteau et al., 2019; Liu et al., 2015; Sun et al., 2019; Turco et al., 2018). While extreme drought conditions are more prevalent in the American West, drier weather conditions across the United States pose a risk for wildfire spread in any climate. For example, the Chimney Tops 2 Fire in Great Smoky Mountains National Park and nearby Gatlinburg, Tennessee, occurred in 2016 when much of the southeastern United States had been under "exceptional drought," or extremely dry, conditions in the months before the fire.

While the short-term effects of climate change, like drought, directly increase fire risk, in the long term, the situation is more complex. Over years or decades, climate can affect factors like fuel buildup, with a drought in one year reducing the amount of fuel available the next. A recent study used a model to integrate ecohydrology, fire spread, and fire effects to simulate 60 years of vegetation, fuel development, and wildfire in an area in the southern Sierra Nevada, California. The study modeled increased temperature and drought, both with and without wildfire. In the short term, high temperature, low precipitation, and high fuel loading led to increased area burned, as expected. However, in the long term, increased temperature led to decreased vegetation and increased fuel decomposition, which reduced fuels and area burned. After the first decade, cycles of large fire years occurred (Kennedy et al., 2021).

Zhong et al. (2021) model how land use and land cover changes will change local atmospheric conditions, including increasing land surface temperature and reducing precipitation, which will in turn lead to longer fire seasons and more extreme fire-weather conditions across the United States. As land use changes exacerbate wildfire risk, climate change and increasing development into the WUI exacerbate the conditions under which a wildfire can ignite and spread. This creates a feedback loop in which fires at the WUI become more likely, frequent, and dangerous.

Land Management at the WUI

Past wildland management practice has been recognized as an additional driver of increasing WUI fire threat. The wildland ecosystem in the western United States mostly evolved with wildfire disruption of the landscape (Hessburg, 2021). However, fire exclusion and suppression have been the emphasis of wildland fire management in the region in the past century (Calkin et al., 2015; Stephens and Ruth, 2005). This land management practice and past logging activities have contributed toward changes in forests in terms of vegetation age, structure, continuity, and density (Calkin et al., 2015; McIntyre et al., 2015; Schoennagel et al., 2017). Fire exclusion has resulted in fuel accumulation and wildlands that are less fire resistant, and evidence indicates that a transition has occurred from a regular, less intense surface fire regime to infrequent, large crown fires in the region (Calkin et al., 2015; Schoennagel et al., 2017). Ironically, the increase in assets at risk due to the growth of the WUI and sociopolitical pressures continues to compel the suppression response to wildfires (Calkin et al., 2015; Kolden, 2019), although there is a growing realization of the need to rectify current wildland management practices (Hessburg, 2021; Kolden, 2019; Schoennagel et al., 2017).

Finding: The threat of severe fires at the WUI is growing rapidly.

THE IMPACTS OF WUI FIRES

Wildfires have major economic, environmental, and social impacts. These include the costs associated with material losses, damage to human health, environmental destruction, and the operational costs of disaster response, such as firefighting efforts. Because of the wide range of impacts, numerous ways exist to measure the incidence

TABLE 2-2 Statistics that Describe Impacts of WUI Fires in the United States

Measure of Impact	Statistic	Year	Reference
Fire-exposure-related civilian deaths	15 deaths*	2017	Thomas et al., 2017
Fire-exposure-related firefighter deaths	18 deaths*	2017	Thomas et al., 2017
Acute-smoke-exposure-related civilian deaths	2,940–21,095 deaths*	2017	Thomas et al., 2017
Civilian injuries	88 injuries*	2017	Thomas et al., 2017
Firefighter injuries	260 injuries*	2017	Thomas et al., 2017
Costs due to injuries	$177,450,535*	2017	Thomas et al., 2017
Total acres burned (from all wildfires)	10,122,336 acres	2020	NIFC, 2020
Structures destroyed by fire	17,904	2020	NIFC, 2020
Residences destroyed by fire	9,630	2020	NIFC, 2020

NOTE: NIFC = National Interagency Fire Center; * estimated values or ranges.

and severity of wildfires. Table 2-2 provides statistics from recent WUI fires, showing several types of impacts and how they were measured. The sources listed in the table are the most recent available, at the time of this report, that are specific to WUI fires. While some impacts, such as number of deaths and costs of property damage, are straightforward to measure, other impacts, like the effects of psychological distress, are not.

The 2021 fire season provides many examples of the growing complexity of impacts of WUI fires as well as the complexity of measuring those impacts. The Caldor Fire was only the second fire (after the Dixie Fire, during the same season) ever to burn across the natural boundary of the Sierra Nevada Mountains (Reinhard and Patel, 2021). As a result, 22,000 residents of Lake Tahoe were told to evacuate, and many lost their homes. The smoke from the fire affected millions of people in several states. In 2020, one in seven Americans (about 47 million people) experienced dangerous air quality from wildfire smoke (Carlsen et al., 2020; based on an analysis of US Environmental Protection Agency AirNow data). Effects of smoke exposure include discomfort (such as headaches, fatigue, and sore throats), severe health effects (for example, particulates exposure can cause cardiac events), and disruption of activities, as people were told to stay indoors. In some cases, the conditions persisted for weeks or even months. Entire communities were lost to fires, such as the town of Greenville, California—a town where people had relocated after losing homes in Paradise, which was destroyed by 2018's Camp Fire. Dry conditions made the fires difficult to control. For example, the Marshall Fire, which destroyed communities in Louisville and Superior, Colorado, in December 2021, burned grasslands in a suburban region that were extremely dry due to months of drought. The US Forest Service stopped performing prescribed burns of federal lands in 2021 because the agency was overwhelmed with fighting wildfires (Sommer, 2021).

In the United States, in many cases, there is a lack of high-quality data specific to WUI fires. It is difficult to extract WUI-specific data for several reasons. First, fires are heterogeneous; they may start as purely wildland fires and later enter populated areas, burning structures and other elements of the built environment for only part of the event timeline. Second, smoke travels long distances, affecting ambient and indoor air quality at regional or even continental scales, which are well beyond the immediate fire zone. When effects are geographically distant from the fire, it can be difficult to confidently attribute environmental impacts, disease, and exacerbation of health conditions to the fire and the emissions that originated at the WUI. Finally, wildfires have both tangible and intangible impacts; while the tangible impacts might be straightforward to measure, "the intangible and potentially long-term wellbeing costs associated with physical injury, pain, psychological distress and behavioral change are much more difficult to evaluate" (Johnston et al., 2021, p. 2).

Fire Incidence by Region

While the expansion of WUI land areas has been more rapid in the eastern United States than in the West, almost half of the WUI in the East is characterized by low-severity fire regimes (Theobald and Romme, 2007).

Only 12 percent of the WUI in the West is characterized as low severity, resulting in a greater risk of wildfires occurring in the western United States.

Some of the greatest densities of WUI communities are in the eastern United States, where intermix communities are widespread (Figure 2-2). As Figure 2-3 shows, WUI areas are growing rapidly in the eastern United States. These denser populations and growing WUI areas mean more people in potential wildfire regions. In 2020, over 33,000 fires occurred in eastern states and burned just under 0.7 million acres (Wibbenmeyer and McDarris, 2021). By contrast, the 26,000 wildfires in the West burned approximately 9.5 million acres. While these larger wildfires in the West attract attention and can have large regional impacts, the size of a fire is not necessarily predictive of economic damage. The Marshall Fire, described later in this chapter, was the most economically destructive fire in Colorado history, yet burned a relatively small area of 6,200 acres. Factors that influence fire severity include land use, vegetation, terrain, weather, and fire history. Although the US Forest Service led the development of Wildfire Risk to Communities (wildfirerisk.org), there are no current federal definitions related to fire risk or fire severity specific to the WUI. California has defined "Fire Hazard Severity Zones" that provide a more nuanced view of this issue at the state level; however, the details of how California determines these zones are not transparently available (OSFM, 2021).

> **Finding:** WUI fires have different characteristics in different regions of the United States that likely lead to different risks, but a comprehensive mapping of communities at risk of fire, and the severity of that risk, is not available.

Demographics of Impacted Populations

Not all populations are equally affected by WUI fires. The limited number of studies that exist examine different regions and draw a variety of conclusions about which populations are most impacted. A 2021 study used wildfire data from the California Fire and Resource Assessment Program and the National Interagency Fire Center and demographic data from the American Community Survey to examine the populations living in areas burned by wildfire in California from 2000 to 2020 (Masri et al., 2021). In addition to a near doubling of the populations in affected areas (largely due to an increase in the area burned by wildfires), the study found that these areas contained lower proportions of racially minoritized groups in general but higher proportions of Native American populations; higher proportions of older residents; and higher proportions of low-income residents, lower median household incomes, and lower home values.

By contrast, a 2020 study of fires in Southern California since 1980 found that WUI neighborhoods in that region are predominantly white and affluent, with increasingly expensive housing (Garrison and Huxman, 2020). Another 2020 study looked specifically at the risks of wildfire to residents of subsidized housing in California (Gabbe et al., 2020). This study was motivated by the difficulties low-income households face in recovering from such disasters, as well as the fact that policy makers can reduce the risk to current and future subsidized housing. The study found that a higher proportion of subsidized housing is located outside the WUI, but that residents of 140,000 subsidized housing units as well as other vulnerable populations in the WUI do live at risk.

These three studies of impacts in California highlight an important point about the availability of information on WUI fires and wildfires more generally: the literature on WUI fires is often focused on the West, particularly the State of California. As discussed earlier in this chapter, WUI fires are not strictly a western phenomenon; however, likely due to a variety of contributing factors, western fires are more strongly represented in the existing research that informed this report. The committee acknowledges the potential bias that this fact introduces to its work. Until broader research can fill in information gaps on the demographics of impacted populations in the East, and elsewhere in the United States, the existing literature, which focuses on the West, can be used to analyze the problem of WUI fires, with an acknowledgment of the limitations.

> **Finding:** Although the majority of the WUI is found in the eastern United States, research on the impacts of WUI fires is more abundant for California than other US states.

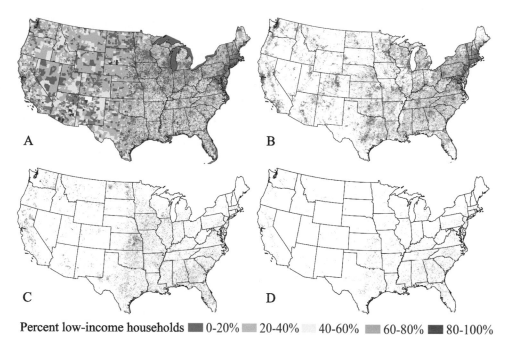

Percent low-income households ■ 0-20% ■ 20-40% 40-60% ■ 60-80% ■ 80-100%

FIGURE 2-4 (A) Map of low-income populations in the contiguous United States. (B) Overlay of low-income populations with WUI areas. (C) Overlay of low-income populations with areas burned by wildfire. (D) Overlay of low-income populations with WUI areas burned by wildfire. Data to create the figures were sourced from the following: low-income populations in the contiguous United States in 2020 (EPA, 2021); WUI areas in the contiguous United States in 2010 (Radeloff et al., 2018; updates to WUI land area mappings are under way using 2020 census data but were unavailable at the time the committee prepared this report); burned areas in the contiguous United States in 2021 (Wiedinmyer et al., 2011).

To begin addressing these information gaps, simple geographic information system analyses could be conducted to combine existing data sets in new ways. In Figure 2-4, national data on the distribution of low-income households has been combined with Radeloff et al.'s WUI maps and a data set that shows the spatial distribution of burned areas from 2021 wildfires (EPA, 2021; Radeloff et al., 2018; Wiedinmyer et al., 2011). The data set on burned areas uses thermal hot spot detection to locate fires; therefore, the data set includes open biomass burning, including wildfires, agricultural fires, and prescribed burning (Wiedinmyer et al., 2011).

This analysis, done by the committee using the above sources, shows that, while 32.5 percent of households in the contiguous United States are categorized as low income (i.e., with income less than two times the federal poverty level; data from the Census Bureau's American Community Survey 2014–2018), 28.5 percent of households in the WUI are categorized as low income, and 36.3 percent of households in WUI areas that burned in 2020 are categorized as low income. This analysis is a simple example of the types of insights that could emerge from combining a precise definition of WUI communities with other mappings. These types of analyses could inform future research or help decision-makers prioritize research.

Finding: Quantitative and widely accepted definitions of WUI communities would enable mappings of risk and help define future research priorities.

USING EXAMPLES TO CONTEXTUALIZE WUI FIRES

The earlier sections of this chapter discuss how pervasive WUI communities and fires are; however, few data are available on how WUI fires differ from other wildfires. For the purposes of this report, the committee chose to explore a set of recent examples of the state of available information. The five WUI fires presented in this section and discussed throughout this report took place in North America within the past 10 years. They were identified by the committee through their own research and through conversations with experts as cases that exhibit distinct characteristics and for which a comparatively large amount of documentation is available. This section provides some situational details for each example fire. A table that contains the results of the committee's literature search for each fire can be found in Appendix C of the report. Later chapters refer to what is known about these fires to enable a deeper discussion of the information needs in WUI fire research.

Waldo Canyon Fire (2012)

The Waldo Canyon Fire was first reported on Saturday, June 23, 2012. It began in the Pike National Forest, approximately three miles west of Colorado Springs, Colorado. The fire burned 18,247 acres over 18 days and was reported as fully contained on July 10, 2012 (City of Colorado Springs, 2013). About 19 percent of the burn was classified as high severity, 40 percent as moderate severity, and 41 percent as low severity. In the severe Colorado wildfire season of 2012, the Waldo Canyon Fire stands out with its high social impact, which was due to its close proximity to Colorado Springs and other communities. The fire forced the evacuation of more than 32,000 people, including 22,000 residents within a two-hour period on June 26, 2012 (Chin et al., 2016). There were two documented fatalities (City of Colorado Springs, 2013).

The Waldo Canyon Fire was, at the time, the most destructive wildfire in Colorado history, destroying more than 300 homes, mostly in the Mountain Shadows neighborhood. These homes were downslope of a major topographic ridge that normally would have kept the fire from spreading. However, erratic winds exceeding 100 km per hour, which were fueled by dry conditions, caused the fire to cross the ridge into the residential neighborhood (Chin et al., 2016). The Colorado Springs Fire Department invited the National Institute of Standards and Technology (NIST) and the US Forest Service Fire and Environmental Research Applications Team to collect post-incident data from the fire, specifically in the Mountain Shadows community. The report found that 1,455 primary structures had existed in the community. Of these, over 300 primary structures, as classified in that report, were completely destroyed, and over 100 more structures were damaged (Maranghides et al., 2015). The Waldo Canyon Fire had losses estimated at US$454 million (FEMA, 2021). An investigation by the US Forest Service and the Federal Bureau of Investigation later determined that the fire had been caused by human activity.

Horse River Fire (2016)

On May 1, 2016, a fire started to the southwest of Fort McMurray, Alberta, Canada, in what became known as the Horse River Fire. Fort McMurray is a city bordered by boreal forest in the Athabasca oil sands region of Alberta and had a population of 78,382 in 2015 (MSB, 2015). The fire quickly spread to populated areas, exacerbated by record-setting high temperatures and low humidity (McGrath, 2016). More than 88,000 residents of the city and surrounding areas were evacuated on May 3, 2016 (McGee, 2019). The fire was declared contained on July 5, 2016, but was not fully extinguished until August of the following year.

Over 1.4 million acres of land were burned (Gabbert, 2017), including 2,579 homes (Wood Buffalo Economic Development, 2017). At the time, the Horse River Fire was the costliest natural disaster in Canadian history, with insurance claims amounting to C$3.58 billion (Suncor Energy Inc., 2018). Today, it remains the most expensive wildfire in Canadian history. There were no fatalities and no injuries. The specific cause of the fire is still unknown, but law enforcement determined that it was most likely the result of human activity.

Chimney Tops 2 Fire (2016)

The 2016 wildfire that engulfed Gatlinburg, Tennessee, became known as the Chimney Tops 2 Fire. First observed around 5:30 p.m. on November 23, it started as a 1.5-acre wildland fire in Great Smoky Mountains National Park, about 5.5 miles away from city limits (Culver, 2016). The steep topography of Chimney Tops prevented firefighters from immediately extinguishing the fire. When high winds picked up four days later on November 27, the fire rapidly grew to 35 acres. The next day, in addition to causing the original fire area to spread, high wind conditions carried firebrands up to one mile and caused new fires to ignite to the north, closer to the city. On November 28, the fire traveled three miles toward Gatlinburg in less than five hours. That evening, the entire Gatlinburg area including nearby Pigeon Forge, which included over 14,000 residents, was placed under a mandatory evacuation order.

A few hours later, around 2:00 a.m. on November 29, the weather shifted, with wind speeds dying down, replaced by steady rain. While the fire was not fully contained until about two weeks later, its rapid spread had ended. There were 14 deaths, 2,500 destroyed or damaged structures, and about 17,000 acres burned (Guthrie et al., 2017). One year after the fire, damages were estimated at over US$500 million (Songer, 2017). The cause of the fire was determined to be arson.

Camp Fire (2018)

The Camp Fire was first reported on November 8, 2018, in Butte County, California. Strongly driven toward populated areas by northeasterly winds, it was 100 percent contained on November 25, 2018, after burning 153,336 acres over 17 days (Mohler, 2019). Burned areas included both interface- and intermix-type WUI (Kramer et al., 2019). Nineteen percent of the fire area was classified as very low soil burn severity, 63 percent as low soil burn severity, 16 percent as moderate soil burn severity, and 2 percent as high soil burn severity (WERT, 2018). The fire destroyed the town of Paradise, California, and damaged the neighboring communities of Magalia, Butte Creek Canyon, and Concow.

The fire resulted in 85 fatalities and three documented injuries (CAL FIRE, 2019). Some 18,804 structures were destroyed, including 13,696 single-family residences, 276 multi-family residences, 4,923 other, minor structures, and 528 commercial structures. An additional 462 single-family residences, 25 multi-family residences, and 102 commercial structures were damaged, but not completely destroyed (Hawks, 2018). In 2019, California Department of Forestry and Fire Protection investigators determined that the cause of the Camp Fire was electrical transmission lines that ignited vegetation at two ignition sites (Mohler, 2019). Pacific Gas and Electric Company, the utility responsible, later pleaded guilty to 84 counts of involuntary manslaughter and paid damages to the wildfire's victims. The Camp Fire was the costliest disaster worldwide in 2018, totaling US$16.5 billion in losses (NIST, 2021; Ruiz-Grossman, 2019).

NIST prepared an extensive case study of the fire's spread based on abundant observations by first responders and other first-hand witnesses (Maranghides et al., 2021). Because this fire is recent, the committee acknowledges that relevant chemistry data and analyses may still be unpublished at this time.

Marshall Fire (2021)

At 10.00 a.m. on December 30, 2021, high winds, which blew over 115 miles per hour in some places, helped drive fast-moving grass fires that rapidly spread through the densely populated suburban areas of Superior and Louisville in Boulder County, Colorado (CPR, 2021; Markus, 2022). The most destructive fire in Colorado history, the Marshall Fire destroyed over 1,000 residential structures and businesses and burned over 6,000 acres (Metzger, 2022). Burn severity information was not available at the time of the publication of this report. High winds created 40-foot-high flames in some areas. At 11:44 a.m. on December 30, residents east of the fire were placed under an evacuation order as firefighters struggled to contain the fire, which was burning in multiple places (Brown and Paul, 2022). Two hours later, a reverse-911 call ordered more residents to evacuate as high winds continued.

The fire was extinguished by heavy snowfall on Friday, December 31, when up to 10 inches of snow blanketed the Denver suburbs. As temperatures dropped below freezing, 7,500 customers of the electric company, Xcel Energy, were without electricity, and 13,000 people in the Boulder area were without gas (CPR, 2022). One person was found dead and another reported missing (CBS Denver, 2022). President Biden approved a major disaster declaration on January 1, 2022. The total value of homes destroyed is estimated at more than $500,000,000, while the value of commercial damage is still being calculated (Cobb, 2022).

This fire occurred only a few months before this report's publication. The cause of the fire is still unknown (Boulder OEM, 2022). Facts about the event will continue to emerge; although many researchers took part in post-fire chemical analyses of air, water, and soil, chemical data will take time to appear in the peer-reviewed literature.

FINDINGS FROM RECENT EXAMPLES OF WUI FIRES

Appendix C describes the literature available on these fires, which the committee used in later chapters, along with additional data from wildfires that the committee found relevant. Despite these studies being relatively widely studied, there are still significant data gaps, especially information on the fuels, combustion, emissions, and atmospheric chemistry.

Finding: There is a paucity of chemical information available about WUI fires.

Where data are available (e.g., from state and local agencies), it often must be drawn from air and water quality monitoring networks, and these networks are not designed specifically for collecting data relevant to wildfires and their emissions. However, supplemental data collection is not generally prioritized during WUI fires, as authorities' first concern is life and property protection. Additionally, the lack of technological and workforce availability limits research efforts. These are some of the challenges facing the research efforts needed to understand the effects of WUI fires.

States like California do more sampling and chemical testing in areas affected by fires than other states do (CalRecycle, 2022). For example, Colorado recommends against testing soil at affected parcels prior to residents returning to rebuild, "as testing can be difficult, expensive and may cause delays in being able to handle the waste" (CDPHE, 2020, p. 1).

Finding: Advancing the understanding of WUI fires will take advance planning and prioritization of data collection during WUI fires.

REFERENCES

Ager, A. A., P. Palaiologou, C. R. Evers, M. A. Day, C. Ringo, and K. Short. 2019. "Wildfire Exposure to the Wildland Urban Interface in the Western US." *Applied Geography* 111: 102059. https://doi.org/10.1016/j.apgeog.2019.102059.

Balch, J. K., B. A. Bradley, J. T. Abatzoglou, R. C. Nagy, E. J. Fusco, and A. L. Mahood. 2017. "Human-Started Wildfires Expand the Fire Niche across the United States." *Proceedings of the National Academy of Sciences of the United States of America* 114 (11): 2946–2951. https://doi.org/10.1073/pnas.1617394114.

Boulder OEM (Boulder Office of Emergency Management). January 6, 2022. "Boulder County Releases Updated List of Structures Damaged and Destroyed in the Marshall Fire." https://www.bouldercounty.org/news/boulder-county-releases-updated-list-of-structures-damaged-and-destroyed-in-the-marshall-fire/ (accessed January 31, 2022).

Bowman, D. M. J. S., C. A. Kolden, J. T. Abatzoglou, F. H. Johnston, G. R. van der Werf, and M. Flannigan. 2020. "Vegetation Fires in the Anthropocene." *Nature Reviews Earth & Environment* 1 (10): 500–515. https://doi.org/10.1038/s43017-020-0085-3.

Brown, J., and J. Paul. January 6, 2022. "The Minute-by-Minute Story of the Marshall Fire's Wind-Fueled Tear through Boulder County." *Colorado Sun.* https://coloradosun.com/2022/01/06/marshall-fire-boulder-county-timeline/.

CAL FIRE (California Department of Forestry and Fire Protection). 2019. "Camp Fire Incident." https://www.fire.ca.gov/incidents/2018/11/8/camp-fire/.

Calkin, D. E., M. P. Thompson, and M. A. Finney. 2015. "Negative Consequences of Positive Feedbacks in US Wildfire Management." *Forest Ecosystems* 2: 9. https://doi.org/10.1186/s40663-015-0033-8.

CalRecycle. 2022. *Wildfire Cleanup Process and Order of Operations.* https://calrecycle.ca.gov/disaster/wildfires/operations/.

Carlsen, A., S. Mcminn, and J. Eng. 2020. "1 in 7 Americans Have Experienced Dangerous Air Quality Due to Wildfires This Year." *National Public Radio.* https://www.npr.org/2020/09/23/915723316/1-in-7-americans-have-experienced-dangerous-air-quality-due-to-wildfires-this-ye (accessed February 25, 2022).

CBS Denver. January 21, 2022. "Marshall Fire: Investigators Consider Underground Mine Fire in Origin." *CBS Denver.* https://denver.cbslocal.com/2022/01/21/marshall-fire-boulder-county-underground-mine-fire/.

CDPHE (Colorado Department of Public Health and the Environment). November 5, 2020. "Important Wildfire Debris Guidance." https://www.larimer.org/sites/default/files/uploads/2020/cdphe_important_wildfire_debris_guidance_0.pdf.

Chin, A., L. An, J. L. Florsheim, L. R. Laurencio, R. A. Marston, A. P. Solverson, G. L. Simon, E. Stinson, and E. Wohl. 2016. "Investigating Feedbacks in Human-Landscape Systems: Lessons Following a Wildfire in Colorado, USA." *Geomorphology* 252: 40–50.

City of Colorado Springs. 2013. *Waldo Canyon Fire Final After Action Report*. Colorado Springs, CO: City of Colorado Springs.

Cobb, E. January 30, 2022. "'A Hurricane of Fire': How the Marshall Fire Tore through Boulder County." *Daily Camera*. https://www.dailycamera.com/2022/01/30/hurricane-of-fire/.

CPR (Colorado Public Radio). December 31, 2021. "Boulder County Fires: More Than 500 Houses Burn, Tens of Thousands Evacuate as Fires Continue to Spread." *CPR News*. https://www.cpr.org/2021/12/30/boulder-county-grass-fires/.

CPR. January 1, 2022. "Boulder County Fires: Three Missing from Marshall, Superior Are Suspected Dead; Thousands without Gas and Electricity in Freezing Temps." *CPR News*. https://www.cpr.org/2022/01/01/boulder-county-fires-water-shut-off-in-superior-thousands-without-gas-and-electricity-in-freezing-temps/.

Culver, A. December 13, 2016. "Timeline: Gatlinburg Wildfires." WATE. https://www.wate.com/news/local-news/timeline-gatlinburg-wildfires/ (accessed December 23, 2021).

EPA (US Environmental Protection Agency). 2021. *EJScreen: Environmental Justice Screening and Mapping Tool*, 2021 version. www.epa.gov/ejscreen (accessed October 19, 2021).

Federal Register. January 4, 2001. "Urban Wildland Interface Communities within the Vicinity of Federal Lands That Are at High Risk from Wildfire." *Federal Register* 66 (01-52): 751–777. https://www.federalregister.gov/documents/2001/01/04/01-52/urban-wildland-interface-communities-within-the-vicinity-of-federal-lands-that-are-at-high-risk-from.

FEMA (Federal Emergency Management Agency). February 11, 2021. "Cedar Heights Saved during Waldo Canyon Fire." https://www.fema.gov/case-study/cedar-heights-saved-during-waldo-canyon-fire.

FMB (Fire Management Board). 2019. *Federal Wildland Fire Policy Terms and Definitions*. https://www.nwcg.gov/sites/default/files/docs/eb-fmb-m-19-004a.pdf.

Gabbe, C. J., G. Pierce, and E. Oxlaj. 2020. "Subsidized Households and Wildfire Hazards in California." *Environmental Management* 66 (5): 873–883.

Gabbert, B. November 21, 2017. "Official Report Shows Fort McMurray Fire Created Lightning-Started Fires 26 Miles Away." *Wildfire Today*. https://wildfiretoday.com/2017/11/21/official-report-shows-fort-mcmurray-fire-created-lightning-started-fires-26-miles-away/.

Garrison, J. D., and T. E. Huxman. 2020. "A Tale of Two Suburbias: Turning Up the Heat in Southern California's Flammable Wildland-Urban Interface." *Cities* 104: 102725.

Guthric, V. H., M. J. Finucane, P. E. Keith, and D. Bart Stinnett. 2017. *After Action Review of the November 28, 2016, Firestorm*. Spring, TX: ABS Group. https://wildfiretoday.com/documents/AAR_ChimneyTops2.pdf.

Hammer, R. B., S. I. Stewart, and V. C. Radeloff. 2009. "Demographic Trends, the Wildland–Urban Interface, and Wildfire Management." *Society & Natural Resources* 22 (8): 777–782. https://doi.org/10.1080/08941920802714042.

Hanberry, B. B. 2020. "Reclassifying the Wildland Urban Interface Using Fire Occurrences for the United States." *Land* 9: 225. https://doi.org/10.3390/land9070225.

Hawks, D. 2018. *Camp Fire*. Carson City, NV: Nevada Division of Forestry.

Hessburg, P. F. 2021. "Wildfire and Climate Change Adaptation of Western North American Forests: A Case for Intentional Management." *Ecological Applications* 31 (8): e02432.

Hurteau, M. D., S. Liang, A. L. Westerling, and C. Wiedinmyer. 2019. "Vegetation-Fire Feedback Reduces Projected Area Burned under Climate Change." *Scientific Reports* 9 (1): 2838. https://doi.org/10.1038/s41598-019-39284-1.

Johnston, D., Y. K. Onder, H. Rahman, and M. Ulubasoglu. 2021. *Evaluating Wildfire Exposure: Using Wellbeing Data to Estimate and Value the Impacts of Wildfire*. https://mpra.ub.uni-muenchen.de/109652/.

Kennedy, M. C., R. R. Bart, C. L. Tague, and J. S. Choate. 2021. "Does Hot and Dry Equal More Wildfire? Contrasting Short- and Long-Term Climate Effects on Fire in the Sierra Nevada, CA." *Ecosphere* 12: e03657. https://doi.org/10.1002/ecs2.3657.

Kolden, C. A. 2019. "We're Not Doing Enough Prescribed Fire in the Western United States to Mitigate Wildfire Risk." *Fire* 2 (2): 30.

Kramer, H. A., M. H. Mockrin, P. M. Alexandre, and V. C. Radeloff. 2019. "High Wildfire Damage in Interface Communities in California." *International Journal of Wildland Fire* 28 (9): 641–650.

Liu, Y., S. Goodrick, and W. Heilman. 2014. "Wildland Fire Emissions, Carbon, and Climate: Wildfire–Climate Interactions." *Forest Ecology and Management* 317: 80–96.

Liu, Z., M. C. Wimberly, A. Lamsal, T. L. Sohl, and T. J. Hawbaker. 2015. "Climate Change and Wildfire Risk in an Expanding Wildland–Urban Interface: A Case Study from the Colorado Front Range Corridor." *Landscape Ecology*, 30 (10): 1943–1957.

Maranghides, A., D. McNamara, R. Vihnanek, J. Restaino, and C. Leland. 2015. *A Case Study of a Community Affected by the Waldo Fire – Event Timeline and Defensive Actions*. NIST Technical Note 1910. Gaithersburg, MD: NIST.

Maranghides, A., E. Link, W. Mell, S. Hawks, M. Wilson, W. Brewer, C. Brown, B. Vihnaneck, and W. D. Walton. 2021. *A Case Study of the Camp Fire – Fire Progression Timeline*. Gaithersburg, MD: NIST. https://nvlpubs.nist.gov/nistpubs/TechnicalNotes/NIST.TN.2135.pdf.

Marks, J. 2021. "The Wildland Urban Interface Is Growing in the United States." *Geography Realm*. https://www.geography-realm.com/the-wildland-urban-interface-is-growing-in-the-united-states/ (accessed February 26, 2022).

Markus, B. January 5, 2022. "Boulder County Firefighters Lost Crucial Early Minutes Because They Couldn't Find the Start of the Marshall Fire." *CPR News*. https://www.cpr.org/2022/01/05/boulder-county-marshall-fire-timeline/.

Masri, S., E. Scaduto, Y. Jin, and J. Wu. 2021. "Disproportionate Impacts of Wildfires among Elderly and Low-Income Communities in California from 2000–2020." *International Journal of Environmental Research and Public Health* 18 (8): 3921. https://doi.org/10.3390/ijerph18083921.

McGee, T. K. 2019. "Preparedness and Experiences of Evacuees from the 2016 Fort McMurray Horse River Wildfire." *Fire* 2 (1): 13.

McGrath, M. May 5, 2016. "'Perfect Storm' of El Niño and Warming Boosted Alberta Fires." *BBC News*. https://www.bbc.com/news/science-environment-36212145.

McIntyre, P. J., J. H. Thorne, C. R. Dolanc, A. L. Flint, L. E. Flint, M. Kelly, and D. D. Ackerly. 2015. "Twentieth-Century Shifts in Forest Structure in California: Denser Forests, Smaller Trees, and Increased Dominance of Oaks." *Proceedings of the National Academy of Sciences* 112 (5): 1458–1463.

Metzger, H. January 3, 2022. "Perimeter of Marshall Fire 100% Contained after Burning over 6,000 Acres." *Colorado Politics*. https://www.coloradopolitics.com/denver/marshall-fire-100-percent-contained/article_6ee3eb84-f49a-5bbe-8ea7-d2a26bec620d.html (accessed March 25, 2022).

Mohler, M. May 15, 2019. "CAL FIRE Investigators Determine the Cause of the Camp Fire." California Department of Forestry and Fire Protection. https://www.fire.ca.gov/media/5121/campfire_cause.pdf.

MSB (Government of Alberta Municipal Services Branch). 2015. *2015 Municipal Affairs Population List*. http://municipalaffairs.alberta.ca/documents/msb/2015_Municipal_Affairs_Population_List.pdf.

NIFC (National Interagency Fire Center). 2020. *Wildland Fire Summary and Statistics Annual Report*. https://www.predictiveservices.nifc.gov/intelligence/2020_statssumm/annual_report_2020.pdf.

NIST (National Institute of Standards and Technology). February 8, 2021. "New Timeline of Deadliest California Wildfire Could Guide Lifesaving Research and Action." https://www.nist.gov/news-events/news/2021/02/new-timeline-deadliest-california-wildfire-could-guide-lifesaving-research.

OSFM (Office of the State Fire Marshal). 2021. "Fire Hazard Severity Zones." https://osfm.fire.ca.gov/divisions/wildfire-planning-engineering/wildfire-prevention-engineering/fire-hazard-severity-zones/.

Peterson, G. C. L., S. E. Prince, and A. G. Rappold. 2021. "Trends in Fire Danger and Population Exposure along the Wildland–Urban Interface." *Environmental Science & Technology* 55: 16257–16265. https://doi.org/10.1021/acs.est.1c03835.

Radeloff, V. C., R. B. Hammer, S. I. Stewart, J. S. Fried, S. S. Holcomb, and J. F. McKeefry. 2005. "The Wildland-Urban Interface in the United States." *Ecological Applications* 15 (3): 799–805.

Radeloff, V. C., D. P. Helmers, H. A. Kramer, M. H. Mockrin, P. M. Alexandre, A. Bar-Massada, V. Butsic, T. J. Hawbaker, S. Martinuzzi, A. D. Syphard, and S. I. Stewart. 2018. "Rapid Growth of the US Wildland-Urban Interface Raises Wildfire Risk." *Proceedings of the National Academy of Sciences* 115 (13): 3314–3319. https://doi.org/10.1073/pnas.1718850115.

Reinhard, S., and J. K. Patel. September 5, 2021. "Caldor Fire's March to the Edge of South Lake Tahoe." *The New York Times*. https://www.nytimes.com/interactive/2021/09/05/us/caldor-fire-lake-tahoe.html (accessed February 24, 2022).

Ruiz-Grossman, S. January 12, 2019. "California's Camp Fire Was the Most Expensive Natural Disaster Worldwide in 2018." *Grist*. https://grist.org/article/californias-camp-fire-was-the-most-expensive-natural-disaster-worldwide-in-2018/.

Schoennagel, T., J. K. Balch, H. Brenkert-Smith, P. E. Dennison, B. J. Harvey, M. A. Krawchuk, N. Mietkiewicz, P. Morgan, M. A. Moritz, R. Rasker, M. G. Turner, and C. Whitlock. 2017. "Adapt to More Wildfire in Western North American Forests as Climate Changes." *Proceedings of the National Academy of Sciences* 114 (18): 4582–4590.

Sommer, L. August 10, 2021. "With Extreme Fires Burning, Forest Service Stops 'Good Fires' Too." *NPR* https://www.npr.org/2021/08/09/1026137249/with-extreme-fires-burning-forest-service-stops-good-fires-too (accessed March 25, 2022).

Songer, J. November 27, 2017. "Gatlinburg One Year after the Devastating Wildfires Then and Now." *AL.com*. https://www.al.com/outdoors/2017/11/gatlinburg_one_year_after_the.html.

Stephens, S. L., and L. W. Ruth. 2005. "Federal Fire Policy in the United States." *Ecological Applications* 15 (2): 532–542.

Stewart, S. I., V. C. Radeloff, R. B. Hammer, and T. J. Hawbaker. 2007. "Defining the Wildland–Urban Interface." *Journal of Forestry* 105 (4): 201–207. https://doi.org/10.1093/jof/105.4.201.

Sun, Q., C. Miao, M. Hanel, A. G. L. Borthwick, Q. Duan, D. Ji, and H. Li. 2019. "Global Heat Stress on Health, Wildfires, and Agricultural Crops under Different Levels of Climate Warming." *Environment International* 128: 125–136. https://doi.org/10.1016/j.envint.2019.04.025.

Suncor Energy Inc. 2018. *Volume 2 – Environmental Impact Assessment*. Lewis In Situ Project. https://open.alberta.ca/dataset/f9e55e85-c164-4227-8be6-b45fb32f812f/resource/c3b4bfd4-c43f-495d-be5d-8e1d80e89554/download/vol2_sec18_socio-economic.pdf (accessed March 25, 2022).

Theobald, D. M., and W. H. Romme. 2007. "Expansion of the US Wildland–Urban Interface." *Landscape and Urban Planning* 83 (4): 340–354. https://doi.org/10.1016/j.landurbplan.2007.06.002.

Thomas, D. S., and D. T. Butry. 2014. "Areas of the U.S. Wildland–Urban Interface Threatened by Wildfire during the 2001–2010 Decade." *Natural Hazards* 71: 1561–1585. https://doi.org/10.1007/s11069-013-0965-7.

Thomas, D., D. Butry, S. Gilbert, D. Webb, and J. Fung. 2017. *The Costs and Losses of Wildfires: A Literature Survey.* NIST Special Publication 1215. https://doi.org/10.6028/NIST.SP.1215.

Turco, M., J. J. Rosa-Canovas, J. Bedia, S. Jerez, J. P. Montavez, M. C. Llasat, and A. Provenzale. 2018. "Exacerbated Fires in Mediterranean Europe Due to Anthropogenic Warming Projected with Non-stationary Climate-Fire Models." *Nature Communications* 9 (1): 3821. https://doi.org/10.1038/s41467-018-06358-z.

USFS (US Forest Service). n.d. "Fire Adapted Communities." https://www.fs.usda.gov/managing-land/fire/fac.

USGCRP (US Global Change Research Program). 2018. *Fourth National Climate Assessment, Volume II: Impacts, Risks, and Adaptation in the United States.* Edited by D. R. Reidmiller, C. W. Avery, D. R. Easterling, K. E. Kunkel, K. L. M. Lewis, T. K. Maycock, and B. C. Stewart. Washington, DC: US Global Change Research Program. https://doi.org/10.7930/NCA4.2018.

WERT (State of California Watershed Emergency Response Team). November 29, 2018. *CAMP FIRE Watershed Emergency Response Team Final Report.* https://ucanr.edu/sites/Rangelands/files/304942.pdf.

Wibbenmeyer, M., and A. McDarris. 2021. "Wildfires in the United States 101: Context and Consequences." *Resources for the Future.* https://www.rff.org/publications/explainers/wildfires-in-the-united-states-101-context-and-consequences/ (accessed February 25, 2022).

Wiedinmyer, C., S. K. Akagi, R. J. Yokelson, L. K. Emmons, J. A. Al-Saadi, J. J. Orlando, and A. J. Soja. 2011. "The Fire INventory from NCAR (FINN): A High Resolution Global Model to Estimate the Emissions from Open Burning." *Geoscientific Model Development* 4 (3): 625–641. https://doi.org/10.5194/gmd-4-625-2011.

Wood Buffalo Economic Development. 2017. *Wood Buffalo Economic Development Regional Data.* http://www.oscaalberta.ca/wp-content/uploads/2015/08/Wood-Buffalo-Regional-Data-2017.pdf.

Zhong, S., T. Wang, P. Sciusco, M. Shen, L. Pei, J. Nikolic, K. McKeehan, H. Kashongwe, P. Hatami-Bahman-Beiglou, K. Camacho, D. Akanga, J. J. Charney, and X. Bian. 2021. "Will Land Use Land Cover Change Drive Atmospheric Conditions to Become More Conducive to Wildfires in the United States?" *International Journal of Climatology* 41 (6): 3578–3597. https://doi.org/10.1002/joc.7036.

3

Materials, Combustion, and Emissions in WUI Fires

Fires in the wildland-urban interface (WUI) differ from wildland fires. Combustible materials in the WUI have different chemical compositions and densities, and are present in different quantities, than the vegetative biomass combusted in wildland fires. The urban materials and their characteristics present in the WUI impact the combustion conditions, the chemical reaction pathways that dominate during combustion, and the emissions and residue released into the environment. Current understanding of the chemistry of WUI fires and their emissions is largely inferred based on information on wildland and urban fires.

Wildland fires are a large source of emissions and have been extensively studied. Over the last several decades, numerous laboratory studies and field efforts have sought to link the characteristics of biomass and fire behavior with the resulting emissions (Jaffe et al., 2020). These studies have made clear that emissions from biomass fires are complex and consist of many chemical species, of which only a limited number have been identified. The composition of the emissions also depends on fire conditions; however, many uncertainties remain in the details of this dependence (Coggon et al., 2019; Jaffe et al., 2020; Permar et al., 2021; Sekimoto et al., 2018).

Data on emissions from the combustion of urban materials is derived largely from enclosure fires (i.e., a fire within a room or compartment inside a building) or from laboratory test methods simulating enclosure fires (Blomqvist and Lönnermark, 2001; Stec and Hull, 2010). Both the specific material elemental composition and the amount of oxygen available for combustion strongly impact released emissions. However, very little is known about the chemical composition of urban materials and the interactions of various mixtures of these materials on combustion processes, the types of species emitted, and these species' interactions under different fire conditions (Purser et al., 2015; Stec and Hull, 2010). The fire emissions produced (per mass of material combusted) from the combustion of human-made materials are expected to be much different than those from biomass.

This chapter summarizes what is known about the materials in the WUI environment, and the combustion conditions, dominant reaction pathways, and resulting emissions from WUI fires. The discussion of materials in the WUI environment and the unique aspects of combustion chemistry described in this chapter provide context to the information needed to describe atmospheric transformations, water and soil impacts, health effects, and measurement systems described in Chapters 4–7.

Box 3-1 defines some of the key terms used in the chapter.

BOX 3-1
Key Terminology

Spatial Scales

Near-field scale: From 1 to 10 km downwind of the fire, where the plume remains quite concentrated, dilution has a major effect on the gas-particle partitioning, and chemistry is driven by fast processes that occur on a timescale of minutes

Local scale: From 10 to 100 km, where chemistry is driven by processes that occur on a timescale of minutes to hours

Regional scale: From 100 to 1,000 km, where chemistry is driven by processes that occur on a timescale of hours to days

Continental scale: Greater than 1,000 km, where chemistry is driven by processes that occur over days

Terms and Definitions

Ash: The solid residue remaining after combustion, generally consisting of minerals and lesser amounts of char; ash may become lofted in the plume to become a component of PM

Biomass: Organic material derived from plants or animals

Black carbon or soot: A component of PM derived from high-temperature flaming processes; black carbon is composed of aggregates of carbon-rich particles typically 20–40 nm in diameter

Char: The solid residue remaining after combustion; char generally refers to carbonaceous residues containing some minerals

Combustion efficiency (CE): The fraction of carbon in the fuel that is emitted as CO_2

Combustion factor (CF): The fraction of combustible material exposed to a fire that was actually consumed or volatilized

Emission factor (EF): The mass of a specific compound (or class of compounds) emitted per kilogram of dry material combusted, for a specified dry material or collection of materials

Emission ratio (ER): The ratio of the mass of a compound emitted to the mass of a reference compound that is conserved in the plume, often CO or CO_2; it is often reported as an "**enhanced**" ER where the background concentrations of the compound and reference compound have been subtracted

Emissions (or effluents): Species emitted into the air, water, soil, or other media, from a process; these are sometimes called releases

Enclosure fire: A fire contained within a room or compartment inside a building in which oxygen supply is typically constrained, contrary to open fires; these are sometimes called compartment fires

Energy content: The amount of energy contained within a mass of fuel; it can be quantified as the higher heating value (or gross calorific value), defined as the amount of heat released from complete combustion of a dry material when the products are returned to 25°C, or the lower heating value (or net calorific value), defined as the amount of heat released from complete combustion of a dry material initially at 25°C when water as a combustion product remains in the vapor state; other initial and final state temperatures may be found in the literature

Equivalence ratio (ϕ): The ratio of the actual fuel/oxidizer ratio to the stoichiometric fuel/oxidizer ratio; the stoichiometric fuel/oxidizer ratio is the ratio that is theoretically required to fully oxidize the fuel

Fine particulate matter ($PM_{2.5}$): Airborne particles with diameters of 2.5 micrometers or less, small enough to enter the lungs and bloodstream, posing risks to human health

Fire plume: Air mass downwind of combustion zone, containing elevated concentrations of combustion products

Flaming combustion: Luminous oxidation of gases evolved from the rapid decomposition of a solid biomass fuel

Glowing combustion: Incandescent heterogeneous oxidation of a solid biomass fuel in which all the volatiles have been driven off

Modified combustion efficiency (MCE): The measured, enhanced emission of CO_2 divided by the sum of the enhanced emission of CO and enhanced emission of CO_2; MCE is typically linearly correlated with CE and used as a proxy for CE since it is easier to measure

Oxidative pyrolysis: The thermal decomposition of a combustible material in the presence of molecular oxygen in the surrounding atmosphere

Particulate matter (PM): A complex mixture of solid particles and liquid droplets found in the air

Pollutant: A chemical or biological substance that harms water, air, or land quality

Plume injection parameters: The initial characteristics of a fire plume, including its injection altitude and multi-phase chemical composition

Pyrolysis: The thermal decomposition of a combustible material in the absence of molecular oxygen; the term "pyrolysis" sometimes appears in wildland fire literature to represent oxidative pyrolysis

Secondary organic aerosol: Organic particulate matter that is formed in the atmosphere from precursor gases

Semi-volatile organic compounds (SVOCs): Organic compounds that, based on their vapor pressure, tend to evaporate from the particle phase within near-field dilution of plumes; SVOCs are of concern because of their abundance in the indoor environment and their ability to accumulate and persist in the human body, the infrastructure of buildings, and environmental dust

Smoldering combustion: Combined processes of thermal decomposition and slow, low-temperature, flameless burning of porous solid biomass fuels; sometimes called glowing combustion

Toxic product yield: The maximum possible mass of a combustion product generated during combustion, per unit mass of test specimen consumed (typically expressed in units of grams per gram or kilograms per kilogram)

Urban fire: Fire that occurs primarily in cities or towns with the potential to rapidly spread to adjoining structures; these fires damage and destroy mostly homes, schools, commercial or industrial buildings, and vehicles

Volatile organic compounds (VOCs): Organic compounds with vapor pressures high enough to exist in the atmosphere primarily in the gas phase, typically excluding methane; VOCs can easily become airborne for inhalation exposure

MATERIALS

Many WUI fires occur in residential areas where the primary fuels are those in and around the home, which can include a diverse array of materials. Figure 3-1 shows some of these potential materials, although many others of different compositions may be present. There are numerous exterior and interior construction materials and furnishings with synthetic chemical compositions including textiles, insulations, paints and coatings, premanufactured woods, added flame retardants, antimicrobial and halogenated finishes, wall coverings, and sealants.

The quantities and compositions, cited in Figure 3-1 and elaborated on in the sections below, are illustrative examples. Structural materials may vary greatly based on the age of construction and geographic region. For example, some substances used in building materials or household goods have been banned or phased out, such as polybrominated diphenyl ether (PBDE) flame retardants, yet they may still be present in older household goods (ATSDR, 2017). Furthermore, the amount and composition of materials within the home, particularly chemicals of concern (such as flame retardants and phthlates), may be linked to race, ethnicity, and socioeconomic factors that are not well captured by the examples presented here. Black, Indigenous, and people of color and low-income groups are more likely to live in older homes in poorer condition. Given the lack of comprehensive surveys of materials in homes and how they may differ for various populations, other data sources, like chemical exposure studies, may provide some information on urban material chemical composition. Researchers have found higher exposures to some chemicals of concern (e.g., volatile organic compounds [VOCs], semi-volatile organic compounds [SVOCs], flame retardants, pesticides, lead) among these groups, demonstrating that the composition of homes and their contents likely vary by race and economic status (Adamkiewicz et al., 2011; Jacobs, 2011; Swope and Hernandez, 2019; Zota et al., 2010), although other factors such as occupational exposure may also play a role.

The unique elemental composition of the materials in the urban environment has a direct impact on the combustion chemistry and emissions described in the other sections of this chapter. Table 3-1 summarizes some of the most common building materials and their fire emissions (Blomqvist et al., 2013; Stec, 2017; Stec and Hull, 2010, 2011;

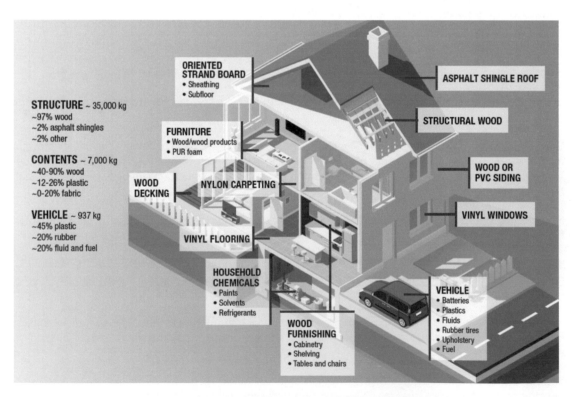

FIGURE 3-1 Examples of the types and quantities of materials that can be found in the home.

TABLE 3-1 Materials Used in Buildings and Their Common Fire Emissions

Material	Most Commonly Released Fire Emissions
Polyurethane foam in insulation	HCN, CO, NO, NO_2, NH_3, HCl, H_3PO_4, PM, PAHs, VOCs, SVOC, TCPP, TCEP, PCDDs, PCDFs, isocyanates
Polyisocyanurate foam in insulation	HCN, CO, NO, NO_2, NH_3, HCl, H_3PO_4, PM, PAHs, VOCs, SVOC, TCPP, TCEP, PCDDs, PCDFs, isocyanates
Phenolic foam in insulation	SO_2, CO, HCl, acrolein, formaldehyde, PM, PAHs, VOCs, SVOCs, TCPP, TCEP, PCDDs, PCDFs
Extruded polystyrene in insulation	HF, HBr, CO, PM, PAHs, VOCs, SVOCs
Glass wool in insulation	HCN, CO, NO_2, HCl, isocyanates
Oriented strand board (OSB)	HCN, CO, NO_2, HCl, acrolein, formaldehyde, PM, PAHs, VOCs, SVOCs, isocyanates
Vinyl siding and/or polyvinyl chloride (PVC) windows	HCl, CO, PCDDs, PCDFs
Upholstery on furniture	HCN, CO, NO, NO_2, NH_3, HCl, H_3PO_4, PM, PAHs, VOCs, SVOC, TCPP, TCEP, PCDDs, PCDFs, isocyanates
Vinyl carpet	HCl, CO, PCDDs, PCDFs
Polyamide carpet	HCN, CO, NO, NO_2, NH_3, PM, PAHs, VOCs, SVOCs, isocyanates
Electrical wiring insulation	HCl, CO, PCDD, PCDFs
Acrylic clothing	HCN, CO, NO, NO_2, NH_3, PM, PAHs, VOCs, SVOCs, isocyanates
Residential furniture	Benzene, toluene, formaldehyde, organophosphate flame retardants

NOTES: PAH = polycyclic aromatic hydrocarbon; TCPP = tris(1-chloro-2-propyl) phosphate; TCEP = tris(2-chloroethyl) phosphate; PCDD = polychlorinated dibenzo-p-dioxin; PCDF = polychlorinated dibenzofuran.

Stec et al., 2013). Table 3-1 relates the materials in the urban environment to the WUI fire emissions that impact the atmosphere, water and soil, and human health as described in Chapters 4–6.

Wildland Materials and Wood Products

The natural woody biomass found in the landscape surrounding WUI structures and wood products used in building construction are of similar macromolecular composition (cellulose, hemicellulose, and lignin). Bain et al. (2003, Appendix 3) list the energy content and chemical composition for a wide range of biomass materials, including wood from numerous tree species and other vegetative types. The higher heating values of different dry woods range from 19 to 20.2 MJ/kg. The heat of combustion per unit mass of biomass material is significantly impacted by the water content, which varies between live and dead woody biomass, biomass species, and season (NWCG, 2021). Wood used in building construction has a lower moisture content (typically less than 12 percent), depending on local ambient conditions (Forest Products Laboratory, 2010).

Biomass also contains 0.1 to 0.4 percent (by mass) nitrogen, 0.01 to 0.09 percent sulfur, up to 0.05 percent chlorine, and traces of other inorganic elements (Tejada et al., 2020). In general, ash levels remaining from the complete combustion of the volatiles and char generated in the pyrolysis/burning of woods are very low (0.01 to 0.07 percent), with only a small fraction of the original sulfur and nitrogen segregated in the char (Bain et al., 2003). Wildland fire ash consists of mineral materials and charred organic components with compositions dependent upon the original fuel type, combustion completeness, and combustion temperatures that the materials experience (Bodi et al., 2014).

In addition to woody biomass (live trees and snags [standing dead trees]), other fuels contribute to the emissions from wildland fires (Atchley et al., 2021). Live foliage in the canopy, dead foliage in the leaf litter, non-woody vegetation, lichen-moss layers, duff layers, and organic soil may be consumed in wildland fires (Ottmar, 2014; Ottmar and Baker, 2007). These other fuel classes like leaf litter, grasses, and shrubs sustain fire spread and are important sources of emissions. For example, in a mixed conifer forest in Oregon, researchers estimated that 33 percent of the biomass consumed in a high-severity scenario could be attributed to the litter and duff layers alone, and 35 percent was attributed to standing trees and snags (Campbell et al., 2007, Table 5). Additionally, duff layers and downed woody debris tend to smolder and can generate far greater emissions per unit mass consumed than litter, shrubs, and tree crowns that are consumed in more intense combustion (Urbanski, 2014).

Vegetative biomass loading in the environment is highly variable in both space and time as vegetation grows, is impacted by insect, disease, and climate conditions, and eventually decays. Biomass loadings can vary greatly among ecosystems; the highest biomass loadings are in wet forests, but only materials that are dry enough to ignite are considered available biomass and thus able to be consumed during a wildland fire. Some of the highest available biomass loadings are in dry forest ecosystems that are common in the western United States (Hessburg et al., 2019). A dry forest can have as much as 70 Mg/ha or about 28,000 kg/acre of available biomass (Ottmar, 2014). The available biomass can also be modified by land management practices, like prescribed fire. About 13 tons/ha (5,000 kg/acre) is consumed in prescribed fire in temperate forests (van Leeuwen et al., 2014).

Urban Materials

The materials in the urban environment differ greatly from the materials in the wildland environment in their amount, elemental composition, energy content, and arrangement in the landscape. Materials unique to urban areas, like polyvinyl chloride or polyurethane, contain much larger amounts of chlorine or nitrogen than biomass and these elements are known to impact combustion chemistry. Though the emissions generated from urban and wildland areas differ, in a WUI fire, they intimately comingle in the plume. In addition, the differing energy density in urban areas compared with wildlands may impact combustion conditions, modifying the timing of emissions and the physical nature of the WUI plume in comparison to the plume of a surrounding wildland fire (Trelles and Pagni, 1997).

In WUI fires, many of the structures burned are residential homes, as seen with the recent WUI fires described in Chapter 2:

- 74 percent of the approximately 18,000 structures destroyed in the Camp Fire (Wallingford, 2018)
- 99 percent of the 344 structures destroyed in the Waldo Canyon Fire (Maranghides et al., 2015)
- 99 percent of the 1,091 structures destroyed in the Marshall Fire (Boulder OEM, 2022)

Large WUI fires—those sweeping through entire communities—may also involve municipal service systems, commercial and industrial buildings, and commercial vehicles and their fuel. These commercial and industrial structures and other components of urban infrastructure likely have material loadings and compositions different from those of residential buildings. They may be composed of more noncombustible materials (e.g., concrete and steel), but may have much higher internal fuel loadings and greater amounts and different chemicals of concern. Thus, they can pose risks different from those of WUI fires confined to residential areas. Specific mappings of commercial and industrial structures and their fuel loadings and chemicals of concern are not broadly available.

Characteristics of Residential Structures

Most houses in the United States are built from wood framing and sheathing covered by an exterior of brick, stone, cement board, or vinyl siding (USCB, 2020). Sheets of fiberglass or foam installations are often added into the building cavities. The roof consists of sheets of plywood to provide the decking for exterior materials such as asphalt tiles, clay tiles, metal sheets, or even fiberglass shingles.

Limited survey data exist on the detailed material composition of residential structures, and even less information exists on elemental composition. Life-cycle analysis tools like the Greenhouse gases, Regulated Emissions, and Energy use in Technologies (GREET) life-cycle model could be used to estimate the mass and elemental composition of buildings (Cai et al., 2021). At present there is no complete accounting of both the fuel load and elemental composition of currently existing structures.

Table 3-2 provides examples of combustible materials loadings and other data for single-family homes built in different years. The committee derived the data in this table from engineering calculations; the assumptions employed in the calculations can impact the results of the analysis and are described in detail in Appendix E. The data represent a snapshot of structure characteristics for a specific location and year of construction, characteristics

TABLE 3-2 Materials Used to Construct Two Single-Family Residences

	The Fuel of Our Homes – from Building Materials to Content (Messerschmidt, 2021)		Analysis of the Lifecycle Impacts and Potential for Avoided Impacts Associated with Single Family Homes (EPA, 2016)	
Year built	2020		1998	
Livable area in ft^2 (m^2)	2,016 (187.3)		2,150 (153.2)	
	Combustible mass in kg (% total combustibles)	Energy content in GJ (% total energy)	Combustible mass in kg (% total combustibles)	Energy content in GJ (% total energy)
Structural wood	15,200 (58%)	281.2 (49%)	13,159 (43%)	243.4 (42%)
Subfloor (OSB or particle board)	3,844 (15%)	71.3 (13%)	10,115 (33%)[a] (OSB panels)	187.5 (33%)
Sheathing	126 (0.5%) (foam board)	3.2 (1%)	-[a]	-
Siding	300 (1%) (PVC siding)	3.0 (1%)	3,485 (12%) (wood siding)	64.5 (11%)
Insulation	990 (4%) (spray polyurethane foam)	25.3 (4%)	183 (1%) (fiberglass)	5.2 (1%)
Roof decking (OSB)	2,108 (8%)	39.1 (7%)	2,393 (8%)	44.4 (8%)
Roof exterior (asphalt shingles)	3,630 (14%)	145.9 (26%)	746 (2%)	30 (5%)
Windows	-		192 (1%)	1.9 (0.3%)
Total combustibles	26,198	569.1	30,272	576.9

NOTE: OSB = oriented strand board.
[a]Includes sheathing.

that vary significantly over time and location. An important limitation of these examples is the lack of information on some components (e.g., air-conditioning fluids, plastic gas piping, plumbing, wire insulation, and electrical materials), which are a small fraction of the mass but may be a concentrated source of species that impact combustion or contribute to hazardous effluents.

Wood-based materials are the largest fraction of combustible mass in the examples shown here; wood and engineered wood products can be derived from a variety of species and can exist in many forms. Engineered wood products, such as oriented strand board (OSB), particle board, and plywood, contain adhesives and fillers that can account for up to 15 percent of the product by mass (Weyerhaueser, 2018). The compositions of the adhesives have changed over time; originally, they contained formaldehyde, but now most adhesives are composed of methylene diphenyl diisocyanate (Blomqvist et al., 2013; Pokhrel et al., 2021; Sandberg, 2016).

Additionally, wood products may be chemically treated to prevent rot, protect against attack by insects or microbes, or confer fire-retardant properties. The use of a treatment depends on the application, with most treatments used in exterior materials. The prevalence of these treatments in structural materials and the particular chemicals used have varied over time. For example, a combination of chromium, copper, and arsenate was commonly used in outdoor wood products (e.g., decks, playgrounds) to protect against microbial attack but was phased out and replaced with alkaline copper quaternary. Flame retardants are increasingly used in exterior and interior structural materials (Lowden and Hull, 2013), with the types of compounds evolving over time (Popescu and Pfriem, 2019). Additives can account for up to 10 percent of the mass of the treated wood, with exact amounts difficult to determine; safety data sheets provide a wide range of mass fractions. Additionally, specific compositional information is not always provided. Different treatments and/or additives can impact combustion properties and ultimate emissions into the environment. For that reason, treated wood cannot be discarded through regular municipal waste in California because of the hazardous nature of the treatments (CalRecycle, 2020).

The composition of the other components of a structure, such as roofing, siding, and insulation, can vary depending on the manufacturer of the product. The amount and type of the materials used also varies by type of home

---

BOX 3-2
How Construction Materials Vary with Region

Building materials vary by geographic location. Some information about construction materials is available from the US Census Bureau, which conducts an annual Survey of Construction and publishes data tables summarizing characteristics of newly constructed housing in each census division.

An analysis of the 2009 data set by the National Association of Home Builders showed that, while vinyl siding was used in a national average of 36.2 percent of homes, regional use of vinyl siding ranged from a low of 8.2 percent in the "West South Central" division (Texas, Oklahoma, Missouri, and Louisiana) to a high of nearly 80 percent in the Middle Atlantic (New York, New Jersey, and Pennsylvania; Emrath, 2010). These regional trends can be strongly impacted by local building codes or environmental conditions. For example, while wood framing is used in most new homes across the United States, masonry or concrete homes account for nearly 20 percent of homes built in parts of the South to protect from floods or high winds and hurricanes (USCB, 2020).

Although Tables 3-1, 3-2, and 3-3 present some examples of the types and quantities of materials found in the built environment, it is important to note that factors such as geographic location and year of construction can introduce variability. In addition to the variability in material types, the elemental composition of construction materials (e.g., structural wood and insulation materials) can vary due to age, location, manufacturer, differences in manufacturing and treatment processes, and other factors.

(e.g., apartment, manufactured home), geographic location, consumer preferences, common building practices, and building codes at the time the residence was built (Box 3-2). For example, exterior finishing has varied substantially over time, with brick and wood exteriors much more common in home construction through the 1970s and 1980s, but with vinyl siding becoming the dominant exterior in new home construction after 1990 (USDOC, 2010).

These changes in consumer and builder preferences can alter the amount of combustible materials, their chemical composition, and the potential for release of chemicals of concern. For example, the transition from wood-based exteriors to asphalt roofing and polyvinyl chloride (PVC) siding changes the combustible mass, elemental composition, and energy content of homes and the nature of any combustion emissions. Similarly, the transition from vermiculite insulation (which may contain asbestos) to blow-in cellulose or polyurethane spray foam changes the amount of combustible materials that will yield emissions of chemicals of concern.

As building codes and practices change over time, the distribution of building ages results in a wide range of housing materials used across the United States, and potentially in WUI areas. Additionally, the frequency and extent of home renovations are not well known (Dixit et al., 2012; Ghattas et al., 2016). Building components like roofing or siding are periodically replaced or repaired, while other parts like the interior framing may remain the same over the lifespan of the building (EPA, 2016). The long lifetimes of houses limit the impact of building codes introduced in the last three decades that mandate the use of ignition-/fire-resistant materials to reduce the risk from wildland fires, leading to regulatory initiatives to promote retrofits for existing homes (State of California, 2019).

Characteristics of Materials in the Home

When exposed to fire, the contents of a home influence combustion conditions and emission composition. The materials inside buildings are typically characterized through fire load surveys, where fire load is the total amount of heat released from complete combustion of the contents of a structure. Physical inventories, questionnaires, and digital surveys have all been used to quantify fire load (Elhami-Khorasani et al., 2020). Fire load assessments date back to the 1940s. They have been done in various countries and do not exhibit trends by country or over time (Xie et al., 2019). However, fire load density (fire load per unit floor area) varies with residence size, with larger areas tending to have lower densities. The average fire load density for residential buildings is 645 ± 212 MJ/m^2 (see Table 2 in Xie et al., 2019).

TABLE 3-3 Typical Composition of the Moveable Fire Load in Residential Buildings

Material	Swedish Model Residences (Blomqvist and McNamee, 2009, Table 5)	Canadian Kitchen (Bwalya et al., 2010)	Canadian Bedroom (Bwalya et al., 2010)
Wood/Paper	79.2%	86.5%	42.3%
Plastics	12.3%	13.5%	26.4%
Fabrics	4.3%	0%	31.4%
Other	4.3%		

Table 3-3 summarizes the bulk composition of a typical home's contents. Only a limited number of studies have evaluated total interior fuel loadings for residences in North America. One study found that the main bedroom has the highest fire loading, followed by the kitchen, living room, basement living space, secondary bedrooms, and dining room (Bwalya et al., 2010). The fire loading density is the greatest in the kitchen and bedrooms, driven by the large amount of wooden cabinetry, furniture, and clothing contained in relatively small spaces. Bwalya et al. (2010) found that wood flooring, cabinets, tables, and display units contributed to the wood loading. Plastics were found in carpeting, foam cushioning, and synthetic fabrics. The fabric or textile content was due to clothing.

Characteristics of Vehicles

Just as it is possible to characterize the main constituents in residential structures as shown in Table 3-3, it is also possible to do so for vehicles. Table 3-4 reports the main components of an average modern light-duty vehicle built in 2017. An average vehicle weighs 1,793 kg and is composed of ~75 percent noncombustible materials and ~25 percent combustible materials by mass. The largest contributions to the combustible mass are plastic components, fluids and lubricants, and rubber. The noncombustible mass consists primarily of various metals and a small amount of glass. The average age of light-duty vehicles in use is 11.9 years, and the combustible composition has remained nearly constant over the past several decades (Davis and Boundy, 2021). The increasing percentage of trucks in the light-duty fleet has led to a slight increase in overall vehicle mass over time (~7 percent greater in 2017 versus 1995). A slight reduction of the noncombustible mass has also occurred, from 81 percent in 1995 to 76 percent in 2017.

Although reliable data exist on the bulk material composition of vehicles, the elemental composition of these materials is not as well known but may be estimated using life-cycle analysis tools (e.g., GREET; Cai et al., 2021). Moreover, the material composition of vehicles will change as electric vehicles become more common. Hybrid electric vehicles accounted for 3.2 percent and battery electric vehicles accounted for 1.7 percent of new light-duty vehicle sales in 2020 (Davis and Boundy, 2021, Table 6.2), with both fractions of these vehicle types expected to grow.

Battery electric vehicles are generally heavier than internal combustion engine vehicles due to the additional battery and associated power systems (Wang et al., 2021). Internal combustion engine and hybrid electric vehicles have gasoline and other flammable liquids. These liquids have high energy densities and may contribute a sizable portion of energy content that is not present in a battery electric vehicle (Table 3-4).

Batteries may be a potent source of compounds of concern that could be liberated during a WUI fire or become a source of toxic combustion products. Most vehicles contain a lead acid battery, which varies from 6 to 16 kg depending on the type of vehicle and is ~70 percent lead by mass (Wang et al., 2021). The nickel metal hydride battery sometimes found in electric vehicles can weigh as much as 1,270 kg and is ~28 percent nickel by mass (Wang et al., 2021). These metals may be released into the environment during a WUI fire. Lithium-ion batteries, also used in electric vehicles, can emit fluorinated gases upon heating and during combustion (Larsson et al., 2017; Sturk et al., 2019). The composition of a lithium-ion battery varies depending on the physical design and the battery chemistry (Elgowainy et al., 2016). The most common lithium-ion battery chemistry uses a lithium cobalt oxide cathode, graphite anode, and lithium hexafluorophosphate electrolyte. However, many different types of battery chemistries are used to achieve various performance features, and it is likely that battery compositions and vehicle designs will continue to evolve, along with their overall contribution to emissions from WUI fires.

TABLE 3-4 Mass and Energy Contributions to an Average Light-Duty Internal Combustion Engine Vehicle, Model Year 2017

Material	Mass in kg (% of combustible mass)	Energy Content in GJ (% of total energy content)
Regular steel	554	
High- and medium-strength steel	347	
Stainless steel	33	
Other steels	14	
Iron casting	110	
Aluminum	189	
Magnesium castings	4	
Copper and brass	31	
Lead	17	
Zinc castings	4	
Powder metal parts	20	
Other metals	2	
Glass	43	
Plastics and plastic composites	155 (34%)	6.7 (43%)
Rubber	93 (20%)	3.1 (19%)
Coatings	13 (3%)	0.3 (2%)
Textiles	21 (5%)	0.6 (4%)
Fluids and lubricants	101 (22%)	2.5 (16%)
Other materials	42 (9%)	1.0 (7%)
Fuel[a]	36 (8%)	1.5 (9%)
Combustible portion	461	15.7
Total mass	3,953	

[a]Amount of gasoline assumed to be 36 kg (13 gallons), corresponding to the default value in the GREET life-cycle analysis (Cai et al., 2021).
SOURCE: Davis and Boundy, 2021.

Persson and Simonson (1998) assessed that 9 percent of the plastic material of an automobile was PVC (containing chlorine) and 17 percent was polyurethane foam (containing nitrogen). Measurements of gas-phase emissions from vehicle fires have shown that compounds with a potentially adverse health impact on humans are produced in significant concentrations: Fent and Evans (2011) carried out a series of vehicle fires and measured air concentrations of formaldehyde, acrolein, and isocyanates. They estimated that personal exposures to these compounds were nearly 10 times the acceptable levels on an additive effects basis. In addition, electric vehicles and Li-ion batteries can be the source of hydrogen chloride (HCl) and hydrogen fluoride (HF), which have more severe health effects than asphyxiant gases (Fent and Evans, 2011). Other fluorine-based compounds, such as POF_3 and COF_2, have not been reported in significant quantities from fire tests with batteries. Table 3-5 presents examples of fire emissions from car components (Larsson et al., 2017; Lönnermark and Blomqvist, 2006; Willstrand et al., 2020).

These studies of vehicle fires also showed that many chemicals (e.g., polycyclic aromatic hydrocarbons [PAHs] and metals) partition into the particle phase, in which form they can be inhaled deep into the lung. Table 3-6 presents an example of the elements identified in PM from a vehicle fire (Lönnermark and Blomqvist, 2006). Such data need to be carefully examined, as metal composition differs among different vehicle and battery types (Larsson et al., 2017; Lönnermark and Blomqvist, 2006; Willstrand et al., 2020).

TABLE 3-5 Fire Emissions from Car Components

Car Component	Fire Emissions
Door panel	CO, HCl, HCN, NO, PCDDs/PCDFs, VOCs, PAHs, isocyanates
Ventilation system	CO, acrolein, formaldehyde, PAHs, VOCs
Floor material	CO, HCN, isocyanates, PAHs, VOCs
Dashboard	CO, HCN, NO, isocyanates, PAHs, VOCs
Upholstery material	CO, HCN, NO, HCl, SO_2, isocyanates, PAHs, PCDDs/PCDFs, VOCs
Electrical wiring	CO, HCl, PAHs, PCDDs/PCDFs, VOCs
Tire	CO, SO_2, PAHs, VOCs

Chemicals in Consumer Products

Various data sets can provide some information on chemicals of concern in consumer products in the urban environment. These data sets have been developed for purposes like chemical exposure assessment, prioritizing of chemical toxicity screening, life-cycle analysis, or green building and design. Some data sources for building products, like Pharos (https://pharosproject.net/) and Building for Environmental and Economic Sustainability (Suh and Lippiatt, 2012), were designed to identify green alternatives or provide information for life-cycle analysis. Some data sources were designed for chemical exposure assessment, for example the Chemical/Product Categories Database (Dionisio et al., 2015) and the OrganoRelease framework (Tao et al., 2018). However, many unknowns about the composition of consumer products still exist. For example, a non-targeted analysis of consumer products by Phillips et al. (2018) measured numerous compounds in consumer products, of which over 85 percent were not listed in any chemical database. These non-targeted analyses along with surveys of consumer products and household exposure measurements demonstrate the difficulty of identifying chemicals of concern in the urban environment (Li and Suh, 2019).

In addition to chemicals of concern in current use, many chemicals have been phased out but still exist in the urban environment due to legacy use (e.g., lead, polychlorinated biphenyls) and varying abatement practices. Weschler (2009) examined major trends in consumer products and building materials since 1950 and noted that usage of chemicals of concern varied considerably over time, and that historical usage was difficult to quantify. Many factors may impact material choices in the home, but the identification of health, environmental, or fire hazards were important drivers of usage trends (Weschler, 2009; Cooper et al., 2016).

TABLE 3-6 Elements in PM Emitted from a Vehicle Fire

Metal	Mass Concentration in PM (mg/kg)
Cadmium	26
Cobalt	5
Chromium	59
Copper	430
Nickel	44
Lead	12,800
Antimony	230
Thallium	80
Zinc	50,300
Fluorine	510
Chlorine	39,000
Bromine	4,000

SOURCE: Lönnermark and Blomqvist, 2006.

Effects of Structure Density on WUI Fire Behavior

In addition to the materials that make up WUI structures, the density of the structures themselves can affect WUI fire behavior. The material loading in the landscape is a key parameter since it can impact the fire spread, fire intensity, and quantity of emissions produced. Additionally, the ratio of urban structures and vehicles to vegetative materials combusted in a WUI fire will impact the combustion chemistry and the resulting amount and type of pollutants released into the atmosphere as well as those impacting air, soil, and water quality (see Chapters 4 and 5 for further details). However, the number of houses involved in a fire can vary greatly from one fire to the next, as can the amount of biomass loading and the wildland acreage burned (CAL FIRE, 2022). And, very little is known about the number of vehicles destroyed in WUI fires, as they are not typically quantified in damage assessments (Kuligowski, 2021).

Interface WUI areas (those at the edge of a large area of wildland) can have a high density of houses, and when these areas are exposed to wildland fire, large numbers of structures can burn and may have fire behavior and combustion emissions very different from WUI fires with less structures. The Marshall Fire in 2021 (estimated 1,081 structures and approximately 6,200 acres burned) and the Tunnel Fire in 1991 (estimated 2,900 structures and 2,000 acres burned; CAL FIRE, 2022) are two such examples, where relatively small grassland fires moved into densely populated areas, and the loading of urban fuels greatly exceeded the loading of wildland vegetation.

Researchers have used models and WUI fire observations to better understand how a high structure density in interface WUI areas impacts the rate of spread and the heat release rate of a fire in an urban area in comparison to the surrounding wildland fire. For example, the high density of combustible materials in a structure in comparison to surrounding wildlands increases the duration of the fire and may influence the spread of the fire (Maranghides et al., 2013, 2015). Burning structures can also generate fire-induced winds that impact the rate of spread, as observed in the Tunnel Fire, where the spread rate slowed when the number of structures involved in the fire increased (Trelles and Pagni, 1997). These examples contrast with WUI fires in intermix areas (alternating areas of housing and wildland), where the structure fuel loading is much lower (it may be comparable with that of vegetative fuels) and may have less of an impact on fire behavior.

Rate of spread is often assessed for WUI fires, but few studies have specifically looked at the impact of structure fuel loading, although it is likely an important factor in fire spread (Masoudvaziri et al., 2021). Additionally, the heat release rate from WUI fires has not been comprehensively evaluated but will likely vary with the differing fuel composition and density in WUI fires compared to wildland fires (Simeoni et al., 2012). Rate of spread and heat release rate can impact combustion conditions, the timing of emissions, and the plume dispersion in the atmosphere. It is currently not known how the loading of urban fuels will impact these characteristics in WUI fires.

Factors Impacting Structure Destruction in a WUI Fire

The number of structures that will burn when exposed to wildland fire is highly variable. Many factors can impact the likelihood of destruction, and the significance of each factor, or combination of factors, is an active area of research.

Environmental conditions coupled with the topography of a WUI area can drive extreme fire behaviors that result in destructive WUI fires (Keeley and Syphard, 2019). The most destructive fires have occurred in the western United States during dry, downslope wind events (e.g., the Santa Ana or Diablo winds) and in terrain that channels fire-induced winds, increasing the probability of spot fires igniting ahead of the main fire front (Keeley and Syphard, 2019). Strong winds and dry conditions are common features of WUI fires across the United States (e.g., Maranghides and McNamara, 2016; the example fires described in Chapter 2).

The number of structures present and their arrangement in the landscape are important factors (Syphard et al., 2019). Areas with a high density of combustible structures may experience greater structure-to-structure spread (Syphard et al., 2021, and references therein), and the increased combustible load may lead to more extreme fire behavior (Keeley and Syphard, 2019). However, fire spread can be mitigated in communities using fire-resistant materials and reduced combustible vegetation around the home. In areas where WUI fires are more frequent, regulations have been implemented to prevent the spread of fire (Philson et al., 2021).

Finally, defensive action by firefighters can reduce destruction and potentially limit the spread of a fire within the urban area. However, it is difficult to quantify the impact of such defensive actions and how important they may be in relation to the other characteristics of WUI structures and their surroundings.

Finding: Limited data exist on the composition of residential building materials and the materials within residences; the composition of these potential WUI fuels will be difficult to ascertain from existing fuel loading records, especially for important species such as halogens, phosphorous, and metals. WUI fire fuel loadings are complex and depend on the density of structures and landscape vegetation, the use of fire-resistant building materials, the number and type of vehicles left behind by evacuees, and defensive actions taken during the fire.

Research need: More information is needed on the amount, type, and chemical composition of consumer products in residences at the WUI, including legacy materials and how these may vary with the age and condition of the home. Data are also needed on the amount and arrangement of urban structures, vehicles, and their surroundings.

Changing Materials at the WUI

As experience with WUI fires grows, some jurisdictions are changing building codes, which in turn will change the composition of structures and the wildland at the WUI. Table 3-7 describes actions that are currently used or proposed for use in WUI areas. For example, California building codes such as Chapter 7A of the California State Building Code (State of California, 2016b) or the National Fire Protection Association *Standard for Reducing Structure Ignition from Wildland Fire* (NFPA, 2018) describe building practices such as hardening of the home using Class A ignition-resistant roofs, enclosed eaves, multipane glass, and the use of noncombustible materials or ignition-resistant materials on the structure exterior. Many areas also have requirements to maintain defensible space by limiting the amount of combustible materials (e.g., landscape vegetation, leaf litter, wood stacks) near the home and reducing the horizontal and vertical distribution of fuels in areas surrounding the house that may promote the spread of a fire.

TABLE 3-7 Current and Potential Approaches to Modify Materials to Reduce Risk of WUI Fire

Applicable Area	Actions
Wildlands	• Use land management approaches to reduce fuel loading or otherwise modify fuel characteristics; approaches include grazing, prescribed fire, thinning, or other biomass reduction programs (USFS, 2022)
Community	• Set the community back from slopes where fire risk may be greater (CDLA, 2016) • Cluster ignition-resistant housing to reduce "collective exposure" (Moritz and Butsic, 2020) • Maintain protective buffers around communities (e.g., agricultural lands, irrigated parks, or other breaks in flammable fuel; Moritz and Butsic, 2020)
External to the structure	• Maintain a defensible space around structures by removing dead or dying vegetation, replacing vegetation with noncombustible materials (i.e., a hardscape), and maintaining space in between trees and between trees and the home (e.g., State of California, 2020; CAL FIRE, n.d.) • "Harden" the home by using fire-resistant materials such as stone, brick, glass, concrete, gypsum, stucco, and cement board, as well as ignition-resistant building practices, such as closing eaves and installing dual-pane windows and exhaust screens (e.g., State of California, 2016a,b)
Inside the structure	• Use interior furnishings with fire-resistant materials and limit the use of synthetic products based on petrochemicals that can be fuel sources • Avoid upholstered furniture, which is often the first item ignited by open flames, and instead use furniture with a fire-barrier material (with no added flame retardants, which can be detrimental to human health) between the cushioning and exterior textile (Zammarano et al., 2020) • Choose materials that are inherently fire resistant so that flame retardants are not used and human exposure is prevented (IFSTA, 2016)

In addition to state and local regulations, initiatives like the Firewise Communities (NFPA https://www.nfpa.org/Public-Education/Fire-causes-and-risks/Wildfire/Firewise-USA) have been established to help WUI communities identify mitigation strategies and encourage residents to take actions to protect their homes. Additionally, organizations such as the National Fire Protection Association and the Insurance Institute for Building Home and Safety have developed policy recommendations aimed at reducing risk (https://www.nfpa.org/outthinkwildfire and https://ibhs.org/wp-content/uploads/2019/05/wildfire-public-policy.pdf, respectively). A summary of the recommended policies and mitigation steps from these resources and others described in subsequent text is provided in Table 3-7.

Prescribed fire is widely used in many parts of the United States, often for ecosystem health as well as reducing the loading of fine combustible biomass (grasses and shrubs). Grazing in shrub- or grassland-dominated areas reduces the fine biomass load and thus fire severity (Colantoni et al., 2020; Davies et al., 2015). Close to urban areas, mechanical treatments like tree and brush removal can reduce the combustibles immediately adjacent to developments. These approaches may have limited effectiveness, especially in shrub and grassland regions where vegetation regrows quickly, so additional actions are needed to reduce WUI fire risk (Schoennagel et al., 2017).

A recent literature survey and compilation of recommendations in California (Moritz and Butsic, 2020) highlighted community-scale risk reduction measures to help protect homes and lives from WUI fires in new developments. One example of a measure that may or may not reduce collective fire risk is clustering structures. The design, maintenance, and use of defensible space for fire protection is improved when neighborhoods are developed more densely and are built with stringent fire-resistant building codes; however, clustering may contribute to high fuel loadings. Wildland fire–resistant construction techniques to harden all the homes in a community can be used to minimize risk and protect homes from wildland fires with minimal additional cost (Quarles and Pohl, 2018); after a community or subdivision has been developed, the responsibility to mitigate risk by modifying the environment falls on homeowners, whose resources for retrofitting structures and maintenance of defensible space may be limited.

In existing communities, the approaches to mitigate risk from WUI fires focus on using ignition-resistant building materials, reducing fuels next to structures, and maintaining defensible space. In some areas, state laws or local ordinances are used to ensure that property owners maintain defensible space around structures. Changes to the building structure are implemented largely through renovations and replacements with ignition-resistant materials to align with current building codes. Proposed California legislation would more widely encourage low-cost retrofits, require disclosure of fire safety retrofits during the sale of existing homes, and provide financial assistance for retrofits in an attempt to protect vulnerable communities that may not otherwise have the means to implement mitigation measures (State of California, 2019).

The effectiveness of any of these mitigation actions at preventing structure loss during a WUI fire has yet to be fully understood, and assessments of previous WUI fires show that many factors may play a role in structure survival (Syphard et al., 2021). A survey of some severe WUI fires in California showed that the impact of mitigation actions was only modest (Syphard and Keeley, 2019) and no single category explained more than 25 percent of the likelihood of survival (Syphard et al., 2021). The most effective actions were using covered eaves, multi-paned windows, and screened vents to reduce ignition risk (Syphard and Keeley, 2019). Maintaining defensible space was less effective, showing the significance of ember ignitions, since embers can travel as far as 1 km in wind-driven fires (Keeley and Syphard, 2019).

Mitigation measures to the structure and its surroundings may not be sufficient to reduce risk in some fire-prone landscapes. Keeley and Syphard (2019) noted landscape features that were common in some destructive fires. For example, developments on the edges of canyons or within a wind corridor may be particularly vulnerable. Additionally, some California coastal areas, despite repeated fire, have rapid regrowth of vegetation and seasonal meteorology that create hazardous conditions (e.g., dry fuels and strong winds) that lead to destructive fires. Some planning ordinances require the siting of developments to avoid areas of high fire risk, although this approach to mitigate risk is rarely used (Moritz and Butsic, 2020).

Finding: A variety of measures, from managing the combustible materials in the urban environment and surrounding landscape to using fire-resistant materials and structural designs, have been pursued to mitigate WUI fire ignition and spread. The collective effectiveness of the approaches and their associated costs remain uncertain.

COMBUSTION

As a result of the fuel loading and material composition of urban dwellings and infrastructure, intermixed with fuel loadings and compositions from wildlands, the chemistry of fires at the WUI is more complex than that of wildland or urban fires. However, the physics of WUI fires, wildland fires, and urban fires remain similar. For solid combustibles, all progress through slow and fast decomposition stages to emit volatile, semi-volatile, and particulate combustibles; wildland, urban, and WUI fires can all experience a variety of different types of combustion conditions, and leave residual char and noncombustible materials. Once in the gas phase, combustible materials can further react in flaming combustion and/or in hot regions without flames that support chemical oxidation reactions producing soot precursors, PAHs, and carbon-containing particulates. Depending on the chemical composition of the materials, collectively and through various reactive extinction and quenching effects, these processes lead to emissions of VOCs, species containing halogens, nitrogen, sulfur, flame-retardant species, organometallics, metals and metal compounds, and carbon-containing particulates.

The spatial and temporal distribution of ignition, evolution, and propagation steps of all large-scale fires depends on the spatial distribution of combustible materials, meteorology, and interactions with fire-induced (i.e., buoyancy-induced) convection and terrain features. Ignition processes and fire spread result from local radiative and convective heat transfer, as well as the spot fires produced by the projection of firebrands (i.e., "glowing" combustion fragments that can drift through the air and cause flaming combustion when they encounter more fuel). Firebrands carried by turbulent fluid dynamics can ignite combustible materials, starting spot fires that are as much as 5 km or more ahead of the major propagation line (e.g., Koo et al., 2010; Maranghides et al., 2021). Predicting and mitigating spot fire ignition from firebrands remains a subject of continuing research (Manzello et al., 2020; Suzuki and Manzello, 2021). The overall heat release and interactions of urban fire emissions with those from surrounding vegetation (which include in-soil, ground level, mid-height, and canopy contributions) will influence the fire plume injection altitude and composition. The fire's burning rate together with these plume injection parameters will affect the downstream plume aging and secondary pollutant generation (Atchley et al., 2021; Hallquist et al., 2009), as discussed in detail in Chapter 4.

This section describes the following:

- The nature of premixed and mixing-limited combustion behavior, which determines the temperatures, chemical kinetics, and heat release associated with fires fueled by condensed phase materials (biomass and human-made products) and liquid combustibles
- The chemical pathways and radical species that dominate combustion chemistry, and how these pathways are influenced by the presence of human-made materials, particularly halogen-, nitrogen-, and metal-containing species
- The factors impacting the interaction of the fire plume with surrounding ambient air, eventually leading to buoyant plume rise and longer characteristic reaction times, changing plume composition and plume injection characteristics.

Premixed and Mixing-Limited Combustion

Flaming combustion occurs under two distinctly different limiting regimes: (1) fuel and air fully mix prior to combustion, termed "premixed" combustion, and (2) fuel and air are initially separate, but react in a region where they simultaneously mix, termed "mixing-limited" or diffusively limited combustion (diffusion flames). Mixing-limited combustion is characteristic of burning solids and liquid fuel sprays that vaporize as they burn, but understanding premixed combustion is important for describing the gas-phase combustion behaviors involved in ignition processes and the "partially premixed" turbulent flame structures typical of fire plumes (Tieszen and Gritzo, 2008).

Mixing-limited flames occur in both wildland and WUI fires. An example is a stabilized, mixing-limited flame held in the air flow around a small-diameter tree branch, from which flammable fuel concentrations are produced as a result of radiative heating from the surrounding fire and surface reactions. Another example is a stabilized, open-air, turbulent, mixing-limited flame generated by the rapid mixing of flammable gases and aerosol exiting from an enclosure fire, such as through a broken window of a burning structure. Premixed and partially premixed flames also occur in both wildland and WUI fires, at times in complex spatial relationships with mixing-limited flames, such as a burning, convecting cloud within a rising fire plume.

For both premixed and mixing-limited combustion, the chemistry will be impacted by the amount of oxidizing species. The fuel-to-oxidizer ratio of any unreacted fuel and oxygen can be referenced against that which is theoretically required to fully oxidize the fuel. The equivalence ratio, ϕ, is defined as

$$\phi = \frac{\frac{[Fuel]}{[Oxidizer]}}{\left(\frac{[Fuel]}{[Oxidizer]}\right)_{Stoich}} \quad \phi > 1 \text{ Fuel Rich}; \ \phi < 1 \text{ Fuel Lean}$$

where the brackets denote molar concentration units. In any flame, the combustion heat release and therefore the maximum theoretical flame temperature correspond to stoichiometric conditions (i.e., $\phi = 1$).

For premixed flames, the maximum theoretical flame temperature is defined by the premixed value of ϕ. Lean and rich flammability-limit values of ϕ exist that bracket the values for which premixed flames can exist. The lean limit for laminar premixed hydrocarbon flames is typically between 0.5 and 0.55 for air, but the rich limits vary substantially with the fuel's chemical composition (Glassman et al., 2014). Chemical reactions can proceed outside these flammability limits, if the gas temperatures are sufficiently high, but no flames will be present. For mixing-limited flames, the mixing of fuel and air as burning occurs results in a flame structure with a range of equivalence ratios spanning from very rich values to very lean values. The peak mixing-limited flame temperature occurs where fuel and oxidizer meet in stoichiometric proportions, and the rate of overall reaction within the flame depends on the rate of fuel/oxidizer mixing.

Flame temperatures achieved in premixed and diffusion flames, in turn, have a complex impact on rates of combustion reactions and on species that result when a quenching or extinction of a flame structure occurs. In premixed flames, dilution of the oxidizer by nitrogen (air) and/or by the potential recirculation of combustion products into the flaming region reduce the peak flame temperature and affect flammability limits. Significant dilution or fluid dynamic interactions of premixed as well as diffusion flames lead to quenching/extinction, resulting in incomplete conversion to final combustion products.

A common belief is that overall chemical oxidation timescales for flammables containing hydrogen, carbon, and oxygen in the gas phase, whether aerosols or vapor, decrease exponentially with increasing temperature. Often this is not the case, especially for species with a large carbon number (carbon number > 4 for alkane structures; carbon number > 8 for aromatic structures). In these cases, the overall rate of oxidation is controlled by the complex interactions of numerous elementary reactions. These reactions each have unique chemical thermodynamic and elementary rate parameters, which in concert change the governing reaction pathways, the likely intermediate species, and the most impactful active radicals (e.g., see Curran, 2019; Kohse-Hoinghaus, 2021; Wang et al., 2019). As a result, the global oxidation reaction rate may increase or decrease with increasing temperature.

Typically, the characteristic reaction times at temperatures below ~500 K are too long to be relevant to the gas phase of the flame and near-field plume but may be important in terms of condensed phase processes (for example, spontaneous heating to ignition; Jones et al., 2015). The overall rate of oxidation increases with increasing temperature from the usual values found in the troposphere (~193 to ~298 K) to around 600 K. As temperature increases through this range, the overall reaction rate eventually "turns over" and decreases with increasing temperature. The decreasing trend continues with increasing temperature until the overall rate abruptly begins increasing with increasing reaction temperature once again. The range of temperatures over which the rate decreases with increasing temperature is termed the "negative-temperature-coefficient regime."

The abrupt transition from the negative-temperature-coefficient regime behavior, frequently termed "hot ignition" in the kinetic literature, results from the rapid thermal decomposition of hydrogen peroxide (H_2O_2). The fast decomposition of this intermediate species, which is much more stable at lower temperatures, leads to the production of OH radicals and autothermal acceleration of the overall rate by the production of water from reactions with fuel species. With further increases in reaction temperature, the thermal decomposition of large hydrocarbon radicals becomes rapid, and eventually chemical chain branching (i.e., reactions for which the consumption of a reactive radical results in the generation of more than one reactive radical) ensues. The turnover temperature, hot ignition temperature, and transition to chemical chain branching behavior are all determined by fuel species, oxygen concentration, and pressure.

Complex behaviors are driven by the range and spatial distributions of temperature found in all fires involving hydrocarbons or oxygenated hydrocarbons with large carbon numbers. The fundamental kinetic details for the oxidation of hydrocarbon and oxygenated hydrocarbon species at temperatures below the transition to chemical chain branching remain a very active area of combustion-related research (Ju, 2021; Ju et al., 2019; Wang et al., 2019). The significance of these kinetic behaviors in the various stages of smoldering combustion and in large-scale free-burning fires has not been explored. However, some of the large-carbon-number hydrocarbon and oxygenated species other than aromatics, known to be present, are likely to react within the fire environment and the plume at temperatures below 800–900 K but above the surrounding ambient temperatures by several hundred degrees Kelvin.

Chemical Pathways and Radical Species

The hydroxyl radical (OH) is a key active species in all oxidation processes occurring in low-, intermediate-, and high-temperature kinetic regimes. Two primary types of reactions involving OH and fuel species occur. Hydroxyl radicals can abstract an H atom directly from a C-H bond site, or OH can add to a molecule, forming an "adduct." In both cases, the result eventually produces other products.

The abstraction reaction can be written generically as $R_iH + OH \Rightarrow R_i + H_2O$, where R_i represents a hydrocarbon or oxygenated hydrocarbon radical with a H atom removed from a C-H bond site. The radical R_i generally proceeds to react with molecular oxygen (i.e., $R_i + O_2 \Rightarrow R_iO_2$); or if R_i is a radical with a large carbon number, it may thermally decompose to form additional radicals ($R_i \Rightarrow R_i', R_i'' + \ldots + $ olefinic species; i.e., species containing carbon double bonds). Reactive radicals such as OH or HO_2 will be regenerated only after a number of additional reactions involving further reactions of R, R', R'', the olefinic species, and intermediates formed from them with oxygen. Many radical and molecular oxygen addition reactions are also possible.

The addition reactions of OH with molecular hydrogen and with carbon monoxide (CO) are unique in that each reaction results in a stable species and a highly reactive H atom (i.e., $H_2 + OH \Rightarrow H + H_2O$ and $CO + OH \Rightarrow HOCO \Rightarrow CO_2 + H$). The adduct HOCO is typically short lived at the reaction temperatures of interest in fires and hot fire plumes. The OH addition reactions represent a class of reactions referred to as "chain carrying" reactions, in which the consumption of a reactive radical leads to the formation of a product and another reactive radical. As noted earlier, both OH and H are highly reactive radicals.

Which type of reaction occurs depends on the reacting species and temperature. Especially at lower temperatures and for olefinic or aromatic species, the addition reaction $R_iH + OH \Rightarrow HOR_iH$ is favored. For ethylene and propene, species often found in wood's oxidative pyrolysis products, the OH addition reaction is also favored over abstraction below ~900 K, but becomes disfavored above ~900 K; for acetylene, addition remains prevalent to temperatures approaching 1050 K (Khaled et al., 2019). Understanding these fundamentals of OH radical reactions is essential to understanding the effects of the relative rates of OH reaction with hydrocarbons and oxygenated hydrocarbons in comparison to OH reaction with CO and hydrogen, at intermediate and high temperatures.

An important characteristic of the homogeneous, gas-phase oxidation of a mixture of R_iH species is that the CO that forms is oxidized to CO_2 (>98 percent) almost entirely *through only one reaction pathway*, $CO + OH \Rightarrow CO_2 + H$. The oxidation reactions of CO with molecular oxygen, oxygen atoms, or hydroperoxyl radicals (HO_2) are all much, much slower than reaction with OH. In comparison to the overall rate constant, the rate constants for all hydrogen abstraction reactions by OH from C-H bonds are typically more than an order of magnitude larger (Atkinson, 2003; Han et al., 2018). This holds for all homogeneous, gas-phase oxidations of mixtures of R_iH species found in wildland and WUI fires.

Phenomenologically, this disparity in rate constants for OH reactions with CO and R_iH leads to a sequential, overall oxidative progress: (1) first hydrocarbons and/or oxygenated hydrocarbons concurrently convert to species with smaller carbon numbers, such as formaldehyde (CH_2O), methane (CH_4), ethylene (C_2H_4), acetylene (C_2H_2), and eventually CO and H_2O, and (2) the oxidation of CO to CO_2 follows (Glassman et al., 2014), only after most of the hydrocarbon species have been substantially depleted. For example, the rate constant for CH_2O (formaldehyde) + OH $\Rightarrow$ HCO + H_2O is larger than that for CH_4 + OH $\Rightarrow$ CH_3 + H_2O, and more than an order of magnitude larger than that for CO + OH. Thus, for mixtures of equal concentrations of CH_2O or CH_4 and CO, the oxidation of CO by OH will be strongly inhibited by the competitive reactions of OH with CH_2O and CH_4. Only after the hydrocarbon species are substantially depleted can CO successfully complete for OH radicals. This means that CO can be used as an indicator species for the overall extent of oxidation of hydrocarbons in plumes, as discussed later in this chapter. On the other hand, however, the addition reactions of OH with C_2H_4 and C_2H_2 do not compete favorably with CO + OH. In fact, for ethylene and acetylene, their co-presence with CO actually promotes CO oxidation (Yetter and Dryer, 1992).

In general, in wildland and WUI fires and their hot plumes, where intermediate and high temperatures are present, the gas-phase oxidation of CO to CO_2 will be negligibly slow in mixtures of larger-carbon-number hydrocarbons, oxygenated hydrocarbons, and CO, as long as the following is true:

$$([CO] \times k_{CO,OH}) \ll \Sigma_j([R_iH] \times k_{i,OH})_j$$

The brackets denote concentration units consistent with the units used in specifying the rate constants, k. The relationship described by the equation has implications in terms of the relative $[R_iH]/[CO]$ ratios found for all the species present, since the OH in the plume is typically the most reactive species with VOCs.

The Effect of Halogens

The introduction of human-made materials as fuels alters the radical concentrations and dominant pathways. Halogens, particularly Cl and Br, play significant roles (Hastie, 1973). The inhibitory effects of halogens, phosphorous, and antimony have inspired their use in formulating flame retardants (Gann and Gilman, 2003; Morgan and Gilman, 2013). Conversely, and as discussed in Chapter 4, chlorine radicals can accelerate the reactions of hydrocarbons under ambient tropospheric conditions (Tanaka, 2003). The gaseous emissions produced in fires with halogens present (whether from flame retardants or the oxidative decomposition of human-made materials containing halogens) have differing compositions and yields than the combustion emissions of materials without halogens (Senkan, 2000). The presence of Cl, which yields HCl (a stable product) during the oxidation process, enhances the rate of radical termination through a number of elementary reactions:

$$Cl + HO_2 \Rightarrow HCl + O_2$$
$$CO + Cl + M \Leftrightarrow COCl + M$$
$$COCl + Cl \Rightarrow CO + Cl_2$$
$$Cl_2 + H \Rightarrow HCl + Cl,$$

The net change produced by these reactions is

$$H + Cl \Rightarrow HCl.$$

Other Cl reactions that yield stable products are

$$HCl + OH \Rightarrow H_2O + Cl$$
$$H + Cl + M \Rightarrow HCl + M.$$

And, at high flame temperatures, the reactions are

$$Cl + Cl + M \Leftrightarrow Cl_2 + M$$
$$H + Cl + M \Leftrightarrow HCl + M.$$

In addition to quenching combustion reactions by enhancing radical termination, halogen chemistry can have a variety of other effects, including the stong inhibition of CO oxidation by CH_3Cl, another species found in emissions from the combustion of plastics containing Cl.

The impacts of halogen chemistry can also vary between premixed and mixing-limited flames. In premixed flames (Babushok et al., 2014), chlorine-containing inhibitors scavenge radicals. Other halogens also enhance termination reactions, with the effectiveness of the inhibition cycles of all the halogens scaling in the order $F < Cl < Br \approx I$, in both laminar premixed flames and perfectly stirred reactor conditions (Schefer and Brown, 1982).

For mixing-limited flames, the impacts are more complex. Halogens have different effects depending on whether the halogenated species is added to the fuel side or to the oxidant side of the mixing-limited flames. Simmons and Wolfhard (1956) noted in their early work on the inhibition of coaxial mixing-limited flames of propane and air by CH_3Br that the inhibitor was much more effective when added to the fuel side rather than the oxidizer side of the flame. Later, Creitz (1961) investigated the inhibition of axial mixing-limited flames of six different fuels with air. Contrary to the work of Simmons and Wolfhard, when adding CH_3Br and CF_3Br to the fuel or oxidizer side, the volume percentage required to extinguish the flame was much greater when the species was added to the fuel side for all fuels studied, except carbon monoxide.

Pitz and Sawyer (1978) investigated the combustion of flammable vapors produced by polyethylene or PVC using an opposed-flow reactor, which combines the flammable vapors with oxidant under mixing-limited conditions. (PVC is essentially polyethylene with one hydrogen atom replaced by one Cl atom.) Pitz (1982) followed this initial work with more refined experiments using chlorinated polyethylene as the solid fuel and numerical modeling of the results. Similar to Simmons and Wolfhard, all the data suggested that halogen inhibits more strongly when added to the fuel side rather than the oxidizer-rich side of mixing-limited flames.

The disparate nature of the work of Pitz with that of Creitz can be to some extent resolved by studies reported by Linteris (1997) on the amount of acid gases generated near the extinction of propane flames in two coaxial diffusion configurations. A wide range of molecular inhibitor candidates containing halogens were compared, each having different chemical structure elemental ratios of carbon/hydrogen/halogen (Cl, Br, F). For the same inhibitor, the amounts of the inhibitor species required for flame extinction in the two configurations were different. In each configuration, the amount also differed whether the inhibitor was added to the fuel or oxidizer side of the flame. The amount of acid gases emitted at extinction was also different for each inhibitor and configuration.

For solid materials burning diffusively, the material C/H/halogen elemental composition is defined, and the flame structure will be one in which the halogen is present on the fuel-rich side. However, once acid gases and other species from the diffusive-limited combustion are produced, these species may be present on both the oxidizer and fuel sides in downstream, partially premixed, turbulent flames, for example, in the case of enclosure fire emissions burning in external, open-flame structures of a building.

The contrast of these studies indicates the complexity of considering the influence of halogens in mixing-limited combustion under different flame configurations, and with chlorinated species of different carbon numbers and halogen contents. In WUI fires, halogens will be present on the fuel side as a halogen-containing solid such as PVC is burned. On the other hand, the HCl and other chlorinated species produced as products may enter other partially premixed and mixing-limited flame structures mixed with air. Interactions of halogens in partially premixed turbulent flames, as are present in wildland and WUI fires, need further investigation. The unknown extent of halogen chemistry in WUI fire plumes limits the applicability to WUI fire plumes of empirical results relating emissions to CO concentrations that are based on wildland fire plumes.

The Effect of Nitrogen Species

Nitrogen species in flames can be produced at relatively low temperatures from the oxidation of organically bound nitrogen species such as nylon, and from the oxidation of plastics (Chaos et al., 2009). At temperatures from about 600 to 1150 K, the presence of even small amounts of NO or NO_2 can provide a rapid conversion of HO_2, a much less reactive radical, to a reactive OH radical. This increase in OH production occurs through a collection of reactions involving very small concentrations of NO and NO_2 present in an oxidizing mixture of R_iH:

$$NO + HO_2 \Rightarrow NO_2 + OH$$
$$NO_2 + H \Rightarrow NO + OH$$
$$H + O_2 + M \Rightarrow HO_2 + M.$$

In combination, these elementary reactions yield the overall reaction result

$$2\,H + O_2 \Rightarrow 2\,OH,$$

thereby transforming the pseudo-termination species HO_2 into reactive OH radicals. A substantial amount of published literature shows the enhancement of oxidation reactions of hydrocarbons and hydrocarbon oxygenates by the presence of NO and/or NO_2 at temperatures from around 600 to 1050 K. Studies also consider other reactions involving NO_2 as an oxidant of hydrocarbon radicals (Alam et al., 2017; Ano and Dryer, 1998; Glarborg et al., 2018; Marrodán et al., 2019). The role of CO_2 and H_2O on NO formation during char-bound N oxidation has also been noted in the literature (Karlström et al., 2020).

As in the case of halogens, interactions of nitrogen species in flames present in wildland, urban, and WUI fires need further investigation. The unknown extent of nitrogen chemistry limits the applicability to WUI fire plumes of empirical results relating emissions to CO concentrations that are based on wildland fire plumes. Finally, both halogenated and nitrogen-containing WUI plume species that evolve from the combustion of synthetic materials are important to initializing subsequent secondary aerosols and emissions generated in plume dispersion and transport (Cai and Griffin, 2006; Choi et al., 2020; Wang and Ruiz, 2017).

All of this discussion is based on chemistries that occur in the presence of oxygen and at relatively high temperatures. Pyrolysis chemistry and smoldering combustion will also be important in some fire environments, and the nature of these chemistries in WUI fires is expected to be different than pyrolysis and smoldering combustion chemistry in wildland and urban fires for many of the same reasons discussed here.

> **Finding:** Both the mass of fuel and the elemental composition of human-made fuel materials are important in assessing the effects of halogens, nitrogen, and other elements that have potential to alter the chemistry and affect the composition of the species found in WUI fire plumes, both in the near field and as the plumes age. Temperatures experienced by these human-made materials during the combustion process will also influence the evolution of WUI fire chemistry by impacting the evolution of key radical species.

> **Finding:** Comprehensive, detailed kinetic models are not available for assessing the individual or combined effects of halogen-, sulfur-, metal-, and nitrogen-containing species, found in biomass or human-made materials, on combustion emissions associated with wildland and WUI fires.

The relative importance of various pathways depends on temperature as well as concentrations of reactants, intermediate products, and products of the combustion chemistry. These conditions are not well understood for combustion at the WUI.

EMISSIONS

The chemistry of WUI plumes evolves in complex ways as temperatures change and as the chemical species in the plume encounter different radical populations. Defining emissions associated with a plume implicitly requires a definition of the point at which a chemically evolving plume is injected into the atmosphere as an emission. There are different ways to define emissions in different technical disciplines. For example, in the area of fire toxicity, the commonly used term "fire emissions" is replaced by the term "fire effluents," which include all gases and aerosols, including suspended particles, created by combustion or pyrolysis (ISO, 2017). In regulatory emissions reporting, the emissions sometimes are defined at the point of release (e.g., in an exhaust stack) and sometimes after some mixing with the ambient atmosphere (e.g., as for flares). There is not yet a precise and widely accepted definition of the term "emission" for WUI fires and, therefore, variable definitions are in use for when and how an emission is injected into a surrounding air mass.

Approaches used to characterize emissions utilize emission factors (EFs) and emission ratios (ERs). Different approaches are employed for wildland fires and enclosure fires, but most approaches use empirical relationships between EFs or ERs and some measure of combustion efficiency.

Wildland Fire Emissions

Andreae and Merlet (2001) summarize a framework of using EFs and the early history of developing methodologies to quantify the impact of biomass burning on air pollution (Akagi et al., 2011; Andreae, 2019; Urbanski, 2013). An EF is typically defined as the mass of a compound emitted per kilogram of dry fuel combusted. Numerous studies on emissions from wildland fires over the past several decades have resulted in a large database of EFs for a wide range of ecosystems and biomass species. Large databases of *average* EFs and their variations over a wide number of fires, derived from plume data, have been generated. Andreae (2019) recently expanded and updated prior efforts, to include additional data and analyses with a global context, and Prichard et al. (2020) have compiled EFs with a North American focus.

EFs are typically defined using the "carbon balance method" as

$$EF_i \left(\frac{g}{kg}\right) = F_C \times 1{,}000 \left(\frac{g}{kg}\right) \times \frac{MW_i(g)}{12(g)} \times \frac{C_i}{C_T}$$

where EF_i is the mass of compound i (in grams) emitted per kilogram of dry fuel burned, F_C is the mass fraction of carbon in the fuel (typically assumed to be 0.45–0.5; Andreae, 2019), MW_i is the molecular weight of compound i, and C_i/C_T is the number of moles of compound i emitted, divided by the total number of moles of carbon emitted. The mass fraction of carbon in the fuel varies little for biomass, but in the case of human-made materials, the value may vary substantially. The value of C_i/C_T is typically determined from an individual fire-averaged measurement, calculated from the averaged ERs using

$$\frac{C_i}{C_T} = \left(\frac{\Delta C_i}{\Delta C_{CO_2}}\right) / \sum_{j=1}^{n} \left(NC_j \times \frac{\Delta C_1}{\Delta C_{CO_2}}\right)$$

where NC_j is the number of carbon atoms in compound j, summed over all carbon-containing compounds, and $\Delta C_i/\Delta C_{CO_2}$ is the fire-averaged ER of compound i to CO_2:

$$\frac{\Delta C_i}{\Delta C_{CO_2}} = \frac{C_{i,plume} - C_{i,background}}{C_{CO_2,plume} - C_{CO_2,background}}.$$

This "carbon balance method" is most accurate when all of the burnt carbon has been volatilized and quantified. Ignoring particulate carbon content and undetermined volatile species typically leads to at most a 10 percent underestimation of the total carbon, even for full-scale fire conditions (Yokelson et al., 1999). Enhanced ERs or "normalized excess ERs" correct for background concentrations and may sometimes use other compounds conserved in the plume other than CO_2. The reference compound chosen is generally selected such that its background concentration in the atmosphere is low and exhibits minimal variation (for example, CO). In this case, the normalized excess ER, $EnR_{i/CO}$, for compound i is defined as

$$EnR_{i/CO} = \frac{\Delta C_i}{\Delta C_{CO}} = \frac{C_{i,plume} - C_{i,background}}{C_{CO,plume} - C_{CO,background}},$$

where ΔC_i is the background-subtracted concentration of species i, $C_{i,plume}$ is the concentration of species i in the plume, and $C_{i,background}$ is the concentration of species i in the background air; similarly, ΔC_{CO} is the background-subtracted CO concentration, $C_{CO,plume}$ is the CO concentration in the plume, and $C_{CO,background}$ is the concentration of CO in the background air.

EFs will depend on the combustion characteristics under which the emissions are generated, and these conditions are frequently characterized by a combustion efficiency (CE) in the plume. The CE determined from plume measurements yields a reasonable estimate of the total carbon consumed in the burning process as well as the relative contributions of smoldering and flaming combustion to the composition of species injected into the plume. The CE is derived from

$$CE = \frac{\Delta C_{CO_2}}{\Delta C_{CO_2} + \Delta C_{CO} + \Delta C_{CH_4} + \Delta C_{NMOC} + \Delta C_{PC}},$$

where the denominator represents all sources of carbon added to the atmosphere by the plume and the subscripts designate CO_2, CO, CH_4, non-methane organic carbon (NMOC), and particulate carbon (PC). The term ΔC_{NMOC} and the term ΔC_{PC} can sometimes be made using direct measurements, for example, methane and total non-methane organic carbon can be estimated using an instrument that is sensitive to total carbon content, although the accuracy of these types of measurements depends on the species being analyzed. Similarly, ΔC_{PC} can be measured using instruments sensitive to total carbon in the particle phase. However, these measurements are often not available, and modified combustion efficiency (MCE) is frequently used as an approximation, ignoring all but CO and CO_2 carbon emissions, as

$$MCE = \frac{\Delta C_{CO_2}}{\Delta C_{CO_2} + \Delta C_{CO}}.$$

Well-ventilated flaming (low smoldering contributions) produces a MCE near 0.99, while the MCE of smoldering itself may vary substantially (~0.65–0.85) but is typically found to be near 0.8. Thus, an overall, fire-integrated, time-averaged MCE near 0.9 might be expected to imply roughly equal contributions from flaming and smoldering. However, some of the emissions from smoldering are likely to be further processed in high-temperature, flaming regions, depending on if the smoldering was initial, concurrent with, or subsequent to the flaming of the wildland fire.

As described in the previous section, the oxidation of CO in WUI fires depends on the temperature history of the fire and the materials combusted in the fire. Historically, the CE and MCE have been used to indicate the contributions of smoldering and flaming combustion in wildfire plumes. CE and MCE might also indicate the contributions of smoldering and flaming combustion in WUI plumes; however, the precise relationship between smoldering and flaming combustion and CE/MCE is uncertain.

The correlation of enhanced ERs for species i as a function of MCE would perhaps allow the determination of a more robust description of the composition of species injected into the plume. MCE is also likely to be time dependent for large-scale fires, resulting in variability of the EF_i values from the averaged value frequently reported over many different fires (Wiggins et al., 2021). Historically, "fresh" EFs from both laboratory and field studies for wildland fires compare favorably for some species with conditions generating the same MCE. Attempts to devolve the averaged measurements in a meaningful manner to obtain a time history of plume composition are complex and contribute to uncertainties in quantifying plume properties. Nonetheless, determining ERs and EFs from plume sampling, along with CE and MCE from plume measurements, has progressed immensely over the last decade.

Data from Recent Measurement Campaigns

Several intense field campaigns and emerging analyses of the resulting data show an improved ability to define the composition of wildland fire and WUI fire plumes. For example, the National Science Foundation and National Center for Atmospheric Research used a C-130 research aircraft during the 2018 Western Wildfire Experiment for

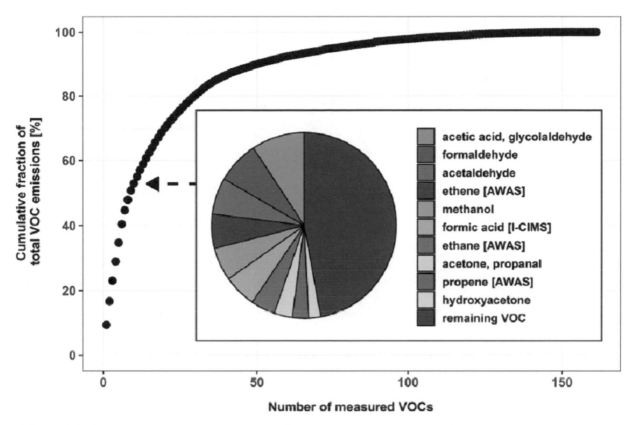

FIGURE 3-2 The cumulative mass fraction of the total measured VOC emissions as a function of measured VOCs during WE-CAN. Measurements were taken by PTR-ToF-MS unless otherwise indicated (AWAS = Advanced Whole Air Sampler; I-CIMS = iodide-adduct, high-resolution time-of-flight chemical ionization mass spectrometry). SOURCE: Permar et al., 2021.

Cloud Chemistry, Aerosol Absorption, and Nitrogen (WE-CAN) field campaign (NCAR-UCAR, 2018) to take comprehensive smoke-characterization measurements; the aircraft traversed fire plumes and sampled and analyzed their compositions using a custom-built proton-transfer-reaction time-of-flight mass spectrometry (PTR-ToF-MS) instrument and other instrumentation (Permar et al., 2021).

Researchers identified and quantified ERs and EFs for 161 VOCs in 31 near-plume flight transects of 24 wildland fires. As shown in Figure 3-2, 10 species accounted for about 53 percent of the measured mass.

Figure 3-3 plots EFs (Permar et al., 2021) for the 20 most abundant emitted VOC species, by mass, as measured in a number of field experiments. Overall, the WE-CAN data campaign increased the number of species that have been identified in wildland fire plumes, found that the emissions of only 98 species accounted for 76 percent of the average total measured VOC mass, and found that the EFs had statistically significant and negative dependences on MCE. VOC mass fractions showed much less MCE dependence, with significant overlap within the observed MCE range, suggesting that a single speciation profile might be able to describe VOC emissions for the western US coniferous forest wildland fires sampled during WE-CAN.

Additional field studies, such as Fire Influence on Regional to Global Environments and Air Quality (FIREX-AQ; Sekimoto et al., 2018), have suggested alternatives to CE and MCE for characterizing combustion consitions (e.g., high- and low-temperature combustion), and have examined the contributions that fire emissions can make to regional photochemistry as the plumes evolve (see Chapter 4). These types of field programs undertaken for wildland fires demonstrate the types of advances that can be made in developing emissions estimates; however, data for WUI fires similar to the wildland fire data from WE-CAN and FIREX-AQ are

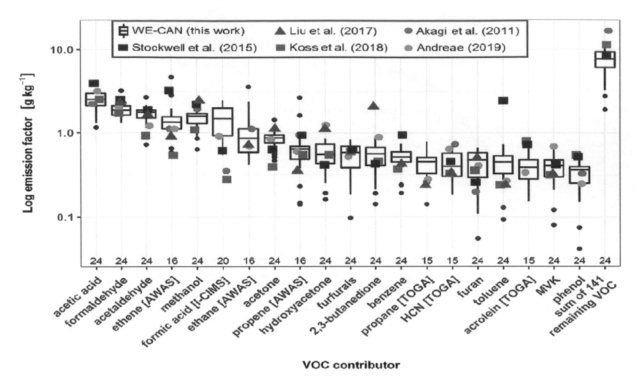

FIGURE 3-3 Box plot of EFs for the 20 most abundantly emitted VOCs, by mass, during WE-CAN, compared with selected data sets from the literature (see Permar et al., 2021, for details). Measurements were taken by PTR-ToF-MS unless otherwise indicated (TOGA = gas chromatography–based trace organic gas analyzer).

not available. EFs for WUI fires would be expected to contain contributions from halogenated compounds, nitrogen-containing compounds, and other species and would likely need to characterize not only combustion characteristics, but also the relative magnitudes and compositions of urban and wildland fuels consumed by the fires.

PM Emissions

Particulate matter (PM), consisting of species contained in a liquid or solid atmospheric aerosol, is one of the most important pollutants associated with fires. While PM accounts for only about 2 percent of the carbon emissions from wildland fires (Urbanski, 2014), its health impacts can be significant (see Chapter 6). Wildland fire PM is primarily composed of organic carbonaceous species, with minor contributions from elemental or "black" carbon (5–10 percent) and less than 5 percent from inorganic compounds (e.g., potassium, chlorine, sulfate), which depend upon the original concentration in the biomass (Jaffe et al., 2020).

Nearly 3,000 different compounds have been detected in PM emitted by fires, but only about half have been identified (Jen et al., 2019). Major classes of organic compounds included sugars, noncyclic aliphatics, organic nitrogen compounds, methoxyphenols, and substituted phenols. Researchers identified a small percentage of emissions from the burning of woody debris that, despite burning under very different conditions (MCE 0.98 vs 0.78), emitted similar compounds, demonstrating that fuel type can be a more important factor than combustion conditions in determining organic speciation (Jen et al., 2019). These organic compounds also span a range of volatilities, with many having the potential to partition between gas and particle phases, depending on the local concentration and temperature (May et al., 2013; Hatch et al., 2018). No comparable analyses of PM generated by WUI fires is available.

Urban Fire Emissions

Emission estimates for WUI fires must involve some parameterization of urban emissions that add to and potentially modify emissions estimates for wildland fires. Current understanding of urban emissions is based on laboratory experiments using configurations and conditions relevant to enclosure fires, (i.e., fires within buildings). The configurations emphasize the pyrolysis, oxidative pyrolysis, and boundary layer conditions that determine the intermediate reaction products (i.e., as the condensed phase reacts), which may be flammable gas-phase species, which can then ignite. In enclosure fires, where heating of the condensed phase is augmented by radiative/convective heat transfer from adjacent regions, the heating leads to the production of excess pyrolyzate in regions of oxidizer deficiency (i.e., incomplete combustion produces prolific amounts of flammable, gas-phase species). In larger enclosures, if ventilation air is abruptly supplied, the pyrolyzate can rapidly ignite and propagate gas-phase combustion via a combination of premixed and mixing-limited flames, referred to as a "flashover" event within the enclosure (Pitts, 1995; Stec and Hull, 2010). Researchers have developed several experimental configurations to characterize the evolution of pyrolyzate species relevant to enclosure fires, from early flaming to flashover.

Full-scale fire simulations reporting comprehensive EFs are limited for urban fires. RISE (Research Institutes of Sweden) has carried out a very limited number of large-scale urban fire simulations to develop emissions data for use in inventories of the fire life cycle (Amon et al., 2019). These studies provide some EFs for a mixture of materials. Table 3-8 provides the EFs of some of the most commonly emitted fire emissions, as measured in these studies.

The EFs in Table 3-8 vary widely, most by more than one order of magnitude. Some EFs vary from two to four orders of magnitude, with the most extreme variation observed in the polychlorinated dibenzo-p-dioxin (PCDD) and polychlorinated dibenzofuran (PCDF) EFs. This is because the full-scale experiments described here had a variety of materials and combustion conditions, with some tests where the ventilation was not controlled, and others in simulated rooms with ventilated, under-ventilated, and flashover conditions. Additionally, there may be even larger variability in the composition of structures, structure density, and combustion conditions both external and internal to the structures. Collectively, the variability in WUI fire emissions may be very large, and effective methods to parameterize emissions given these uncertainties are needed. Nonetheless these data show important distinctions in urban fire emissions from wildland fire emissions that should be evaluated for WUI fires.

Ventilation conditions and the chemical composition of the fuel likely play an important role in the formation of some pollutants like HCl, HF, SO_x, and metals, as these species are emitted only if their precursors are present in the original material consumed in the fire. Additionally, persistent organic pollutants such as PCDD/PCDF are emitted in much greater amounts when Cl is present in the materials and the combustion conditions are favorable for dioxin formation (Gullett et al., 2000; Stec et al., 2013). As shown in fire toxicity testing, the combustion conditions greatly impact the emissions for many different pollutants, and this variation is not consistent across single-component materials.

Full-scale tests with multiple types of materials and their mixtures can provide emissions data and improve upon the methodologies. However, data are limited and smaller-scale apparatuses can be better at controlling the fire conditions, probing the impact of materials on emissions under controlled combustion conditions. The steady-state tube-furnace test (ISO, 2016) produces EFs per unit mass consumption of material that are relevant to the behavior of enclosure fires and immediate human exposure, as a function of the equivalence ratio. Figure 3-4 is a schematic of the ISO/TS 19700 experimental configuration (Stec and Hull, 2010).

As an engineering tool for quantitative assessment of toxic product yields (or EFs), any furnace temperature or ventilation condition can be selected. However, certain conditions relate to ISO fire stages (ISO, 2011), using specified furnace temperatures and primary air flows that approximate those of enclosure fires. A typical analysis may use experiments conducted at three furnace temperatures (650°C, 923 K; 750°C, 1023 K; and 850°C, 1123 K) and stoichiometric, fuel-lean, and fuel-rich equivalence ratios; if necessary, furnace temperatures are increased or decreased to obtain the specified condition.

Most bench-scale fire test methods were developed to assess flammability and burning behavior, and therefore use a well-ventilated fire scenario. However, EFs frequently increase by factors of 10 to 50 in the transition from well-ventilated to under-ventilated flaming. Test apparatuses that are unable to force combustion to occur in oxygen-depleted atmospheres are unsuited to the generation of EFs for enclosure fires. Currently, most existing

TABLE 3-8 EFs from Large-Scale Urban Fire Simulations

Material	CO (g/kg)	NO$_x$/NO (g/kg)	HCN (g/kg)	HCl (g/kg)	PM (g/kg)	C$_6$H$_6$ (g/kg)	CH$_2$O (g/kg)	Pb (mg/kg)	Total PCDD (µg/kg)	Total PCDF (µg/kg)	Reference
Tires (heap or pile)	48.8–58.9			0.41–0.42	16–149				0.009–0.01	0.013–0.07	Lönnermark and Blomqvist, 2005
Scrap tire material					114	2.2		0.47			Lemieux and Ryan, 1993
Vehicle (medium-class 1998 car)	63		1.6	13	64	3.0	1.1	820	2.1	56	Lönnermark and Blomqvist, 2006
Electrical wiring	2.3–22.5			48–53		0.06–1.47	0.1–0.6		0.02–2.28	0.19–60.94	Andersson et al., 2004, 2005; Simonson et al., 2001
Simulated room contents	32–48		1.3–2.2	0.5–1.3		0.17–1.09			0.04–0.20	0.36–1.37	Blomqvist et al., 2004
Sofa with flame retardant in room	24–41	0.9–4.5	0.9–1.4			0.10–1.34	0.5	2.6–25	0.003–1.02	0.2–13.3	Andersson et al., 2003
Simulated room with sofa	14–51		3.5–15	6–18	45–282						Gann et al., 2010
Chemicals in storage room	8–293	0.1–36.3	0.8–44.7	77.8–141	7–76	0.04–7.1					Månsson et al., 1996

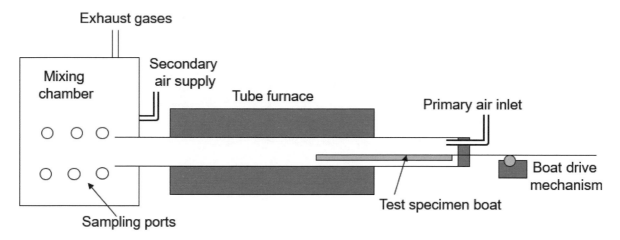

FIGURE 3-4 The steady-state tube-furnace apparatus, ISO/TS 19700.

bench-scale test methods do not represent a particular fire stage; some represent several fire stages separately, while others represent the progress of a fire through an indeterminate number of stages. Furthermore, most test methods produce data that are a function of both the flammability of the specimen and the yield of toxic products. Other methods provide toxic product yield data that are independent of the flammability. Table 3-9 summarizes the attributes of various bench-scale apparatuses. A recent study by RISE evaluated the suitability of four of these five test methods for quantifying the smoke toxicity of polymethylmethacrylate and 11 building insulation products under non-flaming, well-ventilated, and under-ventilated conditions (Blomqvist and Sandinge, 2018, 2021). The study concluded that significant differences in results when testing the same products could be attributed to the differences in the test methods.

Despite this variability, these test methods allow the collection of data that examine the dependence of EFs for commonly used materials, and mixtures of commonly used materials, on combustion conditions. These data can serve as a starting point for more extensive examination of WUI fire emissions under realistic combustion conditions.

Fire Retardants

ISO defines a fire retardant as a substance added, or a treatment applied, to a material in order to delay ignition or to reduce the rate of combustion (ISO, 2017). Fire retardants, which act in the gas phase and interfere with flame reactions, are frequently applied to insulation foams, electrical equipment, and upholstered furniture. The presence or absence of specific flame retardants can dramatically change the emissions associated with a particular material or structure.

A fire retardant should inhibit or even completely suppress the combustion process. Depending on their nature, fire retardants can act chemically and/or physically in the solid, liquid, or gas phase. They interfere with combustion during a particular stage of the combustion process, for example, during heating, decomposition, ignition, or flame spread (Purser, 2009). Several ways exist in which the combustion process can be retarded. It can be done by physical actions (cooling, formation of a protective layer, dilution) or chemical actions, in the condensed or gas phase, as shown in Figure 3-5.

Modern fire retardants prevent fuel molecules from escaping to the gas phase, often by formation of a protective barrier layer, and by inhibiting the flow of fuel and oxygen. This makes the fuel molecules less likely to encounter conditions that could ignite them. While the formulations required for char promotion and intumescence are generally specific to a particular polymer, halogenated flame retardants tend to be nonspecific in their action, and one flame retardant can be incorporated into many polymers. Unfortunately, this ease of use is matched by

TABLE 3-9 Attributes of Five Bench-Scale Testing Apparatuses

Name	The Smoke Density Chamber (ISO 5659)	Controlled Atmosphere Cone Calorimeter	The Non-dynamic Tube Furnace	Steady State Tube Furnace (BS7990 & ISO/TS 19700)	Fire Propagation Apparatus (ISO 12136 & ASTM E-2058)
Heat regimes	25, 50 kW/m²	25, 35, 50 kW/m²	400, 600, 800°C	350, 650, 825°C (15 to 80 kW/m²)	Up to 100 kW/m²
Design	Closed box	Flow-through	Flow-through	Flow-through	Flow-through
Mass loss rate	Variable, sample dependent	Variable, sample dependent	Variable, sample dependent	Fixed	Variable, sample dependent
Sample requirement	75 × 75 mm² results only valid for thickness tested	100 × 100 mm² results only valid for thickness tested	0.1 to 1.0 g	800 × 25 × 25 mm³ (maximum size)	100 × 100 mm² results only valid for thickness tested
Smoldering	Y	Y	Y	Y	Y
Well-ventilated	Y	Y	Y	Y	Y
Under-ventilated	N	N	?	Y	Y
Post-flashover	N	N	?	Y	Y
Advantages	Widely available; currently used for toxicity assessment in mass transport industries	Modification to widely used standard test; measures flammability as well as toxicity	Simple and low cost	Designed for smoke toxicity assessment; ideal for linear or homogeneous products; correlates with large-scale tests for all conditions	Able to replicate high CO yields for under-ventilated flaming
Disadvantages	Poor interlaboratory reproducibility; unable to force under-ventilated flaming; fire stage unknown; sample probe can miss toxic plume	No standard equipment exists; unable to go above 50 kW/m²; toxic product yields do not correlate to large-scale fire tests; equivalence ratio only known after test	Conditions not related to fire stages; does not distinguish between flaming and non-flaming; limited volume of effluent for analysis	Interlaboratory reproducibility not proven for layered materials; separate assessment of flammability required	Expensive; not widely used; capabilities insufficiently demonstrated

SOURCE: Stec and Hull, 2010. ASTM = American Society for Testing and Materials.

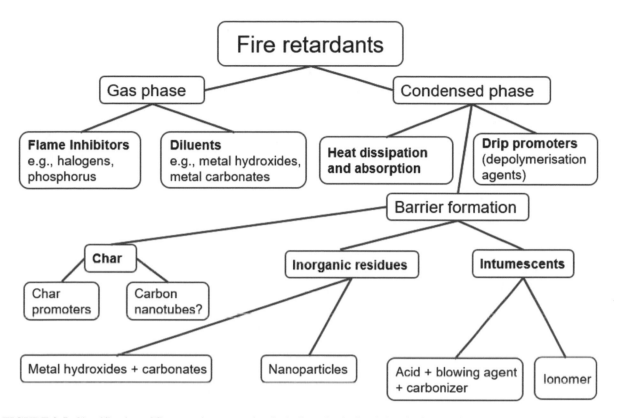

FIGURE 3-5 Classification of fire-retardant strategies, including physical and chemical strategies. SOURCE: Hull et al., 2014.

an ease of release, particularly when operating at elevated temperatures (Hull et al., 2014). Many halogenated flame retardants are persistent, bioaccumulative, and ubiquitous throughout the built and natural environment (Hull et al., 2014).

Halogenated flame retardants, which act in the gas phase, interfere with the free-radical reactions responsible for flaming combustion, as discussed in the earlier section on combustion (and summarized in Babushok et al., 2014). In terms of fire emissions, the presence of hydrogen bromide (HBr) or hydrogen chloride (HCl), produced by flame-retardant combustion, typically produces more carbon monoxide, smoke, and other products of incomplete combustion (e.g., acrolein and formaldehyde), as well as larger cyclic molecules such as PAHs and soot particulates (Molyneux et al., 2014; Schnipper et al., 1995).

Many other types of fire retardants present in condensed media reduce fuel release to the gas phase, frequently by increasing the formation of the surface char layer, decreasing both net radiative heat transfer to the surface and conductive heat transfer into the media. The char may inhibit diffusion of oxygen to the surface and fuel release from it, depending on its porosity. Such barriers have also been deployed in intumescent systems, where gas is released within the molten polymer, causing significant swelling and so increasing the effectiveness of the thermal and physical barrier. Mineral filler flame retardants such as aluminum hydroxide or magnesium hydroxide, char formers based on inorganic phosphates and borates, and intumescents all fall into this category. By keeping the fuel in the condensed phase, they reduce gaseous emissions and simultaneously lower the flammability.

Finding: Limited data exist on emission factors and the sources of uncertainty for WUI fires; no studies of single- or multiple-dwelling, full-scale structures, with realistic contents and realistic WUI fire conditions (e.g., strong winds), have reported emission factors or yields.

Finding: WUI fire emission factors are expected to be different from emission factors for wildland fires and urban fires and to vary widely, depending on structure density and composition, combustion conditions, and other factors.

Finding: Current experimental methods do not capture the dynamic nature of WUI fires; large information gaps lead to large uncertainties in predicting plume composition and structure. Nevertheless, bench-scale reference methods that produce results such as toxic product yields or emission factors are effective tools for applied engineering analyses, material selection, and toxicological evaluations. These methods also allow investigators to control many of the parameters affecting the combustion efficiency and emission factors, often allowing a wider exploration of parameters than possible with enclosure fire tests. These bench-scale analyses could be complemented by field measurements of WUI fire plume composition and structure.

Finding: Halogenated species found in human-made materials and in gas-phase flame retardants change the yields and the types of a wide variety of products of incomplete combustion.

Research need: The interactions between human-made fuels with different elemental compositions (e.g., containing halogens, phosphorous, sulfur, nitrogen, metal salts, and organometallics) and biomass (particularly timber) and other wildland fire fuels as they generate fire plumes need to be modeled mechanistically and explored experimentally at bench, laboratory, and larger scales.

Research need: Experimental and modeling investigations of WUI fire plume compositions should be used to develop emission factors and chemical fingerprints for WUI fire emissions.

REFERENCES

Adamkiewicz, G., A. R. Zota, and M. P. Fabian. 2011. "Moving Environmental Justice Indoors: Understanding Structural Influences on Residential Exposure Patterns in Low-Income Communities." *AJPH*. https://doi.org/10.2105/AJPH.2011.300119.

Akagi, S. K., R. J. Yokelson, C. Wiedinmyer, M. J. Alvarado, J. S. Reid, T. Karl, J. D. Crounse, and P. O. Wennberg. 2011. "Emission Factors for Open and Domestic Biomass Burning for Use in Atmospheric Models." *Atmospheric Chemistry and Physics* 11 (9): 4039–4072. https://doi.org/10.5194/acp-11-4039-2011.

Alam, F. E., F. M. Haas, T. I. Farouk, and F. L. Dryer. 2017. "Influence of Trace Nitrogen Oxides on Natural Gas Oxidation: Flow Reactor Measurements and Kinetic Modeling." *Energy Fuels* 31. https://doi.org/10.1021/acs.energyfuels.6b02369.

Amon, F., J. Gehandler, R. McNamee, M. McNamee, and A. Vilic. 2019. *Measuring the Impact of Fire on the Environment.* Fire Impact Tool, version 1. Gothenburg, Sweden: Research Institutes of Sweden.

Andersson, P., M. Simonson, and L. Rosell. 2003. *Fire-LCA Model: Furniture Study.* SP Report 2003:22. Gothenburg, Sweden: Research Institutes of Sweden.

Andersson, P., L. Rosell, and M. Simonson. 2004. "Small and Large Scale Fire Experiments with Electric Cables under Well-Ventilated and Vitiated Conditions." *Fire Technology* 40 (3). https://doi.org/10.1023/B:FIRE.0000026879.07753.86.

Andersson, P., M. Simonson, L. Rosell, P. Blomqvist, and H. Stripple. 2005. *Fire-LCA Model: Cable Case Study II – NHXMH and NHMH Cable.* Gothenburg, Sweden: Research Institutes of Sweden.

Andreae, M. O. 2019. "Emission of Trace Gases and Aerosols from Biomass Burning – An Updated Assessment." *Atmospheric Chemistry and Physics* 19 (13): 8523–8546. https://doi.org/10.5194/acp-19-8523-2019.

Andreae, M. O., and P. Merlet. 2001. "Emission of Trace Gases and Aerosols from Biomass Burning." *Global Biogeochemical Cycles* 15 (4): 955–966. https://doi.org/10.1029/2000gb001382.

Ano, T. A., and F. L. Dryer. 1998. "Effect of Dimethyl Ether, NO_x, and Ethane on CH_4 Oxidation: High Pressure, Intermediate-Temperature Experiments and Modeling." *Symposium (International) on Combustion* 27 (1). https://doi.org/10.1016/S0082-0784(98)80428-1.

Atchley, A. L., R. Linn, and A. Jonko. 2021. "Effects of Fuel Spatial Distribution on Wildland Fire Behaviour." *International Journal of Wildland Fire* 30 (3). https://doi.org/10.1071/WF20096.

Atkinson, R. 2003. "Kinetics of the Gas-Phase Reactions of OH Radicals with Alkanes and Cycloalkanes." *Atmospheric Chemistry and Physics* 3: 2233–2307. https://doi.org/10.5194/acp-3-2233-2003.

Babushok, V. I., G. T. Linteris, and O. C. Meier. 2014. "Flame Inhibition by CF_3CHCl_2 (HCFC-123)." *Combustion Science and Technology* 186 (6). https://doi.org/10.1080/00102202.2013.878709.

Bain, R. L., W. A. Amos, M. Downing, and R. L. Perlack. 2003. *Biopower Technical Assessment: State of the Industry and Technology*. Golden, CO: National Renewable Energy Laboratory. https://www.nrel.gov/docs/fy03osti/33123.pdf.

Blomqvist, P., and A. Lönnermark. 2001. "Characterization of the Combustion Products in Large-Scale Fire Tests: Comparison of Three Experimental Configurations." *Fire and Materials* 25. https://doi.org/10.1006/fam.761.

Blomqvist, P., and M. S. McNamee. 2009. *Estimation of CO_2-Emissions from Fires in Dwellings, Schools and Cars in the Nordic Countries*. SP Technical Note 2009:13. Borås, Sweden: SP Technical Research Institute of Sweden.

Blomqvist, P., and A. Sandinge. 2018. *Experimental Evaluation of Fire Toxicity Test Methods*. Gothenburg, Sweden: Research Institutes of Sweden.

Blomqvist, P., and A. Sandinge. 2021. "An Experimental Evaluation of the Equivalence Ratios in Tests Apparatus Used for Fire Effluent Toxicity Studies." *Fire and Materials* 45 (8): 1085–1095. https://doi.org/10.1002/fam.2995.

Blomqvist, P., L. Rosell, and M. Simonson. 2004. "Emissions from Fires Part II: Simulated Room Fires." *Fire Technology* 40: 59–73.

Blomqvist, P., M. S. McNamee, A. A. Stec, D. Gylestam, and D. Karlsson. 2013. "Detailed Study of Distribution Patterns of Polycyclic Aromatic Hydrocarbons and Isocyanates under Different Fire Conditions." *Fire and Materials* 38 (1): 125–144. https://doi.org/10.1002/fam.2173.

Bodí, M. B., D. A. Martin, V. N. Balfour, C. Santín, S. H. Doerr, P. Pereira, A. Cerdà, and J. Mataix-Solera. 2014. "Wildland Fire Ash: Production, Composition and Eco-hydro-geomorphic Effects." *Earth-Science Reviews* 130: 103–127. https://doi.org/10.1016/j.earscirev.2013.12.007.

Boulder OEM (Office of Emergency Management). 2022. "Boulder County Releases Updated List of Structures Damaged and Destroyed in the Marshall Fire." https://www.bouldercounty.org/news/boulder-county-releases-updated-list-of-structures-damaged-and-destroyed-in-the-marshall-fire/.

Bwalya, A., G. Lougheed, A. Kashef, and H. Saber. 2010. "Survey Results of Combustible Contents and Floor Areas in Canadian Multi-Family Dwellings." *Fire Technology* 47 (4): 1121–1140. https://doi.org/10.1007/s10694-009-0130-8.

Cai, H., X. Wang, and J. Kelly. 2021. *Building Life-Cycle Analysis with the GREET Building Module: Methodology, Data, and Case Studies*. Lemont, IL: Argonne National Laboratory.

Cai, X., and R. J. Griffin. 2006. "Secondary Aerosol Formation from the Oxidation of Biogenic Hydrocarbons by Chlorine Atoms. *Journal of Geophysical Research: Atmospheres* 111. https://doi.org/ 10.1029/2005JD006857.

CAL FIRE (California Department of Forestry and Fire Protection). n.d. *Wildfire is Coming. Are You . . . Ready?* https://www.readyforwildfire.org/wp-content/uploads/calfire_ready_brochure_LINOweb.pdf (accessed April 1, 2022).

CAL FIRE. 2022. *Stats and Events*. https://www.fire.ca.gov/stats-events/.

CalRecycle. 2020. *Urban Wood Waste*. https://www.calrecycle.ca.gov/ConDemo/Wood/.

Campbell, J., D. Donato, D. Azuma, and B. Law. 2007. "Pyrogenic Carbon Emission from a Large Wildfire in Oregon, United States." *Journal of Geophysical Research: Biogeosciences* 112 (G4). https://doi.org/10.1029/2007jg000451.

CDLA (Colorado Department of Local Affairs). 2016. "Wildland-Urban Interface Code (WUI Code)." In *Planning for Hazards: Land Use Solutions for Colorado*. https://planningforhazards.com/wildland-urban-interface-code-wui-code (accessed March 28, 2022).

Chaos, M., M. P. Burke, Y. Ju, and F. L. Dryer. 2009. "Syngas Chemical Kinetics and Reaction Mechanisms." In *Synthesis Gas Combustion: Fundamentals and Applications* Chapter 2. Edited by Tim C. Lieuwen, Vigor Yang, and Richard Yetter. Boca Raton, FL: CRC Press.

Choi, M. S., X. Qiu, and J. Zhang. 2020. "Study of Secondary Organic Aerosol Formation from Chlorine Radical-Initiated Oxidation of Volatile Organic Compounds in a Polluted Atmosphere Using a 3D Chemical Transport Model." *Environmental Science & Technology* 54 (21). https://doi.org/10.1021/acs.est.0c02958.

Coggon, M. M., C. Y. Lim, A. R. Koss, K. Sekimoto, B. Yuan, J. B. Gilman, D. H. Hagan, V. Selimovic, K. J. Zarzana, S. S. Brown, J. M. Roberts, M. Müller, R. Yokelson, A. Wisthaler, J. E. Krechmer, J. L. Jimenez, C. Cappa, J. H. Kroll, J. de Gouw, and C. Warneke. 2019. "OH Chemistry of Non-methane Organic Gases (NMOGs) Emitted from Laboratory and Ambient Biomass Burning Smoke: Evaluating the Influence of Furans and Oxygenated Aromatics on Ozone and Secondary NMOG Formation." *Atmospheric Chemistry and Physics* 19 (23): 14875–14899. https://doi.org/10.5194/acp-19-14875-2019.

Colantoni, A., G. Egidi, G. Quaranta, R. D'Alessandro, S. Vinci, R. Turco, and L. Salvati. 2020. "Sustainable Land Management, Wildfire Risk and the Role of Grazing in Mediterranean Urban-Rural Interfaces: A Regional Approach from Greece." *Land* 9 (1): 21. https://doi.org/10.3390/land9010021.

Cooper, E. M., G. Kroeger, K. Davis, C. R. Clark, P. L. Ferguson, and H. M. Stapleton. 2016. "Results from Screening Polyurethane Foam Based Consumer Products for Flame Retardant Chemicals: Assessing Impacts on the Change in the Furniture Flammability Standards." *Environmental Science & Technology* 50 (19): 10653–10660. https://doi.org/10.1021/acs.est.6b01602.

Creitz, E. C. 1961. "Inhibition of Diffusion Flames by Methyl Bromide and Trifluoromethyl Bromide Applied to the Fuel and Oxygen Sides of the Reaction Zone." *Journal of Research of the National Bureau of Standards A. Physics and Chemistry* 65A (4): 389–400. https://nvlpubs.nist.gov/nistpubs/jres/65A/jresv65An4p389_A1b.pdf.

Curran, H. J. 2019. "Developing Detailed Chemical Kinetic Mechanisms for Fuel Combustion." *Proceedings of the Combustion Institute* 37 (1). https://doi.org/10.1016/j.proci.2018.06.054.

Davies, K. W., C. S. Boyd, J. D. Bates, and A. Hulet. 2015. "Winter Grazing Can Reduce Wildfire Size, Intensity and Behaviour in a Shrub-Grassland." *International Journal of Wildland Fire* 25: 191–199. https://doi.org/10.1071/WF15055.

Davis, S. C., and R. G. Boundy. 2021. *Transportation Energy Data Book.* ORNL/TM-2020/1770 (Edition 39 of ORNL-5198). Oak Ridge, TN: Oak Ridge National Laboratory. https://tedb.ornl.gov/wp-content/uploads/2021/02/TEDB_Ed_39.pdf.

Dionisio, K. L., A. M. Frame, M. R. Goldsmith, J. F. Wambaugh, A. Liddell, T. Cathey, D. Smith, J. Vail, A. S. Ernstoff, P. Fantke, O. Jolliet, and R. S. Judson. 2015. "Exploring Consumer Exposure Pathways and Patterns of Use for Chemicals in the Environment." *Toxicology Reports* 2: 228–237. https://doi.org/10.1016/j.toxrep.2014.12.009.

Dixit, M. K., J. L. Fernandez-Solis, S. Lavy, and C. H. Culp. 2012. "Need for an Embodied Energy Measurement Protocol for Buildings: A Review Paper." *Renewable and Sustainable Energy Reviews* 16 (6). https://doi.org/10.1016/j.rser.2012.03.021.

Elgowainy, A., J. Han, and J. Ward. 2016. *Cradle-to-Grave Lifecycle Analysis of US Light-Duty Vehicle-Fuel Pathways: A Greenhouse Gas Emissions and Economic Assessment of Current (2015) and Future (2025-2030) Technologies.* Oak Ridge, TN: US Department of Energy, Office of Scientific and Technical Information. https://doi.org/10.2172/1254857.

Elhami-Khorasani, N., J. G. Salado Castillo, E. Saula, T. Josephs, G. Nurlybekova, and T. Gernay. 2020. "Application of a Digitized Fuel Load Surveying Methodology to Office Buildings." *Fire Technology* 57 (1): 101–122. https://doi.org/10.1007/s10694-020-00990-2.

Emrath, P. October 7, 2010. "Characteristics of Single-Family Homes Started in 2009." https://www.nahbclassic.org/generic.aspx?genericContentID=145984 (accessed February 8, 2022).

EPA (US Environmental Protection Agency). 2016. *Analysis of the Lifecycle Impacts and Potential for Avoided Impacts Associated with Single Family Homes.* EPA Report 530-R13-004. https://www.epa.gov/smm/analysis-lifecycle-impacts-and-potential-avoided-impacts-associated-single-family-homes (accessed February 16, 2022).

Fent, K. W., and D. E. Evans. 2011. "Assessing the Risk to Firefighters from Chemical Vapors and Gases during Vehicle Fire Suppression." *Journal of Environmental Monitoring* 13 (3): 536–543. https://doi.org/10.1039/c0em00591f.

Forest Products Laboratory. 2010. *Wood Handbook—Wood as an Engineering Material.* General Technical Report FPL-GTR-190. Madison, WI: US Department of Agriculture, Forest Service.

Gann, R. G., and J. W. Gilman. 2003. "Flame Retardants: An Overview." *Kirk-Othmer Encyclopedia of Chemical Technology.* Hoboken, NJ: John Wiley & Sons Inc. https://doi.org/10.1002/0471238961.1522051807011414.a01.pub2.

Gann, R. G., J. D. Averill, E. L. Johnsson, M. R. Nyden, and R. D. Peacock. 2010. "Fire Effluent Component Yields from Room-Scale Fire Tests." *Fire and Materials* 34 (6): 285–314. https://doi.org/10.1002/fam.1024.

Ghattas, R., J. Gregory, M. Noori, T. R. Miller, E. Olivetti, and S. Greene. 2016. *Life Cycle Assessment for Residential Buildings: A Literature Review and Gap Analysis, Rev. 1.* Cambridge, MA: Massachusetts Institute of Technology Concrete Sustainability Hub. http://hdl.handle.net/1721.1/104794.

Glarborg, P., J. A. Miller, B. Ruscic, and S. J. Klippenstein. 2018. "Modeling Nitrogen Chemistry in Combustion." *Progress in Energy and Combustion Science* 67: 31–68. https://doi.org/10.1016/j.pecs.2018.01.002.

Glassman, I., R. A. Yetter, and N. G. Glumac. 2014. *Combustion* 5th edition. New York, NY: Elsevier.

Gullett, B. K., A. F. Sarofim, K. A. Smith, and C. Procaccini. 2000. "The Role of Chlorine in Dioxin Formation." *Process Safety and Environmental Protection* 78 (1): 47–52. https://doi.org/10.1205/095758200530448.

Hallquist, M., J. C. Wenger, and U. Baltensperger. 2009. "The Formation, Properties and Impact of Secondary Organic Aerosol: Current and Emerging Issues." *Atmospheric Chemistry and Physics* 9 (14): 5155–5236. https://doi.org/10.5194/acp-9-5155-2009.

Han, L., F. Siekmann, and C. Zetzsch. 2018. "Rate Constants for the Reaction of OH Radicals with Hydrocarbons in a Smog Chamber at Low Atmospheric Temperatures." *Atmosphere* 9 (8). https://doi.org/10.3390/atmos9080320.

Hastie, J. W. 1973. "Molecular Basis of Flame Inhibition." *Journal of Research of the National Bureau of Standards A. Physics and Chemistry* 77A (6): 733–754. https://doi.org/10.6028/jres.077A.045.

Hatch, L. E., A. Rivas-Ubach, C. N. Jen, M. Lipton, A. H. Goldstein, and K. C. Barsanti. 2018. "Measurements of I/SVOCs in Biomass-Burning Smoke Using Solid-Phase Extraction Disks and Two-Dimensional Gas Chromatography." *Atmospheric Chemistry and Physics* 18 (24): 17801–17817. https://doi.org/10.5194/acp-18-17801-2018.

Hessburg, P. F., C. L. Miller, S. A. Parks, N. A. Povak, A. H. Taylor, P. E. Higuera, S. J. Prichard, M. P. North, B. M. Collins, M. D. Hurteau, A. J. Larson, C. D. Allen, S. L. Stephens, H. Rivera-Huerta, C. S. Stevens-Rumann, L. D. Daniels, Z. e. Gedalof, R. W. Gray, V. R. Kane, D. J. Churchill, R. K. Hagmann, T. A. Spies, C. A. Cansler, R. T. Belote, T. T. Veblen, M. A. Battaglia, C. Hoffman, C. N. Skinner, H. D. Safford, and R. B. Salter. 2019. "Climate, Environment, and Disturbance History Govern Resilience of Western North American Forests." *Frontiers in Ecology and Evolution* 7. https://doi.org/10.3389/fevo.2019.00239.

Hull, T. R., R. J. Law, and A. Bergman. 2014. "Environmental Drivers for Replacement of Halogenated Flame Retardants." In *Polymer Green Flame Retardants* Chapter 4. Edited by C. D. Papaspyrides and P. Kiliaris. New York, NY: Elsevier. https://doi.org/10.1016/B978-0-444-53808-6.00004-4.

IFSTA (International Fire Service Training Association). 2016. "Interior Finishes and Passive Fire Protection." In *Building Construction Related to the Fire Service* 4th edition, Chapter 5. https://www.ifsta.org/sites/default/files/Chapter%205_BC_4th_edition.pdf (accessed April 1, 2022).

ISO (International Organization for Standardization). 2011. *Guidelines for Assessing the Fire Threat to People.* ISO 19706:2011. Edition 2. https://www.iso.org/standard/56864.html.

ISO. 2016. *Controlled Equivalence Ratio Method for the Determination of Hazardous Components of Fire Effluents—Steady-State Tube Furnace.* ISO/TS 19700:2016. Edition 2. https://www.iso.org/standard/70630.html.

ISO. 2017. *Fire Safety—Vocabulary.* ISO 13943:2017. Edition 3. https://www.iso.org/standard/63321.html.

Jacobs, D. E. 2011. "Environmental Health Disparities in Housing." *American Journal of Public Health* 101 (S1): S115–S122. https://doi.org/10.2105/AJPH.2010.300058.

Jaffe, D. A., S. M. O'Neill, N. K. Larkin, A. L. Holder, D. L. Peterson, J. E. Halofsky, and A. G. Rappold. 2020. "Wildfire and Prescribed Burning Impacts on Air Quality in the United States." *Journal of the Air & Waste Management Association* 70 (6): 583–615. https://doi.org/10.1080/10962247.2020.1749731.

Jen, C. N., L. E. Hatch, V. Selimovic, R. J. Yokelson, R. Weber, A. E. Fernandez, N. M. Kreisberg, K. C. Barsanti, and A. H. Goldstein. 2019. "Speciated and Total Emission Factors of Particulate Organics from Burning Western US Wildland Fuels and Their Dependence on Combustion Efficiency." *Atmospheric Chemistry and Physics* 19 (2): 1013–1026. https://doi.org/10.5194/acp-19-1013-2019.

Jones, J. M., A. Saddawi, B. Dooley, E. J. S. Mitchell, J. Werner, D. J. Waldron, S. Weatherstone, and A. Williams. 2015. "Low Temperature Ignition of Biomass." *Fuel Processing Technology* 134: 372–377. https://doi.org/10.1016/j.fuproc.2015.02.019.

Ju, Y. 2021. "Understanding Cool Flames and Warm Flames." *Proceedings of the Combustion Institute* 38 (1): 83–119. https://doi.org/10.1016/j.proci.2020.09.019.

Ju, Y., C. B. Reuter, O. R. Yehia, and T. I. Farouk. 2019. "Dynamics of Cool Flames." *Progress in Energy and Combustion Science* 75. https://doi.org/10.1016/j.pecs.2019.100787.

Karlström, O., D. Schmid, M. Hupa, and A. Brink. 2020. "Role of CO_2 and H_2O on NO Formation during Biomass Char Oxidation." *Energy & Fuels* 35 (9): 7058–7064. https://doi.org/10.1021/acs.energyfuels.0c03471.

Keeley, J. E., and A. D. Syphard. 2019. "Twenty-First Century California, USA, Wildfires: Fuel-Dominated vs. Wind-Dominated Fires." *Fire Ecology* 15 (1). https://doi.org/10.1186/s42408-019-0041-0.

Khaled, F., B. R. Giri, and A. Farooq. 2019. "On the Reaction of OH Radicals with C2 Hydrocarbons." *Proceedings of the Combustion Institute* 37 (1): 213–219. https://doi.org/10.1016/j.proci.2018.06.052.

Kohse-Hoinghaus, K. 2021. "Combustion in the Future: The Importance of Chemistry." *Proceedings of the Combustion Institute* 38 (1): 1–56. https://doi.org/10.1016/j.proci.2020.06.375.

Koo, E., P. J. Pagni, D. R. Weise, and J. P. Woycheese. 2010. "Firebrands and Spotting Ignition in Large-Scale Fires." *International Journal of Wildland Fire* 19 (7): 818–843. https://doi.org/10.1071/WF07119.

Kuligowski, E. 2021. "Evacuation Decision-Making and Behavior in Wildfires: Past Research, Current Challenges and a Future Research Agenda." *Fire Safety Journal* 120. https://doi.org/10.1016/j.firesaf.2020.103129.

Larsson, F., P. Andersson, P. Blomqvist, and B. E. Mellander. 2017. "Toxic Fluoride Gas Emissions from Lithium-Ion Battery Fires." *Scientific Reports* 7 (1): 10018. https://doi.org/10.1038/s41598-017-09784-z.

Lemieux, P. M., and J. V. Ryan. 1993. "Characterization of Air Pollutants Emitted from a Simulated Scrap Tire Fire." *Air & Waste* 43 (8): 1106–1115. https://doi.org/10.1080/1073161x.1993.10467189.

Li, D., and S. Suh. 2019. "Health Risks of Chemicals in Consumer Products: A Review." *Environment International* 123: 580–587. https://doi.org/10.1016/j.envint.2018.12.033.

Linteris, G. T. 1997. "Acid Gas Production in Inhibited Propane—Air Diffusion Flames." In *Halon Replacements* Chapter 19. Edited by A. W. Miziolek and W. Tsang. Washington, DC: American Chemical Society. https://doi.org/10.1021/bk-1995-0611.ch019.

Lönnermark, A., and P. Blomqvist. 2005. *Emissons from Tyre Fires.* SP Report 2005:43. Borås, Sweden: SP Swedish National Testing and Research Institute. https://www.diva-portal.org/smash/get/diva2:962334/FULLTEXT01.pdf.

Lönnermark, A., and P. Blomqvist. 2006. "Emissions from an Automobile Fire." *Chemosphere* 62 (7): 1043–1056. https://doi.org/10.1016/j.chemosphere.2005.05.002.

Lowden, L. A., and T. R. Hull. 2013. "Flammability Behaviour of Wood and a Review of the Methods for Its Reduction." *Fire Science Reviews* 2: 4. https://doi.org/10.1186/2193-0414-2-4.

Månsson, M., A. Lönnermark, P. Blomqvist, H. Persson, and V. Babrauskas. 1996. *TOXFIRE – Fire Characteristics and Smoke Gas Analyses in Under-Ventilated Large-Scale Combustion Experiments.* SP Report 1996:44. Borås, Sweden: SP Swedish National Testing and Research Institute. https://www.diva-portal.org/smash/get/diva2:962010/FULLTEXT01.pdf.

Manzello, S. L., S. Suzuki, M. J. Gollner, and A. C. Fernandez-Pello. 2020. "Role of Firebrand Combustion in Large Outdoor Fire Spread." *Progress in Energy and Combustion Science* 76. https://doi.org/10.1016/j.pecs.2019.100801.

Maranghides, A., and D. McNamara. 2016. *2011 Wildland Urban Interface Amarillo Fires Report #2 – Assessment of Fire Behavior and WUI Measurement Science.* NIST Technical Note 1909. Gaithersburg, MD: NIST. https://doi.org/10.6028/NIST.TN.1909.

Maranghides, A., D. McNamara, W. Mell, J. Trook, and B. Toman. 2013. *A Case Study of a Community Affected by the Witch and Guejito Fires Report #2 – Evaluating the Effects of Hazard Mitigation Actions on Structure Ignitions.* NIST Technical Note 1796. Gaithersburg, MD: NIST. https://doi.org/10.6028/NIST.TN.1796.

Maranghides, A., D. McNamara, R. Vihnanek, J. Restaino, and C. Leland. 2015. *A Case Study of a Community Affected by the Waldo Fire – Event Timeline and Defensive Actions.* NIST Technical Note 1910. Gaithersburg, MD: NIST. https://doi.org/10.6028/NIST.TN.1910.

Maranghides, A., E. Link, W. R. Mell, S. Hawks, M. Wilson, W. Brewer, C. Brown, B. Vihnaneck, and W. D. Walton. 2021. *A Case Study of the Camp Fire – Fire Progression Timeline.* NIST Technical Note 2135. Gaithersburg, MD: NIST. https://doi.org/10.6028/nist.Tn.2135.

Marrodán, L., Y. Song, M. Lubrano Lavadera, O. Herbinet, M. De Joannon, Y. Ju, M. U. Alzueta, and F. Battin-Leclerc. 2019. "Effects of Bath Gas and NO$_x$ Addition on n-Pentane Low-Temperature Oxidation in a Jet-Stirred Reactor." *Energy & Fuels* 33 (6): 5655–5663. https://doi.org/10.1021/acs.energyfuels.9b00536.

Masoudvaziri, N., F. S. Bardales, and O. K. Keskin. 2021. "Streamlined Wildland-Urban Interface Fire Tracing (SWUIFT): Modeling Wildfire Spread in Communities." *Environmental Modelling & Software* 143. https://doi.org/10.1016/j.envsoft.2021.105097.

May, A. A., E. J. T. Levin, C. J. Hennigan, I. Riipinen, T. Lee, J. L. Collett Jr., J. L. Jimenez, S. M. Kreidenweis, and A. L. Robinson. 2013. "Gas-Particle Partitioning of Primary Organic Aerosol Emissions: 3. Biomass Burning." *Journal of Geophysical Research: Atmospheres* 118 (19): 11327–11338. https://doi.org/10.1002/jgrd.50828.

Messerschmidt, B. 2021. "The Fuel of Our Homes – From Building Materials to Content." Presented at The Chemistry of Urban Wildfires: An Information-Gathering Workshop on June 8, 2021, National Academies of Sciences, Engineering, and Medicine, Washington, DC.

Molyneux, S., A. A. Stec, and T. R. Hull. 2014. "The Effect of Gas Phase Flame Retardants on Fire Effluent Toxicity." *Polymer Degradation and Stability* 106: 36–46. https://doi.org/10.1016/j.polymdegradstab.2013.09.013.

Morgan, A. B., and J. W. Gilman. 2013. "An Overview of Flame Retardancy of Polymeric Materials: Application, Technology, and Future Directions." *Fire and Materials* 37 (4).

Moritz, M., and V. Butsic. 2020. *Building to Coexist with Fire: Community Risk Reduction Measures for New Development in California.* Publication 8680. Davis, CA: University of California, Agriculture and Natural Resources. https://doi.org/10.3733/ucanr.8680.

NCAR-UCAR (National Center for Atmospheric Research–University Corporation for Atmospheric Research). 2018. *WE-CAN: Western Wildfire Experiment for Cloud Chemistry, Aerosol Absorption and Nitrogen.* https://www.eol.ucar.edu/field_projects/we-can.

NFPA (National Fire Protection Association). 2018. *Standard for Reducing Structure Ignition Hazards from Wildland Fire.* NFPA 1144. Quincy, MA: National Fire Protection Association.

NWCG (National Wildfire Coordinating Group). 2021. "Live Fuel Moisture Content." In *Fire Behavior Field Reference Guide.* PMS 437. https://www.nwcg.gov/publications/pms437/fuel-moisture/live-fuel-moisture-content.

Ottmar, R. D. 2014. "Wildland Fire Emissions, Carbon, and Climate: Modeling Fuel Consumption." *Forest Ecology and Management* 317. https://doi.org/10.1016/j.foreco.2013.06.010.

Ottmar, R. D., and S. P. Baker. 2007. *Forest Floor Consumption and Smoke Characterization in Boreal Forested Fuelbed Types of Alaska.* Boise, ID: Joint Fire Science Program. https://www.fs.fed.us/pnw/fera/research/smoke/jfsp_boreal_final_report.pdf.

Permar, W., Q. Wang, and V. Selimovic. 2021. "Emissions of Trace Orgnic Gases From Western US Wildfires Based on WE-CAN Aircraft Measurements." *JGR Atmospheres* 126 (11).

Persson, B., and M. Simonson. 1998. "Fire Emissions into the Atmosphere." *Fire Technology* 34.

Phillips, K. A., A. Yau, K. A. Favela, K. K. Isaacs, A. McEachran, C. Grulke, A. M. Richard, A. J. Williams, J. R. Sobus, R. S. Thomas, and J. F. Wambaugh. 2018. "Suspect Screening Analysis of Chemicals in Consumer Products." *Environmental Science & Technology* 52 (5): 3125–3135. https://doi.org/10.1021/acs.est.7b04781.

Philson, C. S., L. Wagner, and R. Nawathe. 2021. "Mitigating California Wildfire Impact Through Zoning and Housing Policy." *Journal of Science Policy & Governance* 18 (1). https://doi.org/10.38126/JSPG180112.

Pitts, W. M. 1995. "The Global Equivalence Ratio Concept and the Formation Mechanisms of Carbon Monoxide in Enclosure Fires." *Progress in Energy and Combustion Science* 21 (3).

Pitz, W. J. 1982. *Structure, Inhibition and Extinction of Polymer Diffusion Flames*. Report number LBL-15825. Berkeley, CA: Lawrence Berkeley National Laboratory. https://escholarship.org/uc/item/1046z3g2.

Pitz, W. J., and R. F. Sawyer. 1978. "Inhibition Effects on Extinction of Polymer Burning." Presented at Western States Section, The Combustion Institute Spring Meeting, Boulder, CO.

Pokhrel, G., D. J. Gardner, and Y. Han. 2021. "Properties of Wood-Plastic Composites Manufactured from Two Different Wood Feedstocks: Wood Flour and Wood Pellets." *Polymers (Basel)* 13 (16). https://doi.org/10.3390/polym13162769.

Popescu, C. M., and A. Pfriem. 2019. "Treatments and Modification to Improve the Reaction to Fire of Wood and Wood Based Products—An Overview." *Fire and Materials* 44 (1): 100–111. https://doi.org/10.1002/fam.2779.

Prichard, S. J., S. M. O'Neill, and P. Eagle. 2020. "Wildland Fire Emission Factors in North America: Synthesis of Existing Data, Measurement Needs and Management Applications." *International Journal of Wildland Fire* 29 (2). https://doi.org/10.1071/WF19066.

Purser, D. 2009. "Influence of Fire Retardants on Toxic and Environmental Hazards from Fires." In *Fire Retardancy of Polymers: New Strategies and Mechanisms* Chapter 24. London, UK: The Royal Society of Chemistry.

Purser, D. A., R. L. Maynard, and J. C. Wakefield (editors). 2015. *Toxicology, Survival and Health Hazards of Combustion Products*. London: The Royal Society of Chemistry. https://doi.org/10.1039/9781849737487.

Quarles, S. L., and K. Pohl. 2018. *Building a Wildfire-Resistant Home: Codes and Costs*. https://headwaterseconomics.org/wildfire/homes-risk/building-costs-codes/.

Sandberg, D. 2016. "Additives in Wood Products—Today and Future Development." In *Environmental Impacts of Traditional and Innovative Forest-Based Bioproducts* Chapter 4. Edited by A. Kutnar and S. S. Muthu. Singapore: Springer Singapore. https://doi.org/10.1007/978-981-10-0655-5.

Schefer, R. W., and N. J. Brown. 1982. "A Comparative Study of HCl and HBr Combustion Inhibition." *Combustion Science and Technology* 29. https://doi.org/10.1080/00102208208923593.

Schnipper, A., L. Smith-Hansen, and E. S. Thomsen. 1995. "Reduced Combustion Efficiency of Chlorinated Compounds, Resulting in Higher Yields of Carbon Monoxide." *Fire and Materials* 19 (2).

Schoennagel, T., J. K. Balch, and H. Brenkert-Smith. 2017. "Adapt to More Wildfire in Western North American Forests as Climate Changes." *Proceedings of the National Academy of Sciences* 114 (18).

Sekimoto, K., A. R. Koss, J. B. Gilman, V. Selimovic, M. M. Coggon, K. J. Zarzana, B. Yuan, B. M. Lerner, S. S. Brown, C. Warneke, R. J. Yokelson, J. M. Roberts, and J. de Gouw. 2018. "High- and Low-Temperature Pyrolysis Profiles Describe Volatile Organic Compound Emissions from Western US Wildfire Fuels." *Atmospheric Chemistry and Physics* 18 (13): 9263–9281. https://doi.org/10.5194/acp-18-9263-2018.

Senkan, S. M. 2000. "Combustion of Chlorinated Hydrocarbons." *Pollutants from Combustion* 547.

Simeoni, A., J. C. Thomas, and P. Bartoli. 2012. "Flammability Studies for Wildland and Wildland–Urban Interface Fires Applied to Pine Needles and Solid Polymers." *Fire Safety Journal* 54. https://doi.org/10.1016/j.firesaf.2012.08.005.

Simmons, R. F., and H. G. Wolfhard. 1956. "The Influence of Methyl Bromide on Flames. Part 2.—Diffusion Flames." *Transactions of the Faraday Society* 52: 53–59.

Simonson, M., P. Andersson, L. Rosell, V. Emanuelsson, and H. Stripple. 2001. *Fire-LCA Model: Cables Case Study*. SP Report 2001:22. Borås, Sweden: SP Swedish National Testing and Research Institute. https://www.diva-portal.org/smash/get/diva2:962180/FULLTEXT01.pdf.

State of California. 2016a. "Exterior Windows, Skylights and Doors." California Building Code Section 708A. https://up.codes/viewer/california/ca-building-code-2016/chapter/7A/sfm-materials-and-construction-methods-for-exterior-wildfire-exposure#708A (accessed February 16, 2022).

State of California. 2016b. "Materials and Construction Methods for Exterior Wildfire Exposure." California Building Code — Matrix Adoption Table Chapter 7A. https://www.hcd.ca.gov/building-standards/state-housing-law/wildland-urban-interface/docs/2010-part-2-cbc-ch7a.pdf (accessed April 1, 2022).

State of California. 2019. *Fire Safety: Low-Cost Retrofits: Regional Capacity Review: Wildfire Mitigation*. Assembly Bill 38.

State of California. 2020. *Fire Prevention: Wildfire Risk: Defensible Space: Ember-Resistant Zones*. Assembly Bill 3074. https://leginfo.legislature.ca.gov/faces/billTextClient.xhtml?bill_id=201920200AB3074 (accessed April 1, 2022).

Stec, A. A. 2017. "Fire Toxicity – The Elephant in the Room?" *Fire Safety Journal* 91: 79–90. https://doi.org/10.1016/j.firesaf.2017.05.003.

Stec, A. A., and T. R. Hull. 2010. *Fire Toxicity*. Sawston, UK: Woodhead Publishing.

Stec, A. A., and T. R. Hull. 2011. "Assessment of the Fire Toxicity of Building Insulation Materials." *Energy and Buildings* 43 (2–3): 498–506. https://doi.org/10.1016/j.enbuild.2010.10.015.

Stec, A. A., J. Readman, P. Blomqvist, D. Gylestam, D. Karlsson, D. Wojtalewicz, and B. Z. Dlugogorski. 2013. "Analysis of Toxic Effluents Released from PVC Carpet under Different Fire Conditions." *Chemosphere* 90 (1): 65–71. https://doi.org/10.1016/j.chemosphere.2012.07.037.

Sturk, D., L. Rosell, P. Blomqvist, and A. Ahlberg Tidblad. 2019. "Analysis of Li-Ion Battery Gases Vented in an Inert Atmosphere Thermal Test Chamber." *Batteries* 5 (3). https://doi.org/10.3390/batteries5030061.

Suh, S., and B. C. Lippiatt. 2012. "Framework for Hybrid Life Cycle Inventory Databases: A Case Study on the Building for Environmental and Economic Sustainability (BEES) Database." *International Journal of Life Cycle Assessment* 17.

Suzuki, S., and S. L. Manzello. 2021. "Firebrands Generated in Shurijo Castle Fire on October 30th, 2019." *Fire Technology*. https://doi.org/10.1007/s10694-021-01176-0.

Swope, C. B., and D. Hernandez. 2019. "Housing as a Determinant of Health Equity: A Conceptual Model." *Social Science & Medicine* 243.

Syphard, A., and J. Keeley. 2019. "Factors Associated with Structure Loss in the 2013–2018 California Wildfires." *Fire* 2 (3). https://doi.org/10.3390/fire2030049.

Syphard, A. D., H. Rustigian-Romsos, M. Mann, E. Conlisk, M. A. Moritz, and D. Ackerly. 2019. "The Relative Influence of Climate and Housing Development on Current and Projected Future Fire Patterns and Structure Loss across Three California Landscapes." *Global Environmental Change* 56: 41–55. https://doi.org/10.1016/j.gloenvcha.2019.03.007.

Syphard, A. D., H. Rustigian-Romsos, and J. E. Keeley. 2021. "Multiple-Scale Relationships between Vegetation, the Wildland–Urban Interface, and Structure Loss to Wildfire in California." *Fire* 4 (1). https://doi.org/10.3390/fire4010012.

Tanaka, P. L. 2003. "Development of a Chlorine Mechanism for Use in the Carbon Bond IV Chemistry Model." *Journal of Geophysical Research* 108 (D4). https://doi.org/10.1029/2002jd002432.

Tao, M., D. Li, R. Song, S. Suh, and A. A. Keller. 2018. "OrganoRelease – A Framework for Modeling the Release of Organic Chemicals from the Use and Post-use of Consumer Products." *Environmental Pollution* 234: 751–761. https://doi.org/10.1016/j.envpol.2017.11.058.

Tejada, J., J. Wiedenmann, and B. Gall. 2020. "Trace Element Behavior in Wood-Fueled Heat and Power Stations in Terms of an Urban Mining Perspective." *Fuel* 267.

Tieszen, S. R., and L. A. Gritzo. 2008. "Transport Phenomena That Affect Heat Transfer in Fully Turbulent Fires." In *WIT Transactions on State of the Art in Science and Engineering*. Ashurst, Southampton, UK: WIT Press. https://doi.org/10.2495/978-1-84564-160-3/02.

Trelles, J., and P. Pagni. 1997. "Fire-Induced Winds in the 20 October 1991 Oakland Hills Fire." *Fire Safety Science*. https://doi.org/10.3801/IAFSS.FSS.5-911.

Urbanski, S. 2014. "Wildland Fire Emissions, Carbon, and Climate: Emission Factors." *Forest Ecology and Management* 317. https://doi.org/10.1016/j.foreco.2013.05.045.

Urbanski, S. P. 2013. "Combustion Efficiency and Emission Factors for Wildfire-Season Fires in Mixed Conifer Forests of the Northern Rocky Mountains, US." *Atmospheric Chemistry and Physics* 13 (14): 7241–7262. https://doi.org/10.5194/acp-13-7241-2013.

USCB (US Census Bureau). 2020. *Characteristics of New Housing*. https://www.census.gov/construction/chars/.

USDOC (US Department of Commerce). 2010. *2010 Characteristics of New Housing: Single-Family Houses Completed, Units in Multifamily Buildings Completed, Multifamily Buildings Completed, Single-Family Houses Sold, Contractor-Built Houses Started*. https://www2.census.gov/programs-surveys/soc/tables/time-series/chars/c25ann2010.pdf.

USFS (US Forest Service). 2022. *Fuel Treatments*. https://wildfirerisk.org/reduce-risk/fuel-treatments/ (accessed April 1, 2022).

van Leeuwen, T. T., G. R. van der Werf, A. A. Hoffmann, R. G. Detmers, G. Rücker, N. H. F. French, S. Archibald, J. A. Carvalho Jr., G. D. Cook, W. J. de Groot, C. Hély, E. S. Kasischke, S. Kloster, J. L. McCarty, M. L. Pettinari, P. Savadogo, E. C. Alvarado, L. Boschetti, S. Manuri, C. P. Meyer, F. Siegert, L. A. Trollope, and W. S. W. Trollope. 2014. "Biomass Burning Fuel Consumption Rates: A Field Measurement Database." *Biogeosciences* 11 (24): 7305–7329. https://doi.org/10.5194/bg-11-7305-2014.

Wallingford, N. 2018. *Camp Incident Damage Inspection Report, CABTU 016737*. https://www.nist.gov/system/files/documents/2020/11/16/2018%20Camp%20Incident%20DINS%20Final%20Report.pdf.

Wang, D. S., and L. H. Ruiz. 2017. "Secondary Organic Aerosol from Chlorine-Initiated Oxidation of Isoprene." *Atmospheric Chemistry and Physics* 17. https://doi.org/10.5194/acp-17-13491-2017.

Wang, M., A. Elgowainy, Z. Lu, A. Bafana, S. Banerjee, P. T. Benavides, P. Bobba, A. Burnham, H. Cai, U. Gracida, T. R. Hawkins, R. K. Iyer, J. C. Kelly, T. Kim, K. Kingsbury, H. Kwon, U. Lee, Y. Li, X. Liu, L. Ou, N. Siddique, P. Sun, P. Vyawahare, O. Winjobi, M. Wu, H. Xu, E. Yoo, G. G. Zaimes, and G. Zang. 2021. *Greenhouse Gases, Regulated Emissions, and Energy Use in Technologies Model*. Computer software. Washington, DC: US Department of Energy, Office of Energy Efficiency and Renewable Energy. https://doi.org/10.11578/GREET-Net-2021/dc.20210903.1.

Wang, Z., O. Herbinet, N. Hansen, and F. Battin-Leclerc. 2019. "Exploring Hydroperoxides in Combustion: History, Recent Advances and Perspectives." *Progress in Energy and Combustion Science* 73: 132–181. https://doi.org/10.1016/j.pecs.2019.02.003.

Weschler, C. J. 2009. "Changes in Indoor Pollutants Since the 1950s." *Atmospheric Environment* 43 (1).

Weyerhaueser. 2018. *Safety Data Sheet (SDS): Weyerhaeuser Oriented Strand Board (OSB) Products including: Sheathing, Edge™, Edge Gold™, RBS™ (Radiant Barrier Sheathing), Rim Board, SturdiStep™*. Longview, WA: Weyerhaueser.

Wiggins, E. B., A. Andrews, C. Sweeney, J. B. Miller, C. E. Miller, S. Veraverbeke, R. Commane, S. Wofsy, J. M. Henderson, and J. T. Randerson. 2021. "Boreal Forest Fire CO and CH_4 Emission Factors Derived from Tower Observations in Alaska during the Extreme Fire Season of 2015." *Atmospheric Chemistry and Physics* 21 (11): 8557–8574. https://doi.org/10.5194/acp-21-8557-2021.

Willstrand, O., R. Bisschop, P. Blomqvist, A. Temple, and J. Anderson. 2020. *Toxic Gases from Fire in Electric Vehicles*. RISE Report 2020:90. Gothenburg, Sweden: Research Institutes of Sweden. http://ri.diva-portal.org/smash/get/diva2:1522149/FULLTEXT01.pdf.

Xie, Q., J. Xiao, P. Gardoni, and K. Hu. 2019. "Probabilistic Analysis of Building Fire Severity Based on Fire Load Density Models." *Fire Technology* 55 (4): 1349–1375. https://doi.org/10.1007/s10694-018-0716-0.

Yetter, R. A., and F. L. Dryer. 1992. "Inhibition of Moist Carbon Monoxide Oxidation by Trace Amounts of Hydrocarbons." Presented at 24th International Symposium on Combustion, The University of Sydney, Sydney, Australia. https://doi.org/10.1016/S0082-0784(06)80093-7.

Yokelson, R. J., J. G. Goode, D. E. Ward, R. A. Susott, R. E. Babbitt, D. D. Wade, I. Bertschi, D. W. T. Griffith, and W. M. Hao. 1999. "Emissions of Formaldehyde, Acetic Acid, Methanol, and Other Trace Gases from Biomass Fires in North Carolina Measured by Airborne Fourier Transform Infrared Spectroscopy." *Journal of Geophysical Research: Atmospheres* 104 (D23): 30109–30125. https://doi.org/10.1029/1999jd900817.

Zammarano, M., M. S. Hoehler, J. R. Shields, A. L. Thompson, I. Kim, I. T. Leventon, and M. F. Bundy. 2020. *Full-Scale Experiments to Demonstrate Flammability Risk of Residential Upholstered Furniture and Mitigation Using Barrier Fabric*. NIST Technical Note 2129. Gaithersburg, MD: NIST. https://doi.org/10.6028/NIST.TN.2129.

Zota, A. R., G. Adamkiewicz, and R. A. Morello-Frosch. 2010. "Are PBDEs an Environmental Equity Concern? Exposure Disparities by Socioeconomic Status." *Environmental Science and Technology* 44 (15): 5691–5692. https://doi.org/10.1021/es101723d.

4

Atmospheric Transport and Chemical Transformations

Wildland-urban interface (WUI) fires can have substantial negative impacts on human health, visibility, and quality of life, not only in the vicinity of the fire, but also hundreds of kilometers downwind. For example, even 240 km downwind of the Camp Fire, in the Bay Area of northern California, daily fine particulate matter ($PM_{2.5}$) concentrations remained elevated (70–200 $\mu g/m^3$) for over 2 weeks, well above the National Ambient Air Quality Standard of 35 $\mu g/m^3$ (Rooney et al., 2020). The population experiencing this exposure was in excess of seven million people. Thus, this WUI fire resulted in substantial adverse particulate matter exposures to large populations beyond the immediate zone of the fire. Smoke from major fires sometimes affects air quality and visibility on a continental scale, and seasonally averaged summertime contributions from wildland fire smoke $PM_{2.5}$ can be comparable to contributions from non-smoke $PM_{2.5}$ in some US regions (e.g., in the Pacific Northwest; O'Dell et al., 2019).

Finding: A large number of people are exposed to WUI fire emissions over broad geographic areas.

While national monitoring networks provide data for assessing exposures of routinely monitored air pollutants, such as fine particulate matter, the gas- and particle-phase smoke composition specifically associated with WUI fires is not well understood. WUI smoke composition differs from wildland fire smoke because of direct emissions from the combustion and volatilization of human-made materials and the chemical interaction of these emissions with wildland fire emissions (see Chapter 3). Atmospheric chemistry and physical changes (e.g., dilution; phase changes; changes in partitioning of compounds between the gas phase, particles, and cloud droplets; deposition), which are dynamic on timescales of minutes to days, lead to dramatic changes in pollutant composition downwind of fires. However, the impact of WUI emissions and chemistry on regional exposures is not well understood. This chapter explores this impact, emphasizing atmospheric chemistry that informs inhalation exposures and the resulting health effects associated with emissions from WUI fires.

It is important to recognize that, in addition to chemical processes, the physical processes of wet and dry atmospheric deposition can also be a source of contaminated water and soil and thus impact exposures through ingestion. Organic nitrogen (Yu et al., 2020), isocyanic acid (Roberts and Liu, 2019; Roberts et al., 2011) and per- and polyfluoroalkyl substances (PFASs; Pike et al., 2021; Shimizu et al., 2021) are examples of water-soluble compounds found in emissions from burning biomass or in emissions from burning human-made materials. Wet deposition of water-soluble compounds can contribute to water and/or soil contamination and exposure through WUI fire–associated wet deposition. The topic of water and soil contamination is examined in Chapter 5.

Finally, while this chapter will not cover visibility, it is worth noting that the impacts of WUI fires on visibility can be substantial and can impact transportation safety, property values, mental health, and quality of life.

In describing chemical processes that affect the concentrations and composition of air pollutants from WUI fires, the committee defines four zones beyond the fire: the near-field, local, regional, and continental scales. Box 4-1 defines these spatial scales and other key terminology used in the chapter.

PRIMARY SPECIES WITH TOXIC POTENTIAL DOWNWIND OF WUI FIRES

Table 4-1 lists some of the toxicants expected to be directly emitted from WUI fires (i.e., the primary pollutants) and processes that impact their transport, their partitioning between the gas and particle phases in the atmosphere, their transformation, and their degradation. This list draws from laboratory and field measurements focusing on wildland fires, structure fires, and WUI fires, and information about materials (see also Chapter 3). An estimate of species' atmospheric lifetimes is provided. This estimate is based on the primary degradation pathways being photolysis, reaction with hydroxyl radical, and reaction with ozone in the gas phase. Light intensities, hydroxyl and other radical concentrations, and ozone concentrations that the toxicants are exposed to will vary considerably, so these lifetime estimates are approximate and relative indicators of the persistence of the individual chemical species. The possibility that other loss processes dominate in WUI fire plumes (e.g., oxidation by chlorine radical, multiphase chemistry, cloud chemistry) is discussed in the next section, "Atmospheric Transformations."

ATMOSPHERIC TRANSFORMATIONS

The Downwind Fate of WUI Fire Emissions

A wide range of plume studies including studies of wildland fire plumes provide confidence that WUI smoke concentrations decrease with distance downwind through dilution and surface deposition and undergo chemical and phase changes in the atmosphere (Figure 4-1); however, the influence of various chemical processes on the downwind composition of WUI fire plumes is poorly understood. Exposures (i.e., who is exposed, to what, and at what concentrations) depend most on whether the emissions are lofted above the mixed layer and transported long distances before being downmixed to ground level or are trapped near the surface in the near field. Plume rise (i.e., plume injection height) depends on the overall heat released by the fire, as discussed in Chapter 3, and is a major factor determining transport altitude. Meteorology, including transport altitude—the location of the plume within or above the layer of the atmosphere that is well mixed (mixed layer)—also affects toxicant concentrations, gas-particle partitioning, photochemistry, cloud processing, and multiphase chemistry (e.g., through its influence on dilution, temperature and saturation vapor pressure, solar radiation, aerosol liquid water content, and phase separation) as described for wildland fires by Jaffe et al. (2020). Photochemical oxidation in a wildland fire plume can produce elevated ground-level ozone (Dreessen et al., 2016; Lu et al., 2016) and oxidized aerosol on a regional scale (Garofalo et al., 2019; Junghenn Noyes et al., 2020). This is also expected in WUI fire plumes.

Finding: Atmospheric chemistry transforms emissions with distance from the fire, so that downwind communities are exposed to more oxidized mixtures than near-field communities.

A variety of chemical and physical processes that take place on timescales of minutes to days reduce the concentrations of primary species with toxic potential and can also form secondary species with toxic potential. A distinguishing feature of wildland fires, as an emission source, is the prevalence of incomplete combustion and thermal degradation of oxygenated fuel components, such as cellulose, when compared to most primary emission sources such as automotive exhaust. This results in an abundance of primary oxygenated, water-soluble compounds in wildland fire emissions (e.g., HONO, HNO_3, HCl, formaldehyde, glycolaldehyde, acetic acid, phenols, furfurals, isocyanic acid, and amines; Hatch et al., 2015; Friedli et al., 2001; Juncosa Calahorrano et al., 2021; Koss et al., 2018; Roberts et al., 2011; Selimovic et al., 2018; Tomaz et al., 2018). Comprehensive volatile organic compound (VOC) measurements from controlled burns of 18 US wildland fire fuels suggest

BOX 4-1
Key Terminology

Spatial Scales

Near-field scale: From 1 to 10 km downwind of the fire, where the plume remains quite concentrated, dilution has a major effect on the gas-particle partitioning, and chemistry is driven by fast processes that occur on a timescale of minutes

Local scale: From 10 to 100 km, where chemistry is driven by processes that occur on a timescale of minutes to hours

Regional scale: From 100 to 1,000 km, where chemistry is driven by processes that occur on a timescale of hours to days

Continental scale: Greater than 1,000 km, where chemistry is driven by processes that occur over days

Terms and Definitions

Atmospheric lifetime: Average time a molecule of species *i* remains in the atmosphere

Dry atmospheric deposition: The removal of particles and gases from the atmosphere to surfaces in the absence of precipitation (i.e., through gravitational settling, impaction, interception, and diffusion)

Fine particulate matter ($PM_{2.5}$): Airborne particles with diameters of 2.5 micrometers or less, small enough to enter the lungs and bloodstream, posing risks to human health

Fire plume: Air mass downwind of combustion zone, containing elevated concentrations of combustion products

Intermediate-volatility organic compounds (IVOCs): Compounds with vapor pressures between those of VOCs and semi-VOCs; a suite of compounds that, based on their vapor pressure, tend to evaporate from the particle phase with near-field dilution of plumes

Partitioning: The distribution of a species *i* between two media, such as air and particles, or air and water

Pollutant: A chemical or biological substance that harms water, air, or land quality

Prescribed burn: Fire set intentionally for forest or farmland management

Primary organic aerosol: Organic particulate matter that is emitted from the source (e.g., WUI fire) in particulate form

Primary species with toxic potential: Toxic substance that is emitted directly from the source (e.g., WUI fire)

Secondary organic aerosol: Organic particulate matter that is formed in the atmosphere from precursor gases

Secondary species with toxic potential: Toxic substances that are formed through atmospheric chemistry

(continued)

Volatile organic compounds (VOCs): Organic compounds with vapor pressures high enough to exist in the atmosphere primarily in the gas phase, typically excluding methane; VOCs can easily become airborne for inhalation exposure

Wet atmospheric deposition: The removal of particles and gases to the earth's surface via scavenging by precipitation

VOC emissions from wildland fires are roughly 60 percent oxygenated (Gilman et al., 2015). The atmospheric fate of water-soluble compounds is determined by the competition between the rates of gas-phase photochemical reactions, reactive uptake in clouds and wet aerosols, and wet deposition; understanding the fate of water-soluble compounds in wildland fire plumes is an active area of research.

The impacts of fire emissions from human-made materials and firefighting activities on the chemistry and fate of WUI fire plumes is not well understood. Relative to wildland fire emissions, WUI emissions have increased levels of several types of compounds, including reactive halogenated compounds (HCl, HBr, HF), dioxins, phenols from the degradation of polymers, aldehydes, nitrogen-containing organics such as isocyanates, brominated and fluorinated organics, and metals (see Chapter 3 and Table 4-1). These species and radicals derived from their atmospheric chemistry could alter the rate of chemical transformations in WUI plumes, changing the lifetime of primary species with toxic potential and the formation of secondary species with toxic potential. The abundance, atmospheric dynamics, and fate of WUI-enriched species are of interest for several reasons, as discussed below.

Near-Field Changes in Partitioning with Dilution

At the near-field scale, downwind of the immediate zone of the fire, dilution of emissions with clean air results in near-field smoke concentrations that are substantially lower than concentrations in the immediate zone. It also results in substantial evaporation from primary organic aerosols, shifting some compounds from the particle to the gas phase. This alters the fate of the mixture and the dose inhaled by exposed populations. For example, laboratory studies of a primary diesel combustion aerosol suggest that dilution to ambient conditions results in evaporation of roughly three quarters of the primary organic aerosol (Robinson et al., 2007, Figure 1A).

Several intermediate-volatility organic compounds (IVOCs), which evaporate during dilution, can subsequently undergo gas-phase oxidation over minutes to hours (i.e., at the near-field and local scales), changing the composition and increasing the oxidation state of organics in both the gas and particle phases (Lambe et al., 2012; Robinson et al., 2007). Phenol and guaiacol, formed through the thermal degradation of lignin, are examples of IVOC wildland fire emissions (Alves et al., 2011; Hatch et al., 2018; Yee et al., 2013).

WUI fire emissions also have increased levels of dioxins and synthetic polymer degradation products (see Chapter 3; Reisch, 2018; Ruokojärvi et al., 2000), some of which are IVOCs and could evaporate from emitted particles with near-field dilution. Some PFASs are also IVOCs. While they are only a small portion of the total mass of human-made materials associated with structures, PFASs are ubiquitous in indoor environments because of their use in consumer products (Wang et al., 2017). They are also found in Class B firefighting foams stored in fire departments for use on chemical fires in some states. The shift of IVOCs from the particle to the gas phase with dilution can have a large influence on their chemistry and persistence in the atmosphere. For example, in the particle phase, polychlorinated dibenzo-p-dioxins (PCDDs) persist for 1–2 weeks in the atmosphere, whereas in the gas phase, some PCDDs have atmospheric lifetimes as short as 0.5 to 2.0 days against oxidation (Table 4-1). Note that accounting for phase changes with near-field dilution in atmospheric models requires that care is taken to document dilution and temperature conditions during emissions measurements.

TABLE 4-1 Selected Major Primary Species with Toxic Potential Emitted from WUI Fires

	Transport and Partitioning	Transformation and Degradation	Atmospheric Lifetime with Respect to OH and O_3
Carbon monoxide (CO)	Partitions to the atmosphere and is distributed globally via wind. CO in the troposphere is slowly transported to the mesosphere and stratosphere.	Stable under environmental conditions. Reactions with molecular oxygen or water vapor are very slow at ambient temperatures and pressure. Primary degradation pathway of tropospheric CO is via its reaction with photochemically produced hydroxyl radicals, resulting in formation of CO_2.[a] In the stratosphere, it reacts with atomic oxygen generated by the photodissociation of O_2 to form CO_2.[a]	CO remains in the atmosphere for an average of about 2 months.[b]
Formaldehyde (HCHO)	Released to the atmosphere in large amounts and is also formed in the atmosphere by the oxidation of hydrocarbons. Efficiently transferred into clouds, rain, and surface water. Dry deposition and wet removal half-lives are estimated to be 19 and 50 hours, respectively.[a]	Breaks down in the gas phase to formic acid and CO. Reacts via direct photolysis and hydroxyl radical oxidation. Oxidation can also occur in cloud droplets—producing formic acid (a component of acid rain).[a]	The HCHO half-life due to reaction with the hydroxyl radical is approximately 16 hours in clean air and much less in polluted air and wildland fire plumes with elevated reactive oxidant concentrations.[b]
Benzene (C_6H_6)	Gas-phase aromatic compound released into the atmosphere in large amounts from wildfires and from fossil fuels.[c]	Reacts with ozone and hydroxyl radicals.[a]	Benzene's half-life due to reaction with the hydroxyl radical is approximately 9 days.[b]
Toluene (C_7H_8)	Gas-phase aromatic compound released into the atmosphere in large amounts from wildfires and from fossil fuels.[c]	Reacts with ozone and hydroxyl radicals.[a]	Toluene's half-life due to reaction with the hydroxyl radical is approximately 2 days.[b]
Acrolein (C_3H_4O)	Limited atmospheric transport. No partitioning from vapor phase to atmospheric particles. Removed from atmosphere by wet deposition.[a]	Reacts with hydroxyl radicals. Products include CO, HCHO, glyoxal, and glycolaldehyde. In the presence of nitrogen oxides, products include peroxynitrate, NO, glycolaldehyde, malonaldehyde, and 3-hydroxypropanaldehyde. Direct photolysis is minor under ambient conditions.[a]	Acrolein's half-life due to reaction with the hydroxyl radical is approximately 5 hours in clean air and about half that time in polluted air.[b]
Hydrogen cyanide (HCN)	Slow degradation in air; thus, it can be transported over long distances. Largely remains in lower altitudes (troposphere); only 2% of tropospheric hydrogen cyanide is transferred to the stratosphere. Water-soluble cyanide particles are expected to be removed by both wet and dry deposition.[a]	Resistant to photolysis. Reacts with hydroxyl radicals to form CO and NO. This photooxidation occurs at least an order of magnitude faster at lower altitudes (0–8 km) than at upper tropospheric altitudes (10–12 km).[a]	The residence time of HCN in the atmosphere has been estimated to be approximately 1.3–5.0 years, depending on the hydroxyl radical concentration.

continued

TABLE 4-1 Continued

	Transport and Partitioning	Transformation and Degradation	Atmospheric Lifetime with Respect to OH and O_3
Ammonia (NH_3)	The pH and temperature influence transport and partitioning. Dry deposition of NH_3 dominates when there are large NH_3 emissions. Where NH_3 emissions are lower, wet deposition of neutralized ammonium aerosols dominates.[a]	Neutralizes acidic air pollutants (e.g., H_2SO_4, HNO_3, or HCl) to form ammonium salts in aerosols.	The half-life for ammonia in the atmosphere is estimated to be a few days; however, it will dynamically partition very quickly between the gas phase (as ammonia) and the particle phase (as ammonium), depending on the acidity of the particles in the atmosphere.
Nitrogen oxides (NO_x)	Dominant source of NO_x in the air is combustion; 90–95% of NO_x molecules are emitted as NO and 5–10% as NO_2; however, NO is rapidly oxidized to NO_2. Wet precipitation and dry deposition remove NO_x from the atmosphere (US Environmental Protection Agency, 2021).	NO_x contributes to the formation of photochemical smog and acid rain (US Environmental Protection Agency, 2021). Oxidation products of NO_x in the troposphere include HNO_3, HO_2NO_2, HNO_2, peroxyacylnitrates, N_2O_5, nitrate radical (NO_3), and organic nitrates. Dominant form of NO_x (NO, NO_2, HNO_3, etc.) in the lower atmosphere varies, depending on sunlight intensity, temperature, pollutant emissions, pollutant emission lag time, and meteorological factors.	This group of compounds is characterized by relatively short atmospheric lifetimes of hours to days; NO and NO_2 react very quickly with ozone, reaching a dynamic equilibrium.
Sulfur dioxide (SO_2)	Oxidized rapidly in atmospheric waters and removed from the atmosphere by precipitation or dry deposition, mainly as sulfuric acid (acid rain).	Oxidized by OH radicals in the gas phase and in clouds/fogs by peroxides, NO_2, and transition metals, or catalytically on surfaces.[a] Reaction forms SO_3 and subsequently sulfate.[a]	Depending on the hydroxyl radical concentration, sulfur dioxide may rapidly be converted to sulfuric acid, and partitions into the particle phase. This process has a half-life of hours to days at typical hydroxyl radical concentrations. When conditions are favorable, aqueous oxidation can be considerably faster.
Hydrogen sulfide (H_2S)	Partitions to surface water, groundwater, or moist soil via sorption.[a]	Oxidized by oxygen to give sulfur dioxide (SO_2) and ultimately sulfate compounds. Not expected to undergo significant photolysis.[a]	H_2S has lifetimes in air ranging from approximately 1 day in the summer to 42 days in the winter.

Compound			
Hydrogen chloride (HCl)	Acts as a reservoir species, temporarily removing chlorine radicals from a catalytic ozone destruction cycle. Can be found in the gas phase, clouds, and aerosols, dependent on temperature, humidity, and ammonia level. Lower ambient temperature and higher relative humidity favor aerosol formation.[a]	Can react with hydroxyl radicals to form chloride radicals and water. Chloride radicals can deplete O_3 and react rapidly with VOCs. Oxidation of VOCs by chloride radicals can lead to secondary aerosol production.[a]	The half-life for HCl in the atmosphere is estimated to be a few days; however, it will dynamically partition very quickly between the gas phase and the particle phase, depending on the concentration of ammonia and available cations such as sodium.
Hydrogen bromide (HBr)	Limited data on HBr in the atmosphere. Likely to be removed by both wet and dry deposition. Both acidic bromine and particulate bromine are observed in the troposphere.[a]	Does not undergo photolysis in the troposphere. Can react with hydroxyl radicals to form bromide radicals and water. Like chlorine radicals, bromide radicals (Br·) can deplete O_3 and react rapidly with VOCs. Experimental data show HBr may be indirectly formed in the upper atmosphere from organic bromine compounds (such as CH_3Br) via the reaction of BrO with hydroxyl, OH, which then contributes to the depletion of ozone.[a]	The half-life for HBr in the atmosphere is estimated to be a few hours; however, it will dynamically partition very quickly between the gas phase and the particle phase, depending on the concentration of ammonia and available cations such as sodium.
Hydrogen fluoride (HF)	Absorbed by atmospheric water (rain, clouds, fog, snow, aerosol liquid water), forming aqueous hydrofluoric acid. Removed by dry and wet deposit on (i.e., acid precipitation). Can be neutralized (e.g., by sodium or other cation) and taken up by particulate matter or dust.[a]	HF is much too reactive to reach the upper atmosphere, and tropospheric sources therefore do not interfere with the ozone layer. Polymerization or depolymerization reaction of HF does not destroy or remove HF or its oligomers (dimer $(HF)_2$, hexamer $(HF)_6$, and octamer $(HF)_8$) from the air.[a]	The atmospheric lifetime of HF is less than 4 days. Deposition is the dominant mechanism driving removal from the atmosphere (Cheng, 2018).
Per- and polyfluoroalkyl substances (PFASs, e.g., perfluorooctane sulfonate, perfluorooctanoic acid)	Found in both the gas and particle phases. Removed via wet and dry deposition.[a]	PFASs include thousands of compounds across several compound classes. Perfluorinated carboxylic acids and sulfonates are resistant to photolysis and atmospheric photooxidation, but fluorotelomer and fluorosulfamido alcohols can be oxidized by OH radicals to form carboxylic acids and sulfonates (Ellis et al., 2004; Young et al., 2007)	The lifetimes are days to weeks (Ellis et al., 2003).

continued

76

TABLE 4-1 Continued

	Transport and Partitioning	Transformation and Degradation	Atmospheric Lifetime with Respect to OH and O₃
Polycyclic aromatic hydrocarbons (PAHs)	Subject to short- and long-range transport and are removed by wet and dry deposition onto soil, water, and vegetation. Partitioning of PAHs between the gas and the condensed phases depends on their volatility, the temperature, and the concentration of particles in the atmosphere. PAHs having two to three rings are present in air predominantly in the vapor phase; those with four rings exist both in the vapor and particulate phase; and those having five or more rings are found predominantly in the particle phase. Atmospheric residence time and transport distance depend on the size of the particles to which the PAHs are sorbed and on climatic conditions (which will determine rates of wet and dry deposition). Particles with a diameter range of 0.1–3.0 μm, with which airborne PAHs are principally associated, remain airborne for 5–10 days and can be transported long distances.	Two types of chemical reactions transform PAHs: (1) reactions between PAHs adsorbed on particle surfaces and oxidant gases like NO_2, O_3, and SO_3, and (2) gas-phase photooxidation producing oxidized derivatives such as quinones, ketones, or acids. Photochemical reactions involve NO_x, N_2O_5, OH, O_3, SO_2, and peroxyacetyl nitrate. Reactions of PAHs, including fluoranthene and pyrene, with the OH radical (in the presence of NO_x) and with N_2O_5 lead to the formation of nitroarenes in the ambient air, which are more mutagenic than their parent PAHs. Reaction with ozone or peroxyacetylnitrate yields diones; nitrogen oxide reactions yield nitro- and dinitro-PAHs. Compounds adsorbed to particles are more resistant to photochemical reactions than gas-phase PAHs.[a] Variations in chemical composition of different types of particles such as diesel exhaust and wood smoke might strongly affect the reactivity of PAHs.[a]	The atmospheric lifetimes of naphthalene and phenanthrene, due to reaction with the hydroxyl radical in relatively clean air, are about 6 hours and 10 hours, respectively; PAHs may be more resistant to oxidation if sorbed onto particles.[b]
Polychlorinated biphenyls (PCBs)	Enter the atmosphere when volatilized from soil and water. In the atmosphere, PCBs are present both in the vapor phase and sorbed to aerosol particles. Transport and partitioning behavior of PCBs depends on the number of chlorines on the biphenyl molecule, because of the influence on compound vapor pressure. Biphenyls with 0–1 chlorine atoms remain in the atmosphere, while those with 1–4 chlorines gradually migrate toward polar latitudes in a series of volatilization/deposition cycles; those with 4–8 chlorines remain in the mid-latitudes, and those with 8–9 chlorines remain close to the source of contamination. PCBs in the vapor phase are more mobile and transported farther than particle-bound PCBs.	The ability of PCBs to be degraded or transformed in the environment depends on the degree of chlorination of the biphenyl molecule as well as on the isomeric substitution pattern. The vapor-phase reaction of PCBs with hydroxyl radicals is the dominant transformation process in the atmosphere. Possible reaction schemes involve the formation of a 2-hydroxybiphenyl intermediate, which quickly degrades to chlorinated benzoic acid. Insufficient data are available on the importance of photolysis and/or chemical reactions of particle-phase congeners.[a]	The calculated tropospheric lifetime for the reaction of PCBs increases as the number of chlorine substitutions increases. The tropospheric half-lives are 2 days for monochlorobiphenyls, 4 days for dichlorobiphenyls, 10 days for trichlorobiphenyls, and 15 days for tetrachlorobiphenyls; PCBs may be more resistant to oxidation if sorbed onto particles.[b]

	Heavier and coplanar PCBs tend to be particle bound and/or more readily degraded in the atmosphere. PCBs are removed by wet deposition; by dry deposition of aerosols; and by vapor adsorption or partitioning to water, soil, and plant interfaces.[a]		
Polychlorinated dibenzo-p-dioxins (PCDDs)	Combustion-generated PCDDs associated with particulate matter can be distributed over large areas. During transport, PCDDs partition between the vapor phase and particle-bound phase. However, partitioning is largely to the particulate phase. Removed from the atmosphere via wet deposition, particle dry deposition, and gas-phase dry deposition. The less chlorinated PCDD congeners (tetra-CDD and penta-CDD) occur in greater proportion in the vapor and dissolved phases of air and rain, whereas the more chlorinated congeners (hepta-CDD and octa-CDD) are associated with particles.[a]	PCDDs' dominant transformation processes are photolysis and gas-phase diffusion, and volatilization with subsequent photolysis. Transformation reactions depend on whether the PCDD is in the vapor or particulate phase. Vapor-phase PCDDs are not likely to undergo reactions with atmospheric ozone, nitrate, or hydroperoxy radicals; however, reactions with hydroxyl radicals may be significant, particularly for the less-chlorinated congeners (up to tetra-CDD). Based on the photolysis lifetimes of PCDDs in solution, it is expected that vapor-phase PCDDs will also undergo photolysis in the atmosphere, although reactions with hydroxyl radicals will predominate. In the air, the low vapor pressure of octa-CDD results in its partitioning primarily to the particulate phase rather than the vapor phase; therefore, atmospheric photodegradation is less likely to occur for this tightly bound congener.[a]	The atmospheric lifetimes of PCDDs are estimated to range from 0.5 days for mono-CDD to 9.6 days for octa-CDD, with tetra-CDD having a lifetime of 0.8–2 days. Particulate-bound PCDDs are removed by wet or dry deposition with an atmospheric lifetime of approximately 10 days and, to a lesser extent, by photolysis.
Polybrominated diphenyl ethers (PBDEs)	Exist in both gas and particle phases in the atmosphere. Particulate-phase PBDEs are removed from the atmosphere by wet and dry deposition.[a]	Vapor-phase PBDEs may be degraded via oxidation with hydroxyl radicals or direct photolysis, with photolysis being the dominant method. Limited data are available on the rate and extent of photolysis of PBDEs in air.[a]	The half-lives for this reaction in air are estimated to be 29, 140, and 476 days, respectively, for penta-, octa-, and deca-BDE homologs, calculated using a structure estimation method.
Phosphate ester flame retardants	Can be present in the gas or particle phase. Subject to wet and dry deposition.[a]	May be degraded through reaction with hydroxyl radicals. Semi-volatile phosphate esters have the potential to hydrolyze to diesters, monoesters, and phosphoric acid.[a]	Atmospheric half-lives are on the order of 1–12 hours for the selected phosphate esters. Particle-bound phosphate esters are more resistant to OH oxidation with lifetimes of 4–14 days (Liu et al., 2014).

NOTES: Based on laboratory and field measurements from wildland fires, WUI fires, and structure fires, as well as information about materials. HCl, HBr, HF, PFASs, PCBs, PCDDs, PDBEs, and phosphate ester flame retardants are expected to be found in higher concentrations in WUI fires than in wildland fires.

[a]US Agency for Toxic Substances and Disease Registry. 2021. https://www.atsdr.ccc.gov/ (accessed February 26, 2022).

[b]US Environmental Protection Agency. 2021. *EPI Suite™ Estimation Program Interface.* https://www.epa.gov/tsca-screening-tools/epi-suitetm-estimation-program-interface (accessed December 22, 2021).

[c]Jaffe et al., 2020; Evtyugina et al., 2013.

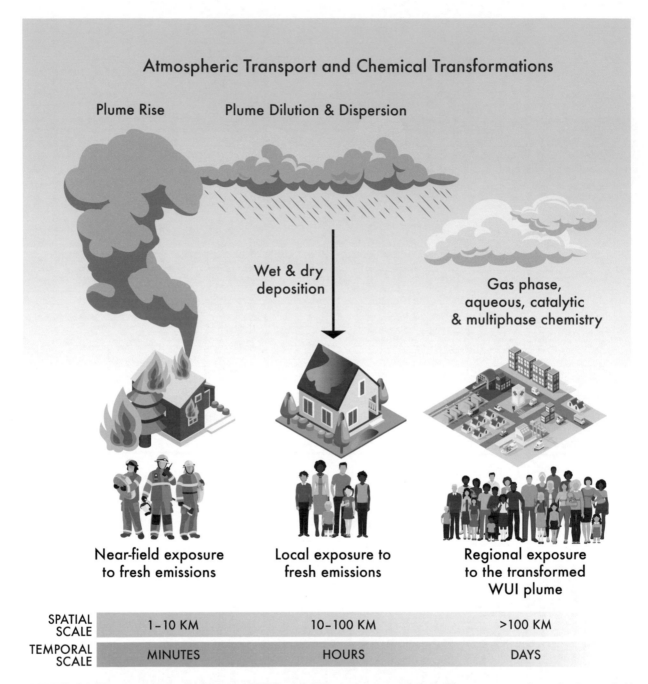

FIGURE 4-1 The movement and changes of WUI emissions over space and time. The processes, shown in the top half above the structures, are discussed in this chapter. The elements showing exposure, in the bottom half below the structures, are discussed in Chapter 6.

Recent progress has been made in quantifying IVOCs in particulate emissions from wildland fires and predicting their downwind chemistry in models (Hatch et al., 2017; Theodoritsi et al., 2021; Yee et al., 2013). However, the characterization of IVOC emissions from burning structures in WUI fires is limited, and predictive modeling has not yet explored the implications of the near-field phase changes and subsequent local- to regional-scale chemical processing of these species. Measurements provide limited evidence for the enhancement of structure-related tracer species above the ambient background in regional measurements; however, some of these species or their reaction products may persist over regional scales (Table 4-1).

Formation of Secondary Species with Toxic Potential

While concentrations of primary species with toxic potential decrease with distance downwind as a result of dilution with ambient air, wet and dry deposition, and oxidation (Table 4-1), oxidation can also produce secondary species with toxic potential (Table 4-2). Measurement of tracer compounds that are specific to WUI fire emissions (primary tracers) and WUI chemistry (secondary tracers) could be used to identify and understand the influence of WUI fires on downwind communities. While little is known about secondary species with toxic potential from WUI fires, elevated concentrations of the hydroxyl radical, reactive nitrogen, reactive chlorine, and transition metals are likely to exist in WUI plumes (see "Major Atmospheric Oxidants Driving Gas-Phase Chemistry" below) and will react with a variety of unsaturated organic compounds to form more oxygenated, chlorinated, and nitrated products that may have increased toxicity (Tuet et al., 2017; Wong et al., 2019). For example,

- Nitrophenols are likely to form in WUI fire plumes since phenols are components of wildland fire plumes and nitrophenols are known to form through daytime and nighttime gas-phase radical chemistry and through reactive nitrogen (e.g., N_2O_5, $ClNO_2$) chemistry in aqueous or multiphase reactions in the atmosphere (Harrison et al., 2005).
- Chlorinated and oxygenated aromatics are likely to form in WUI fire plumes, since aromatic compounds are substantial components of wildland fire plumes, chlorine radicals and hydroxyl radicals are expected to be present in WUI fire plumes, and chlorinated aromatics and oxygenated aromatics are known to form in the atmosphere (e.g., from gas-phase chlorine radical reaction with PAHs; Ohura et al., 2005; Riva et al., 2015).

TABLE 4-2 Examples of Secondary Species with Toxic Potential Observed or Proposed to Occur in WUI Fires

Compound or Compound Class	Precursor	Formation Pathway
Ozone	NO_x and VOCs	Gas-phase oxidation of VOCs in the presence of NO_x and ultraviolet light
Secondary $PM_{2.5}$	SO_2, H_2S, NO_x, NH_3, VOCs, and IVOCs	(1) Gas-phase oxidation followed by vapor pressure–driven partitioning into the particle phase; (2) aqueous or multiphase reaction in aerosol, clouds, or fogs followed by water evaporation and retention of lower volatility products in the particle phase; (3) heterogeneous or catalytic surface reactions on aerosol particles
Formaldehyde, other aldehydes	Most VOCs	Gas phase oxidation of VOCs (Luecken et al., 2018)
Nitroaromatics	Aromatics	Gas-phase and aqueous or multiphase chemistry with reactive nitrogen species (e.g., nitrophenols; Harrison et al., 2005; Pang et al., 2019)
Chlorinated aromatics	Aromatics	From chlorine radical reaction (e.g., chlorinated PAHs; Ohura et al., 2005; Riva et al., 2015)
Perfluorocarboxylic acids	Fluorotelomer alcohols	OH oxidation in gas phase (e.g., perfluorooctanoic acid from 8:2 fluorotelomer alcohol; Wallington et al., 2006)

NOTE: Based on observation in wildland fires or WUI fires, or knowledge about WUI materials or emissions and fundamental atmospheric chemistry knowledge.

- To the extent that PFASs are emitted from WUI fires, oxidation may yield perfluorocarboxylic acids. Photochemical hydroxyl radical oxidation of fluorotelomer alcohols (a type of PFAS found in homes; Hall et al., 2020; Sha et al., 2018), where present, yields perfluorocarboxylic acids such as perfluorooctanoic acid (Wallington et al., 2006).

Stable unique products of these chemistries (e.g., products of chlorine radical chemistry; tracers of synthetic polymer degradation products) could be used as chemical tracers to identify the influence of secondary formation, understand how human-made materials alter the chemistry, and determine the dominant daytime and nighttime oxidants, and potentially to identify dominant formation pathways (e.g., gas-phase chemistry vs aqueous or multi-phase chemistry).

Wildland fire emissions are known to increase formation of ozone, a secondary species with toxic potential formed through photochemical reactions involving oxides of nitrogen and VOCs (Xu et al., 2021); insights pertaining to ozone formation in WUI fires are provided in the section called "Ozone Formation," below. Like-wise, secondary $PM_{2.5}$ will also form downwind of WUI fires. Particulate sulfate forms from the oxidation of SO_2 and H_2S in the gas phase, in clouds and fogs, or in some cases on catalytic surfaces to form sulfuric acid. Sulfuric acid or its neutralized salt (e.g., ammonium sulfate) has a low vapor pressure and thus condenses on existing particles or forms new particles (Seinfeld and Pandis, 2016). Particulate nitrate forms from the oxidation of oxides of nitrogen to nitric acid and the subsequent neutralization of nitric acid by ammonia (Seinfeld and Pandis, 2016). As with sulfate, secondary organic aerosol can form through gas-phase and aqueous reactions; its formation is described in detail in the "Secondary Organic Aerosol Formation" section below. As the aero-sol ages, primary and secondary $PM_{2.5}$ components are frequently found in the same particles (i.e., internally mixed). However, the components are often phase separated, with sulfate, nitrate, and associated water in the core and organics in the shell, except at high relative humidities when the water content of the aerosol particles is substantial and liquid-like components can become homogeneously mixed (Ciobanu et al., 2009; Li et al., 2021; Zuend and Seinfeld, 2012).

Much is left to learn about the formation of secondary species with toxic potential downwind of WUI fires. An unresolved question is whether the "per unit mass" toxicity of the WUI fire plume mixture increases or decreases with distance downwind. O'Dell et al. (2020) found that hazard ratios and cancer risks associated with gas-phase hazardous air pollutants derived from wildland fires decreased with distance downwind during the Western Wildfire Experiment for Cloud Chemistry, Aerosol Absorption, and Nitrogen (WE-CAN) campaign. The largest hazardous air pollutant contributors were acrolein, hydrogen cyanide, and formaldehyde. Other investigators observed an increase in the oxidative potential (per unit mass) of organic aerosol when smoke from wildland fire fuels was oxidized (aged), suggesting that the potential of organic aerosol to play a role in oxidative stress may also increase with atmospheric aging (Wong et al., 2019). Little understanding exists regarding how far from the immediate fire zone human-made materials alter the toxicity of the air pollution mixture, and a more comprehensive assessment of the evolution of primary and secondary WUI fire plume toxicants is needed to understand the evolution of health impacts with atmospheric aging.

Finding: The dominant health-relevant secondary species (secondary species with toxic potential) associated with WUI fires have not been definitively identified, and more work is needed to characterize species with toxic potential.

Finding: It is not well understood how and how far downwind human-made materials influence the composition of WUI fire plumes.

Research need: Research is needed to identify compounds that can serve as tracers of (1) a WUI fire's influence on an air mass (e.g., aerosol tracers), (2) the influence of human-made materials on the chemistry, (3) the influence of particular oxidants, and (4) the influence of specific types of chemistry (e.g., multiphase chemistry) in WUI fire plumes.

Major Atmospheric Oxidants Driving Gas-Phase Chemistry

OH radicals, ozone, nitrate radicals, and chlorine radicals are expected to be the main atmospheric gas-phase oxidants in WUI fire plumes, although other halogens may also play a role. Much of this expectation comes from the current knowledge of WUI fire emissions, fundamental atmospheric chemistry, wildland fire studies (e.g., the recent WE-CAN and Fire Influence on Regional to Global Environments and Air Quality [FIREX-AQ] studies), and studies of atmospheric halogen chemistry in industrial areas. Coordinated measurements, including the use of tracers of chemical processes, have played an important role in improving our understanding of atmospheric chemistry. Atmospheric oxidants transform the smoke mixture as it travels downwind, forming more oxidized and functionalized gases and secondary organic aerosols (e.g., wildland fire aerosol; Zhou et al., 2017).

OH radicals (daytime), nitrate radicals (nighttime), and ozone are mainly responsible for oxidation of organic gases in wildland fire plumes; the implications for oxidation by compound class are provided for wildfires in the FIREX-AQ study by Decker et al. (2021). HONO and HCHO photolysis provides a rapid daytime source of OH radicals (Theys et al., 2020; Veres et al., 2010). In fact, Theys et al. (2020) argue, based on satellite-derived concentrations, that globally, HONO emissions are responsible for two-thirds of OH production in fresh wildland fire plumes, although their importance drops off rapidly as the plume ages (Xu et al., 2021). OH radical concentrations that are 5–20 times higher than those in background air have been measured in wildland fire smoke (Hobbs et al., 2003; Yokelson et al., 2009). Nitrate radical, produced from the oxidation of NO_2 by O_3, is a major nighttime oxidant in wildland fire plumes and may be enhanced in WUI fires due to high emissions of oxides of nitrogen formed as a result of nitrogen in the fire's fuel (fuel NO_x). However, the NO_3 radical is rapidly lost to photolysis during the day (Ng et al., 2017). Oxygenated aromatics in wildland fires are highly reactive with the NO_3 radical (Akherati et al., 2020; Decker et al., 2019). A wildland fire oxidant reactivity budget quantifying the roles of OH (daytime), NO_3 (nighttime), and O_3 (both) through the use of aircraft measurements has been provided by Decker et al. (2021).

The role of halogens in WUI fire plume chemistry warrants additional consideration because of the increase in both halogen and fuel NO_x emissions in these fires (Chapter 3). HCl and particulate chlorine are emitted from wildland fires (McMeeking et al., 2009), but HCl production is increased by the decomposition of polymers from the burning of structures (Chapter 3). Cl radicals are formed at the near-field scale through mechanisms described in Chapter 3. Under atmospheric conditions that lead to cycling between reactive and nonreactive forms of nitrogen (e.g., conditions with high concentrations of N_2O_5), the impact of Cl radicals on chemistry persists with regional transport. Multiphase or aqueous reactions with N_2O_5 release chlorine (as $ClNO_2$) from aerosols during transport, and Cl radicals are subsequently formed via daytime photolysis (Ahern et al., 2018; Goldberger et al., 2019; Thornton et al., 2010).

The Cl radical is roughly an order of magnitude more reactive with many VOCs than the OH radical (Faxon and Allen, 2013). It fragments, oxidizes, and sometimes forms chlorinated organics, and it produces OH radicals (Faxon and Allen, 2013). The OH-to-Cl radical ratio is high enough that OH is the dominant oxidant in most cities (e.g., 9 percent of the primary radical source in Los Angeles; Young et al., 2014). However, the chlorine radical can be an important oxidant in several continental locations (e.g., the Houston, Texas, ship channel; Riemer et al., 2008; Tanaka et al., 2003). Young et al. (2014) conclude based on chemical modeling that the direct influence of the Cl radical on VOC oxidation is evident (e.g., in VOC tracer ratios) when OH-to-Cl radical ratios are less than 200, but that the true impact of the Cl radical is greater because of secondary reactions and the recycling of reactive chlorine.

It is logical that chlorine and potentially other halogens would play an enhanced role in WUI fire plumes. A chlorine radical reactivity budget for WUI fires is needed to better assess the impact of this potentially important radical species. In addition to HCl, HBr is also emitted from structure fires (Chapter 3), yet the contribution of reactive bromine chemistry is not understood.

Finding: Coordinated measurements, including the use of tracers of chemical processes, have proven useful in improving our understanding of the chemical evolution in aging wildland fire plumes but are much more limited for WUI fires.

Finding: Current understanding of the dominant daytime and nighttime oxidant species, including an assessment of any enhanced role of halogen chemistry, is incomplete; this information is important to estimate the atmospheric lifetime of primary species with toxic potential and predict the formation of secondary species with toxic potential downwind of WUI fires.

Research need: Field measurements and controlled experiments are needed to identify the dominant daytime and nighttime atmospheric oxidants, their concentration ranges, and the influence of emissions from human-made materials on oxidation; this information is important to determine typical atmospheric lifetimes for primary species with toxic potential and better predict concentrations of secondary species with toxic potential in WUI fire plumes.

Ozone Formation

Wildland fires have a substantial impact on regional summertime ozone concentrations (and exposures) in the western United States due to their substantial emissions of ozone precursors (i.e., VOCs and NO_x; Friedli et al., 2001; Koss et al., 2018; Xu et al., 2021). Lu et al. (2016) attribute one-third of summer days exceeding the National Ambient Air Quality Standard for ozone in the western United States to wildland fires (daily maximum 8-hour average, MDA8, ozone concentrations exceeding 70 ppbv). Wildland fire MDA8 ozone increases can be as high as 5–40 ppbv on some days (Dreessen et al., 2016; Gong et al., 2017; Lu et al., 2016), and the majority of observations suggest at least some ozone increase (Jaffe and Wigder, 2012). VOC-to-NO_x ratios are approximately 10–30 for temperate wildland fires, suggesting that NO_x is typically the limiting precursor after the first few hours of rapid formation and transport (Akagi et al., 2011; Andreae, 2019; Jaffe and Wigder, 2012; Jaffe et al., 2020; Xu et al., 2021). Thus, the partitioning of reactive nitrogen between NO_x, NO_3, peroxyacetyl nitrate, and other N species will regulate ozone production, and ozone production may increase where the wildland fire plumes interact with NO_x-rich pollution (e.g., when they encounter cities and powerplant plumes). Peroxyacetyl nitrate acts as a long-range NO_x reservoir species, enabling ozone formation to continue over hundreds of kilometers (Alvarado et al., 2010; Briggs et al., 2017).

It is possible that increased reactive nitrogen and chlorine emissions in WUI fires may drive faster ozone formation (e.g., by increasing OH) and more ozone production (e.g., via NO_x and Cl reservoir species) at a regional scale. However, the magnitude of these effects is not known. Ozone formation is a photochemical process, and thus it also depends on the photolysis rate, which can be reduced by aerosol extinction (e.g., the rate can be reduced by 10–20 percent; Real et al., 2007). Likewise, clouds and wet aerosols can reduce ozone formation by removing water-soluble ozone precursors (Yokelson et al., 2003).

Finding: The increase in reactive nitrogen and halogen emissions in WUI fires relative to wildland fires, and the VOC-to-NO_x ratios of WUI fire emissions, are not well characterized; this information could improve prediction of ozone from WUI fires.

Secondary Organic Aerosol Formation

Aircraft, remote sensing, field, and laboratory studies provide insights into secondary organic aerosol (SOA) formation from wildland fire emissions (Hodshire et al., 2019). For example,

- Garofalo et al. (2019) did not see a change in the total organic aerosol mass concentration (relative to CO) with distance downwind but did see an increase in the degree of oxygenation, as indicated by an increase in oxygenation markers and a concurrent decrease in the biomass burning marker.
- Palm et al. (2020) simulated the wildfire chemistry, constrained by aircraft measurements, and concluded that 87 percent of the organic aerosol formed in the atmosphere was composed of evaporated IVOCs. Phenolic SOA precursors accounted for the majority of the remaining SOA. At 3–6 hours of daytime oxidation, two-thirds of the aerosol mass was primary, and one-third was secondary.

- Through remote sensing, validated in part by aircraft measurements, Junghenn Noyes et al. (2020) documented an increasing ratio of brown carbon to black carbon aerosol with distance downwind of the Government Flats Complex Fire in Oregon, suggesting the atmospheric formation of light-absorbing organic aerosol in the plume.

SOA Formation via Vapor Pressure–Driven Partitioning

Based on fundamental principles of atmospheric chemistry, the committee expects that gas-phase oxidation via OH and Cl (daytime), NO_3 (nighttime), and O_3 results in fragmentation of some WUI fire organics and functionalization of others. Thus, gas-phase oxidation can lead to the formation of smaller, more oxygenated (water-soluble) gas-phase compounds. However, when oxidation leads to functionalization and the products have a low enough vapor pressure, they condense (sorb into preexisting aerosol) and form a SOA (Kroll and Seinfeld, 2008). SOA formation through vapor pressure–driven partitioning happens on a timescale of minutes to hours (Palm et al., 2020). SOA formed from wildland fires via vapor pressure–driven partitioning typically has a lower volatility and higher O-to-C ratios than primary organic aerosol emitted by wildland fires (Garofalo et al., 2019; Palm et al., 2020).

Oxygenated aromatic compounds and IVOCs are understood to be the main SOA precursors through gas-phase oxidation and vapor pressure–driven partitioning (Akherati et al., 2020; Chan et al., 2009; Coggon et al., 2019; Patoulias et al., 2021; Posner et al., 2019; Tkacik et al., 2012; Yee et al., 2013; Zhao et al., 2014b). These are among the SOA precursors in WUI fire plumes. However, model prediction of the magnitude of SOA formation downwind of wildland fires (Theodoritsi et al., 2021) and WUI fires remains uncertain due in part to inadequate chemical characterization of IVOCs and other precursor emissions, the partial understanding of oxidation mechanisms, and a lack of thermodynamic data for the oxidation products. Phenol, guaiacol, dioxins, and synthetic polymer degradation products are among the potential WUI precursors to SOA formation through gas-phase oxidation and vapor pressure–based partitioning.

SOA Formation via Aqueous, Heterogeneous, or Multiphase Chemistry

Chemistry in WUI fire plumes can also occur on the surfaces of aerosol particles (heterogeneous) and in the liquid water contained in clouds, fogs, and aerosol particles (multiphase or aqueous). Atmospheric liquid water and water-soluble compounds are abundant in wildland fire plumes. Formaldehyde, isocyanic acid, glycolaldehyde, acetic acid, phenols, furfurals, and amines, which are found in wildland fire plumes (Friedli et al., 2001; Hatch et al., 2015; Koss et al., 2018; Leslie et al., 2019; McFall et al., 2020; Permar et al., 2021; Schauer et al., 2001; Selimovic et al., 2018; Tomaz et al., 2018), for example, will partition into clouds and wet aerosols, where they can react further through radical or non-radical reactions (Bianco et al., 2020). In some cases, products of aqueous organic chemistry remain in the particle phase after water evaporation (e.g., Brégonzio-Rozier et al., 2016; Lee et al., 2011), and water evaporation itself can drive accretion reactions that enhance particle-phase retention (e.g., for aldehydes and amines; Hawkins et al., 2014; Loeffler et al., 2006). Thus, chemistry in clouds and wet aerosols can lead to "aqueous" SOA formation (Blando and Turpin, 2000; Lamkaddam et al., 2021; McNeill, 2015; Tilgner et al., 2013; Volkamer et al., 2009).

Because the precursors are small and oxidized, with high O-to-C ratios, the SOA produced also has high O-to-C ratios. In clouds, radical reactions dominate, whereas in wet aerosols, non-radical reactions can be important (Lim et al., 2010; McNeill, 2015). For example, peroxides and epoxides from isoprene oxidation undergo rapid reactive uptake via acid catalysis in wet aerosols (Surratt et al., 2010); this multiphase isoprene chemistry is a major source of SOA in the southeastern United States (Budisulistiorini et al., 2013; Ying et al., 2015). The aqueous oxidation of *particulate* organics in clouds and wet aerosols can also result in volatile losses of oxidation products—a reduction in organic aerosol mass.

Several studies provide evidence for the influence of atmospheric aqueous chemistry on wildland fire smoke evolution:

- Decreasing concentrations of methanol and increasing concentrations of formaldehyde above clouds affected by savanna fires have been attributed to in-cloud oxidation of savanna fire emissions (Tabazadeh et al., 2004; Yokelson et al., 2003).

- Zhang et al. (2017) measured an increase in the aerosol brown-carbon-to-black-carbon ratio in the outflow, compared to inflow, of convective clouds impacted by a wildland fire plume, suggesting aqueous SOA formation.
- Gilardoni et al. (2016) observed the formation of a light-absorbing SOA in fog and wet aerosol impacted by residential woodburning and estimated that aqueous chemistry accounts for approximately one-half of the SOA mass in the Po Valley, Italy, in a typical winter (Gilardoni et al., 2016; Paglione et al., 2020).
- Lin et al. (2010) concluded that cloud processing of straw-burning emissions could explain high correlations between cloud chemistry tracers (i.e., oxalate and sulfate) and humic-like substances in ambient aerosols affected by the field burning of rice straw.
- Cook et al. (2017) observed light-absorbing aqueous-phase products of syringol and guaiacol oxidation in wildland fire–influenced cloud water samples.
- Tomaz et al. (2018) observed the formation of known SOA components upon aqueous OH radical oxidation of emissions from burning mixtures of forest materials.
- The aqueous SOA formation potential of several wildland fire emissions (e.g., acetic, glycolic, pyruvic, succinic, glutaric, adipic, and lactic acids; glycolaldehyde; phenol; catechol; guaiacol; vanillin; levoglucosan) and the aqueous formation of nitrophenols among other products have been studied in the laboratory (Carlton et al., 2006; Chang and Thompson, 2010; Charbouillot et al., 2012; Li et al., 2014; Lim et al., 2013; Pang et al., 2019; Perri et al., 2009; Sun et al., 2010; Tan et al., 2012; Vidović et al., 2020; Yu et al., 2014, 2016; Zhao et al., 2014a).
- Models estimate in-cloud SOA from biomass burning at 20–30 Tg/yr globally (Liu et al., 2012; Lin et al., 2014).

Predictions of SOA formation through aqueous chemistry are highly uncertain (Barth et al., 2021), in part because of uncertainties in aqueous-phase oxidant concentrations. It is generally accepted that OH concentrations are depleted in clouds in comparison to equilibrium (Arakaki et al., 2013), although there is some indication that concentrations could be much higher under certain circumstances (Lamkaddam et al., 2021; Paulson et al., 2019) and that reactions at air-water interfaces can be orders of magnitude higher than in bulk water (Enami et al., 2014). Photosensitized reactions, HONO, halogens, reactive nitrogen, H_2O_2, and metals (Fenton chemistry) are all possible sources of aqueous-phase radicals (Bianco et al., 2020). Organics could also plausibly recycle radicals through autocatalysis (Rossignol et al., 2014). Acidity is also a critical factor in determining the partitioning of aqueous-phase products between gas and particle phases (Ortiz-Montalvo et al., 2014; Tilgner et al., 2021).

Overall, significant progress has been made in understanding the gas-phase chemistry, gas-to-particle partitioning, and SOA formation associated with the burning of biomass from fields and forests. Many of the most important processes are controlled by radical chemistry. Because WUI fires have the potential to have significantly different radical reactivities than wildland fires, due to the increased presence of halogen radicals, metals, and other species, critical chemical pathways have the potential to have different rates and to be qualitatively different than in wildland fires. More work is needed to understand WUI plume chemistry, especially the multiphase, aqueous, and catalytic chemistry. Issues include identifying the critical WUI species and processes. A combination of aircraft and other field measurements (see Chapter 7), controlled laboratory experiments, and modeling (see below) will be needed to address these questions.

Finding: Multiphase, aqueous, and catalytic chemistry in WUI fire plumes is poorly understood.

Research need: An improved understanding of key processes and formation pathways such as the roles of multiphase, aqueous, and catalytic chemistry in WUI fire plumes is needed.

CURRENT PRACTICE IN MODELING FAR PLUMES

A critical role of smoke models is to provide forecasts for public health guidance; coordinated measurements and modeling, used in conjunction with controlled experiments, also test our understanding of emissions and atmospheric processing, and enable model refinements. The physical movement of wildland fire plumes over length scales of hundreds to thousands of kilometers is typically modeled using trajectory or dispersion models. These types of models typically model few or none of the chemical transformations that occur in wildland fire and WUI plumes. They do, however, predict dilution, some physical processes such as deposition, and the direction of movement of plumes. These features make transport and dispersion models useful for predicting likely atmospheric concentrations of particulate matter and are used in public notification of risk (activities to avoid). For example, the smoke forecasting system from the National Oceanic and Atmospheric Administration uses National Weather Service inputs from the North American Mesoscale model and smoke dispersion simulations from the Hybrid Single-Particle Lagrangian Integrated Trajectory model (Stein et al., 2015) to produce a daily prediction of smoke transport and concentration (NOAA ARL, 2012). As an example, Figure 4-2 shows fire location (top) and fire plume projections (bottom) for September 16, 2021, when fire plumes from Idaho and California extended east as far as Minnesota (NOAA ARL, 2021).

The accuracy of these models of plume trajectories is dependent on the ability to accurately assess the vertical height of the plume, since wind speeds and direction can vary significantly with height. Paugam et al. (2016) provide a review of plume rise performance in chemical transport models along with the atmospheric and fire parameters governing plume rise and assess the sensitivity of dispersion predictions to the predicted plume rise. More recent tools (https://hwp-viz.gsd.esrl.noaa.gov/smoke/; https://rapidrefresh.noaa.gov/hrrr/HRRRsmoke/) that make use of fire radiative power data to estimate plume rise have provided significant advances in trajectory modeling of wildfires (Ahmadov et al., 2017).

In an evaluation of the forecasting ability of 12 smoke models for a case study on a single wildland fire, Ye et al. (2021) found that all models underpredicted the amount and extent of the plume, while predictions of $PM_{2.5}$ were biased in both directions, depending on the model. The authors attributed biases to uncertainties in emissions and plume rise, although atmospheric chemistry could certainly also play a role. This study demonstrates challenges involved in smoke forecasting and provides some insight into where additional effort is needed to improve prognostic modeling for WUI fires. Comprehensive summaries of smoke models can be found in the literature (Jaffe et al., 2020; Ye et al., 2021).

Finding: Models are currently available and can be useful for predicting likely impacts, public notification of risk (activities to avoid), and forecasting of particulate matter concentrations.

Research need: Continued improvement is needed in smoke dispersion models for use in both retrospective and prospective modes.

Different types of models are used to characterize chemical transformations in wildland fire and WUI plumes over local, regional, and continental scales. The complex gas-phase chemical transformations, particle-phase chemical transformations, and gas-to-particle partitioning associated with wildland fires and WUI fires, outlined in this chapter, are tracked using computationally intensive models that couple chemical transport and transformation. These modeling tools are broadly referred to as chemical transport models, and include modeling frameworks such as the Community Multiscale Air Quality (CMAQ) Modeling System (https://www.epa.gov/cmaq) and the Comprehensive Air Quality Model with Extensions (https://www.camx.com/). Because these models are so computationally intensive, simplified gas-phase chemical mechanisms, particle-phase chemical mechanisms, and models of gas-to-particle partitioning processes are used in the models. These "reduced" mechanisms track a limited number of molecular species and compound classes explicitly. The majority of primary and secondary pollutants are grouped into very broad "lumped" chemical classes such as inorganic compounds, VOCs, semi-VOCs, water-soluble organic compounds, and particulate matter. The general relationships between lumped and explicitly tracked molecular species in predictive chemical transport models, and unresolved species, is illustrated in Figure 4-3.

Chemical transport models are used to couple the modeling of transport and chemical transformation and are widely used in developing and evaluating air quality management plans for reducing ozone, particulate matter, and other pollutants in urban and regional atmospheres. Chemical transport models have also been applied to examine the coupled transport and chemical transformation of fire plumes, most commonly in retrospective analyses of wildland fire and WUI fire events, rather than in forecast mode (Goodrick et al., 2012).

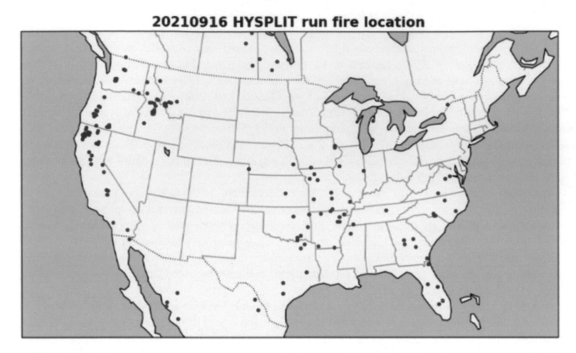

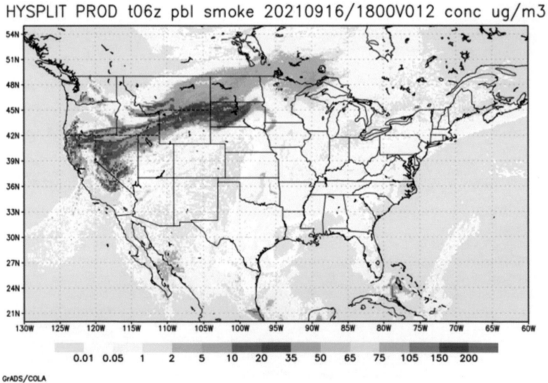

FIGURE 4-2 Example of daily fire locations (top) and fire plume projections (bottom) for September 16, 2021, from the National Oceanic and Atmospheric Administration.

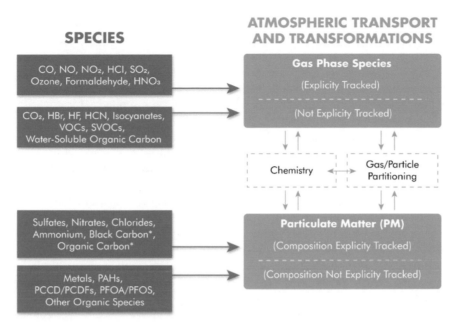

*Defined based on thermal-optical carbon analysis protocols described by Spada and Hyslop 2018; Chow et al, 2007).

FIGURE 4-3 This schematic overview indicates which emissions are explicitly tracked and which are lumped (not explicitly tracked) by models. *Black carbon and organic carbon are defined based on thermal-optical carbon analysis protocols described by Spada and Hyslop (2018) and Chow et al. (2007).

Chemical transport models have generally not been optimized to address the unique chemistries and transport characteristics of wildland fire plumes. Nevertheless, a number of fire episodes that have impacted air quality over regional to continental scales in the United States have been modeled (Fann et al., 2018; Huang et al., 2021; Jiang and Yoo, 2019; Wilkins et al., 2018). These simulations, when compared to ambient observations, have tended to underestimate peak ozone and particulate matter concentrations associated with fires. In addition, the simplified chemistries used in the models for computational efficiency simulate only a limited number of chemical species, and most model applications are not formulated to track toxicants generated by fires. Limitations of current-generation chemical transport modeling tools in simulating regional- to continental-scale fire events are described in more detail below.

Limitations of Condensed Chemistries

Chemical transport models use simplified representations of chemical transformations because models of the full chemistry, even if all reactions and rates were known, would be too computationally intensive to use within the context of the models. One commonly used simplification is to represent classes of compounds by a single chemical species. For example, the reactions of all substituted aromatic compounds might be represented by the reactions of one or two "lumped" species. Another commonly used simplification is to represent collections of similar functional groups in molecules by a single lumped species. For example, the reactions of all paraffinic carbons in a mixture might be represented by a single chemical species.

These "lumped" or condensed mechanisms pose multiple problems for modeling fire emissions. First, in lumping chemical species, information about specific molecular concentrations, which is needed for estimating concentrations of toxicants, is lost. New condensed mechanisms with the ability to track key individual chemical species could be developed.

A second, and related, problem is that changing individual reactions in condensed mechanisms is problematic. As a simple example, consider a condensed mechanism that seeks to model specific aromatic species and therefore separates a particular aromatic species (e.g., toluene) from the lumped aromatics category. This can change the average reactivity of the "lumped" species, because the reactivity of toluene may be different than that of the xylenes and ethyl benzene, which toluene may have been lumped with. This means that to update the toluene chemistry, the entire mechanism has to be updated (Whitten et al., 2010). Since even modest changes made to condensed mechanisms require that the entire mechanism be reevaluated, the development of condensed mechanisms suitable for specialized applications such as wildland or WUI fires has been limited. This difficulty in updating the chemical lumping is not limited to the chemical mechanisms.

A third issue is that condensed mechanisms have been designed with gas-phase chemistry in mind, without consideration for partitioning between the gas phase and clouds or aerosols. In some cases, species are "lumped" together across a wide range of Henry's law constants, which makes it difficult to incorporate cloud and multiphase aerosol chemistry into the models, to predict the concentrations of water-soluble toxicants (e.g., isocyanic acid) and to predict deposition. This limitation affects the prediction of SOA and could affect the accuracy of gas-phase chemistry prediction, especially in the near field, by making it difficult to accurately account for the loss of reactive water-soluble gas-phase species to clouds and wet aerosols (e.g., formaldehyde, formic acid, HONO, HCl).

A fourth issue is the omission of classes of compounds (e.g., Coggon et al., 2019). For example, condensed mechanisms are not designed to track IVOCs, which are emitted as particulate matter but volatilize in the near field. Nor do they typically track furans, which have been reported to account for roughly 10 percent of ozone formation in the first 4 hours of wildfire smoke aging (Coggon et al., 2019).

These limitations are of particular concern for wildland fire and WUI fire model applications, given the importance of tracking toxicants for exposure assessment and the abundance of water-soluble species and IVOCs emitted by WUI fires.

Finding: Current models lack the chemical specificity needed to track in detail the types of toxicants associated with wildland fires and WUI fires.

Research need: Development of condensed chemical mechanisms is needed for use in applying chemical transport models to WUI fires.

Sub-Grid-Cell Chemical Processing of Emissions

For industrial point sources, near-source limitations of chemical transport models are frequently overcome through the use of sub-grid-cell (i.e., plume-in-grid [PiG]) chemical modeling tools; however, such tools are not designed for wildland fire or WUI fire emissions. Chemical transport modeling (e.g., CMAQ; Appel et al., 2021) is typically performed using gridded domains with a horizontal resolution of 12 km, 36 km, or more. The vertical depth of the model grid cells increases with height above ground level, and ranges from depths as small as tens of meters to depths as large as hundreds of meters at higher elevations. When emissions are modeled, they are assumed to immediately and completely mix throughout the grid cell into which they are emitted. Thus, the horizontal extent of the grid cell means that spatial variability in emissions over roughly kilometer scales is lost. Since fire burn fronts are often at smaller spatial scales (Freitas et al., 2007; Sokolik et al., 2019), chemical transport models will unrealistically mix fire plumes.

For sources with sub-grid scales, the unrealistic, immediate dilution of emissions leads chemical transport models to underestimate the rates of reactions. The rate of a chemical reaction scales directly with concentration, and therefore dilution of an emission that might be occurring over a 1 km^2 area to a grid cell that is 36 km by 36 km in extent will reduce concentrations of the emitted species by a factor of more than 1,000. When the reactants are mixed over a large volume in a simulation, the reactions also are simulated to occur over a large volume. For unimolecular chemical reactions, using a large grid-cell size may have only a modest effect on accuracy. Many atmospheric chemical reactions are bimolecular, however. Emission plumes, occurring over a 1 km^2 area, with multiple reactants participating in bimolecular reactions, that are immediately mixed over an area of 1,000 km^2, will have the rates of the bimolecular reactions reduced by a factor of 10^6. Thus, even if these reactions are assumed to occur over a volume that is a factor of 10^3 larger, the volume-integrated rate of reaction will still be orders of magnitude too low. Thus, modeling of chemical transformations at sub-grid-cell levels is frequently needed.

The usefulness of sub-grid-cell processing is well known, and tools that model sub-grid-cell plumes within the context of a gridded model (PiG modeling tools) have been developed (Karamchandani et al., 2011; Wei et al., 2021). These PiG tools have their own reduced chemical mechanisms, and those mechanisms have been developed primarily for modeling industrial plumes, especially plumes of power plants that are rich in NO_x (Karamchandani et al., 2011). The PiG tools designed to model industrial point sources model plume mixing with the surrounding atmosphere based on an industrial point-source plume structure, which differs from the typical wildland fire or WUI fire plume structure. Because the chemistry and plume dispersion currently available within PiG tools are not well suited for fire plumes, some wildland fire modeling studies have made use of box models to perform initial sub-grid-scale chemistry (Peng et al., 2021). However, it should be noted that the general framework of PiG tools could be useful for fire modeling.

Finding: Current models lack detailed physical modeling of the unique features of fire plume structures.

Finding: Models fail to capture key chemical interactions within the plume, including the interactions of emission sources with ambient air, clouds, and aerosols.

Research need: Development of sub-grid-cell processing techniques suitable for fire plumes is needed.

Research need: Regional chemical transport models need substantial enhancements in order to adequately describe WUI fire plumes, and research support is needed to drive these improvements.

Overall, chemical transport models, when used in combination with field measurements, are powerful tools for understanding and predicting air quality impacts of fires. Enhancements to chemical transport models, coupled with measurements to evaluate the performance of the improvements, could be used to develop new capabilities, including but not limited to the following:

- Modeling of individual chemical species that are either toxicants or surrogates for the emission of classes of WUI emissions, with consideration for properties that affect partitioning between the gas phase, clouds, and aerosols (e.g., Henry's law constant, vapor pressure)
- Development of sub-grid-cell models that could be used to model the mixing of fire plumes with ambient air and the structure of single-structure or neighborhood-scale plumes
- Development of chemical transport models that could be used in prognostic mode to facilitate first-responder activity during environmental crises, and communication/decision-making coordination (a technology link to communicate predictions at the different scales of responders)

Research need: A combination of research approaches is needed to understand and predict toxicant concentrations downwind of WUI fires and ultimately mitigate their health risks.

Research need: There is a need to identify a subset of WUI fire–associated species (based on their toxicity or importance in chemistry) that are prioritized in both measurement and modeling efforts; attention is needed toward selecting priority species that span a range of physicochemical properties (e.g., vapor pressure, Henry's law constant, oxidation state) and reactivity.

REFERENCES

Ahern, A. T., L. Goldberger, L. Jahl, J. Thornton, and R. C. Sullivan. 2018. "Production of N_2O_5 and $ClNO_2$ through Nocturnal Processing of Biomass-Burning Aerosol." *Environmental Science & Technology* 52(2):550–559. https://doi.org/10.1021/acs.est.7b04386.

Ahmadov, R., G. Grell, E. James, I. Csiszar, M. Tsidulko, B. Pierce, et al. 2017. "Using VIIRS Fire Radiative Power Data to Simulate Biomass Burning Emissions, Plume Rise and Smoke Transport in a Real-Time Air Quality Modeling System." 2017. *IEEE International Geoscience and Remote Sensing Symposium* 2806–2808. New York, NY: IEEE.

Akagi, S. K., R. J. Yokelson, C. Wiedinmyer, M. J. Alvarado, J. S. Reid, T. Karl, J. D. Crounse, and P. O. Wennberg. 2011. "Emission Factors for Open and Domestic Biomass Burning for Use in Atmospheric Models." *Atmospheric Chemistry and Physics* 11(9):4039–4072. https://doi.org/10.5194/acp-11-4039-2011.

Akherati, A., Y. He, M. M. Coggon, A. R. Koss, A. L. Hodshire, K. Sekimoto, C. Warneke, J. de Gouw, L. Yee, J. H. Seinfeld, T. B. Onasch, S. C. Herndon, W. B. Knighton, C. D. Cappa, M. J. Kleeman, C. Y. Lim, J. H. Kroll, J. R. Pierce, and S. H. Jathar. 2020. "Oxygenated Aromatic Compounds are Important Precursors of Secondary Organic Aerosol in Biomass-Burning Emissions." *Environmental Science & Technology* 54(14):8568–8579. https://doi.org/10.1021/acs.est.0c01345.

Alvarado, M. J., J. A. Logan, J. Mao, E. Apel, D. Riemer, D. Blake, R. C. Cohen, K. E. Min, A. E. Perring, E. C. Browne, P. J. Wooldridge, G. S. Diskin, G. W. Sachse, H. Fuelberg, W. R. Sessions, D. L. Harrigan, G. Huey, J. Liao, A. Case-Hanks, J. L. Jimenez, M. J. Cubison, S. A. Vay, A. J. Weinheimer, D. J. Knapp, D. D. Montzka, F. M. Flocke, I. B. Pollack, P. O. Wennberg, A. Kurten, J. Crounse, J. M. S. Clair, A. Wisthaler, T. Mikoviny, R. M. Yantosca, C. C. Carouge, and P. Le Sager. 2010. "Nitrogen Oxides and PAN in Plumes from Boreal Fires during ARCTAS-B and Their Impact on Ozone: An Integrated Analysis of Aircraft and Satellite Observations." *Atmospheric Chemistry and Physics* 10(20):9739–9760. https://doi.org/10.5194/acp-10-9739-2010.

Alves, C. A., A. Vicente, C. Monteiro, C. Goncalves, M. Evtyugina, and C. Pio. 2011. "Emission of Trace Gases and Organic Components in Smoke Particles from a Wildfire in a Mixed-Evergreen Forest in Portugal." *Science of the Total Environment* 409(8):1466–1475. https://doi.org/10.1016/j.scitotenv.2010.12.025.

Andreae, M. O. 2019. "Emission of Trace Gases and Aerosols from Biomass Burning – An Updated Assessment." *Atmospheric Chemistry and Physics* 19(13):8523–8546. https://doi.org/10.5194/acp-19-8523-2019.

Appel, K. W., O. Bash, K. M. Fahey, M. Foley, R. C. Gilliam, C. Hogrefe, . . . and D. C. Wong. 2021. "The Community Multiscale Air Quality (CMAQ) Model Versions 5.3 and 5.3.1: System Updates and Evaluation." *Geoscientific Model Development* 14(5):2867–2897.

Arakaki, T., C. Anastasio, Y. Kuroki, H. Nakajima, K. Okada, Y. Kotani, D. Handa, S. Azechi, T. Kimura, A. Tsuhako, and Y. Miyagi. 2013. "A General Scavenging Rate Constant for Reaction of Hydroxyl Radical with Organic Carbon in Atmospheric Waters." *Environmental Science & Technology* 47(15):8196–8203. https://doi.org/10.1021/es401927b.

Barth, M. C., B. Ervens, H. Herrmann, A. Tilgner, V. F. McNeill, W. G. Tsui, L. Deguillaume, N. Chaumerliac, A. Carlton, and S. M. Lance. 2021. "Box Model Intercomparison of Cloud Chemistry." *Journal of Geophysical Research: Atmospheres* 126(21):e2021JD035486. https://doi.org/10.1029/2021JD035486.

Bianco, A., M. Passananti, M. Brigante, and G. Mailhot. 2020. "Photochemistry of the Cloud Aqueous Phase: A Review." *Molecules* 25(2):423. https://doi.org/10.3390/molecules25020423.

Blando, J. D., and B. J. Turpin. 2000. "Secondary Organic Aerosol Formation in Cloud and Fog Droplets: A Literature Evaluation of Plausibility." *Atmospheric Environment* 34(10):1623–1632. https://doi.org/10.1016/S1352-2310(99)00392-1.

Brégonzio-Rozier, L., C. Giorio, F. Siekmann, E. Pangui, S. B. Morales, B. Temime-Roussel, A. Gratien, V. Michoud, M. Cazaunau, H. L. DeWitt, A. Tapparo, A. Monod, and J. F. Doussin. 2016. "Secondary Organic Aerosol Formation from Isoprene Photooxidation during Cloud Condensation–Evaporation Cycles." *Atmospheric Chemistry and Physics* 16(3):1747–1760. https://doi.org/10.5194/acp-16-1747-2016.

Briggs, N. L., D. A. Jaffe, H. Gao, J. R. Hee, P. M. Baylon, Q. Zhang, S. Zhou, S. C. Collier, P. D. Sampson, and R. A. Cary. 2017. "Particulate Matter, Ozone, and Nitrogen Species in Aged Wildfire Plumes Observed at the Mount Bachelor Observatory." *Aerosol and Air Quality Research* 16(12):3075–3087. https://doi.org/10.4209/aaqr.2016.03.0120.

Budisulistiorini, S. H., M. R. Canagaratna, P. L. Croteau, W. J. Marth, K. Baumann, E. S. Edgerton, S. L. Shaw, E. M. Knipping, D. R. Worsnop, J. T. Jayne, A. Gold, and J. D. Surratt. 2013. "Real-Time Continuous Characterization of Secondary Organic Aerosol Derived from Isoprene Epoxydiols in Downtown Atlanta, Georgia, Using the Aerodyne Aerosol Chemical Speciation Monitor." *Environmental Science & Technology* 47(11):5686–5694. https://doi.org/10.1021/es400023n.

Carlton, A. G., B. J. Turpin, H. J. Lim, K. E. Altieri, and S. Seitzinger. 2006. "Link between Isoprene and Secondary Organic Aerosol (SOA): Pyruvic Acid Oxidation Yields Low Volatility Organic Acids in Clouds." *Geophysical Research Letters* 33(6). https://doi.org/10.1029/2005GL025374.

Chan, A. W. H., K. Kautzman, P. Chhabra, J. Surratt, M. Chan, J. Crounse, A. Kürten, P. Wennberg, R. Flagan, and J. Seinfeld. 2009. "Secondary Organic Aerosol Formation from Photooxidation of Naphthalene and Alkylnaphthalenes: Implications for Oxidation of Intermediate Volatility Organic Compounds (IVOCs)." *Atmospheric Chemistry and Physics* 9(9): 3049–3060. https://doi.org/10.5194/acp-9-3049-2009.

Chang, J. L., and J. E. Thompson. 2010. "Characterization of Colored Products Formed during Irradiation of Aqueous Solutions Containing H_2O_2 and Phenolic Compounds." *Atmospheric Environment* 44(4):541–551. https://doi.org/10.1016/j.atmosenv.2009.10.042.

Charbouillot, T., S. Gorini, G. Voyard, M. Parazols, M. Brigante, L. Deguillaume, A.-M. Delort, and G. Mailhot. 2012. "Mechanism of Carboxylic Acid Photooxidation in Atmospheric Aqueous Phase: Formation, Fate and Reactivity." *Atmospheric Environment* 56:1–8. https://doi.org/10.1016/j.atmosenv.2012.03.079.

Cheng, M. D. 2018. "Atmospheric Chemistry of Hydrogen Fluoride." *Journal of Atmospheric Chemistry* 75:1–16. https://doi.org/10.1007/s10874-017-9359-7.

Chow, J. C., J. G. Watson, L.-W.A. Chen, M. Chang, N. F. Robinson, D. L. Trimble, and S. D. Kohl. 2007. "The IMPROVE: A temperature protocol for thermal/optical carbon analysis: Maintaining consistency with a long-term database." *Journal of the Air & Waste Management Association* 57(9):1014–1023. https://doi.org/10.3155/1047-3289.57.9.1014.

Ciobanu, V. G., C. Marcolli, U. K. Krieger, U. Weers, and T. Peter. 2009. "Liquid–Liquid Phase Separation in Mixed Organic/Inorganic Aerosol Particles." *The Journal of Physical Chemistry A* 113(41):10966–10978. https://doi.org/10.1021/jp905054d.

Coggon, M. M., C. Y. Lim, A. R. Koss, K. Sekimoto, B. Yuan, J. B. Gilman, D. H. Hagan, V. Selimovic, K. J. Zarzana, S. S. Brown, J. M. Roberts, M. Müller, R. Yokelson, A. Wisthaler, J. E. Krechmer, J. L. Jimenez, C. Cappa, J. H. Kroll, J. de Gouw, and C. Warneke. 2019. "OH Chemistry of Non-methane Organic Gases (NMOGs) Emitted from Laboratory and Ambient Biomass Burning Smoke: Evaluating the Influence of Furans and Oxygenated Aromatics on Ozone and Secondary NMOG Formation." *Atmospheric Chemistry and Physics* 19(23):14875–14899. https://doi.org/10.5194/acp-19-14875-2019.

Cook, R. D., Y.-H. Lin, Z. Peng, E. Boone, R. K. Chu, J. E. Dukett, M. J. Gunsch, W. Zhang, N. Tolic, and A. Laskin. 2017. "Biogenic, Urban, and Wildfire Influences on the Molecular Composition of Dissolved Organic Compounds in Cloud Water." *Atmospheric Chemistry and Physics* 17(24):15167–15180. https://doi.org/10.5194/acp-17-15167-2017.

Decker, Z. C. J., K. J. Zarzana, M. Coggon, K. E. Min, I. Pollack, T. B. Ryerson, J. Peischl, P. Edwards, W. P. Dube, M. Z. Markovic, J. M. Roberts, P. R. Veres, M. Graus, C. Warneke, J. de Gouw, L. E. Hatch, K. C. Barsanti, and S. S. Brown. 2019. "Nighttime Chemical Transformation in Biomass Burning Plumes: A Box Model Analysis Initialized with Aircraft Observations." *Environmental Science & Technology* 53(5):2529–2538. https://doi.org/10.1021/acs.est.8b05359.

Decker, Z. C., M. A. Robinson, K. C. Barsanti, I. Bourgeois, M. M. Coggon, J. P. DiGangi, G. S. Diskin, F. M. Flocke, A. Franchin, and C. D. Fredrickson. 2021. "Nighttime and Daytime Dark Oxidation Chemistry in Wildfire Plumes: An Observation and Model Analysis of FIREX-AQ Aircraft Data." *Atmospheric Chemistry and Physics* 21(21):16293–16317. https://doi.org/10.5194/acp-21-16293-2021.

Dreessen, J., J. Sullivan, and R. Delgado. 2016. "Observations and Impacts of Transported Canadian Wildfire Smoke on Ozone and Aerosol Air Quality in the Maryland Region on June 9–12, 2015." *Journal of the Air & Waste Management Association* 66(9):842–862. https://doi.org/10.1080/10962247.2016.1161674.

Ellis, D. A., J. W. Martin, S. A. Mabury, M. D. Hurley, M. P. Sulbaek Andersen, and T. J. Wallington. 2003. "Atmospheric Lifetime of Fluorotelomer Alcohols." *Environmental Science & Technology* 37(17):3816–3820. https://doi.org/10.1021/es034136j.

Ellis, D. A., J. W. Martin, A. O. De Silva, S. A. Mabury, M. D. Hurley, M. P. Sulbaek Andersen, and T. J. Wallington. 2004. "Degradation of Fluorotelomer Alcohols: A Likely Atmospheric Source of Perfluorinated Carboxylic Acids." *Environmental Science & Technology* 38(12):3316–3321. https://doi.org/10.1021/es049860w.

Enami, S., Y. Sakamoto, and A. J. Colussi. 2014. "Fenton Chemistry at Aqueous Interfaces." *Proceedings of the National Academy of Sciences* 111(2):623–628. https://doi.org/10.1073/pnas.1314885111.

Evtyugina, M., A. I. Calvo, T. Nunes, C. Alves, A. P. Fernandes, L. Tarelho, . . . and C. Pio. 2013. "VOC Emissions of Smouldering Combustion from Mediterranean Wildfires in Central Portugal." *Atmospheric Environment* 64:339–348.

Fann, N., B. Alman, R. A. Broome, G. G. Morgan, F. H. Johnston, G. Pouliot, and A. G. Rappold. 2018. "The Health Impacts and Economic Value of Wildland Fire Episodes in the U.S.: 2008–2012." *Science of the Total Environment* 610–611:802–809. https://doi.org/10.1016/j.scitotenv.2017.08.024.

Faxon, C. B., and D. T. Allen. 2013. "Chlorine Chemistry in Urban Atmospheres: A Review." *Environmental Chemistry* 10(3):221–233. https://doi.org/10.1071/EN13026.

Freitas, S. R., K. M. Longo, R. Chatfield, D. Latham, M. A. F. Silva Dias, M. O. Andreae, . . . and J. A. Carvalho Jr. 2007. "Including the Sub-grid Scale Plume Rise of Vegetation Fires in Low Resolution Atmospheric Transport Models." *Atmospheric Chemistry and Physics* 7(13):3385–3398.

Friedli, H., E. Atlas, V. Stroud, L. Giovanni, T. Campos, and L. Radke. 2001. "Volatile Organic Trace Gases Emitted from North American Wildfires." *Global Biogeochemical Cycles* 15(2):435–452. https://doi.org/10.1029/2000GB001328.

Garofalo, L. A., M. A. Pothier, E. J. Levin, T. Campos, S. M. Kreidenweis, and D. K. Farmer. 2019. "Emission and Evolution of Submicron Organic Aerosol in Smoke from Wildfires in the Western United States." *ACS Earth and Space Chemistry* 3(7):1237–1247. https://doi.org/10.1021/acsearthspacechem.9b00125.

Gilardoni, S., P. Massoli, M. Paglione, L. Giulianelli, C. Carbone, M. Rinaldi, S. Decesari, S. Sandrini, F. Costabile, G. P. Gobbi, M. C. Pietrogrande, M. Visentin, F. Scotto, S. Fuzzi, and M. C. Facchini. 2016. "Direct Observation of Aqueous Secondary Organic Aerosol from Biomass-Burning Emissions." *Proceedings of the National Academy of Sciences* 113(36):10013–10018. https://doi.org/10.1073/pnas.1602212113.

Gilman, J., B. Lerner, W. Kuster, P. D. Goldan, C. Warneke, P. Veres, J. Roberts, J. De Gouw, I. Burling, and R. Yokelson. 2015. "Biomass Burning Emissions and Potential Air Quality Impacts of Volatile Organic Compounds and Other Trace Gases from Fuels Common in the US." *Atmospheric Chemistry and Physics* 15(24):13915–13938. https://doi.org/10.5194/acpd-15-21713-2015.

Goldberger, L. A., L. G. Jahl, J. A. Thornton, and R. C. Sullivan. 2019. "N_2O_5 Reactive Uptake Kinetics and Chlorine Activation on Authentic Biomass-Burning Aerosol." *Environmental Science: Processes & Impacts* 21(10):1684–1698. https://doi.org/10.1039/c9em00330d.

Gong, X., A. S. Kaulfus, U. S. Nair, and D. A. Jaffe. 2017. "Quantifying O_3 Impacts in Urban Areas Due to Wildfires Using a Generalized Additive Model." *Environmental Science & Technology* 51(22):13216–13223. https://doi.org/10.1021/acs.est.7b03130.

Goodrick, S. L., G. L. Achtemeier, N. K. Larkin, Y. Liu, and T. Strand. 2012. "Modelling Smoke Transport from Wildland Fires: A Review." *International Journal of Wildland Fire* 22(1):83–94.

Hall, S. M., S. Patton, M. Petreas, S. Zhang, A. L. Phillips, K. Hoffman, and H. M. Stapleton. 2020. "Per- and Polyfluoroalkyl Substances in Dust Collected from Residential Homes and Fire Stations in North America." *Environmental Science & Technology* 54(22):14558–14567. https://doi.org/10.1021/acs.est.0c04869.

Harrison, M. A. J., S. Barra, D. Borghesi, D. Vione, C. Arsene, and R. Iulian Olariu. 2005. "Nitrated Phenols in the Atmosphere: A Review." *Atmospheric Environment* 39(2):231–248. https://doi.org/10.1016/j.atmosenv.2004.09.044.

Hatch, L. E., W. Luo, J. F. Pankow, R. J. Yokelson, C. E. Stockwell, and K. Barsanti. 2015. "Identification and Quantification of Gaseous Organic Compounds Emitted from Biomass Burning Using Two-Dimensional Gas Chromatography–Time-of-Flight Mass Spectrometry." *Atmospheric Chemistry and Physics* 15(4):1865–1899. https://doi.org/10.5194/acp-15-1865-2015.

Hatch, L. E., R. J. Yokelson, C. E. Stockwell, P. R. Veres, I. J. Simpson, D. R. Blake, J. J. Orlando, and K. C. Barsanti. 2017. "Multi-instrument Comparison and Compilation of Non-methane Organic Gas Emissions from Biomass Burning and Implications for Smoke-Derived Secondary Organic Aerosol Precursors." *Atmospheric Chemistry and Physics* 17(2):1471–1489. https://doi.org/10.5194/acp-17-1471-2017.

Hatch, L. E., A. Rivas-Ubach, C. N. Jen, M. Lipton, A. H. Goldstein, and K. C. Barsanti. 2018. "Measurements of I/SVOCs in Biomass-Burning Smoke Using Solid-Phase Extraction Disks and Two-Dimensional Gas Chromatography." *Atmospheric Chemistry and Physics* 18(24):17801–17817. https://doi.org/10.5194/acp-18-17801-2018.

Hawkins, L. N., M. J. Baril, N. Sedehi, M. M. Galloway, D. O. De Haan, G. P. Schill, and M. A. Tolbert. 2014. "Formation of Semisolid, Oligomerized Aqueous SOA: Lab Simulations of Cloud Processing." *Environmental Science & Technology* 48(4):2273–2280. https://doi.org/10.1021/es4049626.

Hobbs, P. V., P. Sinha, R. J. Yokelson, T. J. Christian, D. R. Blake, S. Gao, T. W. Kirchstetter, T. Novakov, and P. Pilewskie. 2003. "Evolution of Gases and Particles from a Savanna Fire in South Africa." *Journal of Geophysical Research: Atmospheres* 108(D13). https://doi.org/10.1029/2002JD002352.

Hodshire, A. L., A. Akherati, M. J. Alvarado, B. Brown-Steiner, S. H. Jathar, J. L. Jimenez, S. M. Kreidenweis, C. R. Lonsdale, T. B. Onasch, A. M. Ortega, and J. R. Pierce. 2019. "Aging Effects on Biomass Burning Aerosol Mass and Composition: A Critical Review of Field and Laboratory Studies." *Environmental Science & Technology* 53(17):10007–10022. https://doi.org/10.1021/acs.est.9b02588.

Huang, R., R. Lal, M. Qin, Y. Hu, A. G. Russell, M. T. Odman, S. Afrin, F. Garcia-Menendez, and S. M. O'Neill. 2021. "Application and Evaluation of a Low-Cost PM Sensor and Data Fusion with CMAQ Simulations to Quantify the Impacts of Prescribed Burning on Air Quality in Southwestern Georgia, USA." *Journal of the Air & Waste Management Association* 71(7):815–829. https://doi.org/10.1080/10962247.2021.1924311.

Jaffe, D. A., and N. L. Wigder. 2012. "Ozone Production from Wildfires: A Critical Review." *Atmospheric Environment* 51:1–10. https://doi.org/10.1016/j.atmosenv.2011.11.063.

Jaffe, D. A., S. M. O'Neill, N. K. Larkin, A. L. Holder, D. L. Peterson, J. E. Halofsky, and A. G. Rappold. 2020. "Wildfire and Prescribed Burning Impacts on Air Quality in the United States." *Journal of the Air & Waste Management Association* 70(6):583–615. https://doi.org/10.1080/10962247.2020.1749731.

Jiang, X., and E. H. Enki Yoo. 2019. "Modeling Wildland Fire-Specific $PM_{2.5}$ Concentrations for Uncertainty-Aware Health Impact Assessments." *Environmental Science & Technology* 53(20):11828–11839. https://doi.org/10.1021/acs.est.9b02660.

Juncosa Calahorrano, J. F., J. Lindaas, K. O'Dell, B. B. Palm, Q. Peng, F. Flocke, I. B. Pollack, L. A. Garofalo, D. K. Farmer, and J. R. Pierce. 2021. "Daytime Oxidized Reactive Nitrogen Partitioning in Western US Wildfire Smoke Plumes." *Journal of Geophysical Research: Atmospheres* 126(4):e2020JD033484. https://doi.org/10.1029/2020JD033484.

Junghenn Noyes, K., R. Kahn, A. Sedlacek, L. Kleinman, J. Limbacher, and Z. Li. 2020. "Wildfire Smoke Particle Properties and Evolution, from Space-Based Multi-Angle Imaging." *Remote Sensing* 12(5):769. https://doi.org/10.3390/rs12050769.

Karamchandani, P., K. Vijayaraghavan, and G. Yarwood. 2011. "Sub-grid Scale Plume Modeling." *Atmosphere* 2(3):389–406.

Koss, A. R., K. Sekimoto, J. B. Gilman, V. Selimovic, M. M. Coggon, K. J. Zarzana, B. Yuan, B. M. Lerner, S. S. Brown, and J. L. Jimenez. 2018. "Non-methane Organic Gas Emissions from Biomass Burning: Identification, Quantification, and Emission Factors from PTR-ToF during the FIREX 2016 Laboratory Experiment." *Atmospheric Chemistry and Physics* 18(5):3299–3319. https://doi.org/10.5194/acp-18-3299-2018.

Kroll, J. H., and J. H. Seinfeld. 2008. "Chemistry of Secondary Organic Aerosol: Formation and Evolution of Low-Volatility Organics in the Atmosphere." *Atmospheric Environment* 42(16):3593–3624. https://doi.org/10.1016/j.atmosenv.2008.01.003.

Lambe, A. T., T. B. Onasch, D. R. Croasdale, J. P. Wright, A. T. Martin, J. P. Franklin, P. Massoli, J. H. Kroll, M. R. Canagaratna, W. H. Brune, D. R. Worsnop, and P. Davidovits. 2012. "Transitions from Functionalization to Fragmentation Reactions of Laboratory Secondary Organic Aerosol (SOA) Generated from the OH Oxidation of Alkane Precursors." *Environmental Science & Technology* 46(10):5430–5437. https://doi.org/10.1021/es300274t.

Lamkaddam, H., J. Dommen, A. Ranjithkumar, H. Gordon, G. Wehrle, J. Krechmer, F. Majluf, D. Salionov, J. Schmale, S. Bjelić, K. S. Carslaw, I. E. Haddad, and U. Baltensperger. 2021. "Large Contribution to Secondary Organic Aerosol from Isoprene Cloud Chemistry." *Science Advances* 7(13):eabe2952. https://doi.org/10.1126/sciadv.abe2952.

Lee, A. K. Y., P. Herckes, W. R. Leaitch, A. M. Macdonald, and J. P. D. Abbatt. 2011. "Aqueous OH Oxidation of Ambient Organic Aerosol and Cloud Water Organics: Formation of Highly Oxidized Products." *Geophysical Research Letters* 38(11). https://doi.org/10.1029/2011GL047439.

Leslie, M. D., M. Ridoli, J. G. Murphy, and N. Borduas-Dedekind. 2019. "Isocyanic Acid (HNCO) and Its Fate in the Atmosphere: A Review." *Environmental Science: Processes & Impacts* 21(5):793–808. https://doi.org/10.1039/c9em00003h.

Li, W., L. Liu, J. Zhang, L. Xu, Y. Wang, Y. Sun, and Z. Shi. 2021. "Microscopic Evidence for Phase Separation of Organic Species and Inorganic Salts in Fine Ambient Aerosol Particles." *Environmental Science & Technology* 55(4):2234–2242. https://doi.org/10.1021/acs.est.0c02333.

Li, Y. J., D. D. Huang, H. Y. Cheung, A. K. Y. Lee, and C. K. Chan. 2014. "Aqueous-Phase Photochemical Oxidation and Direct Photolysis of Vanillin – A Model Compound of Methoxy Phenols from Biomass Burning." *Atmospheric Chemistry and Physics* 14(6):2871–2885. https://doi.org/10.5194/acp-14-2871-2014.

Lim, Y. B., Y. Tan, M. J. Perri, S. P. Seitzinger, and B. J. Turpin. 2010. "Aqueous Chemistry and Its Role in Secondary Organic Aerosol (SOA) Formation." *Atmospheric Chemistry and Physics* 10(21):10521–10539. https://doi.org/10.5194/acp-10-10521-2010.

Lim, Y., Y. Tan, and B. Turpin. 2013. "Chemical Insights, Explicit Chemistry, and Yields of Secondary Organic Aerosol from OH Radical Oxidation of Methylglyoxal and Glyoxal in the Aqueous Phase." *Atmospheric Chemistry and Physics* 13(17):8651–8667. https://doi.org/10.5194/acp-13-8651-2013.

Lin, G., S. Sillman, J. Penner, and A. Ito. 2014. "Global Modeling of SOA: The Use of Different Mechanisms for Aqueous-Phase Formation." *Atmospheric Chemistry and Physics* 14(11):5451–5475. https://doi.org/10.5194/acp-14-5451-2014.

Lin, P., G. Engling, and J. Yu. 2010. "Humic-Like Substances in Fresh Emissions of Rice Straw Burning and in Ambient Aerosols in the Pearl River Delta Region, China." *Atmospheric Chemistry and Physics* 10(14):6487–6500. https://doi.org/10.5194/acp-10-6487-2010.

Liu, J., L. W. Horowitz, S. Fan, A. G. Carlton, and H. Levy. 2012. "Global In-Cloud Production of Secondary Organic Aerosols: Implementation of a Detailed Chemical Mechanism in the GFDL Atmospheric Model AM3." *Journal of Geophysical Research: Atmospheres* 117(D15). https://doi.org/10.1029/2012JD017838.

Liu, Y., J. Liggio, T. Harner, L. Jantunen, M. Shoeib, and S. M. Li. 2014. "Heterogeneous OH Initiated Oxidation: A Possible Explanation for the Persistence of Organophosphate Flame Retardants in Air." *Environmental Science & Technology* 48(2):1041–1048. https://doi.org/10.1021/es404515k.

Loeffler, K. W., C. A. Koehler, N. M. Paul, and D. O. De Haan. 2006. "Oligomer Formation in Evaporating Aqueous Glyoxal and Methyl Glyoxal Solutions." *Environmental Science & Technology* 40(20):6318–6323. https://doi.org/10.1021/es060810w.

Lu, X., L. Zhang, X. Yue, J. Zhang, D. A. Jaffe, A. Stohl, Y. Zhao, and J. Shao. 2016. "Wildfire Influences on the Variability and Trend of Summer Surface Ozone in the Mountainous Western United States." *Atmospheric Chemistry and Physics* 16(22):14687–14702. https://doi.org/10.5194/acp-16-14687-2016.

Luecken, D. J., S. L. Napelenok, M. Strum, R. Scheffe, and S. Phillips. 2018. "Sensitivity of Ambient Atmospheric Formaldehyde and Ozone to Precursor Species and Source Types across the United States." *Environmental Science & Technology* 52(8):4668–4675. https://doi.org/10.1021/acs.est.7b05509.

McFall, A. S., A. W. Johnson, and C. Anastasio. 2020. "Air–Water Partitioning of Biomass-Burning Phenols and the Effects of Temperature and Salinity." *Environmental Science & Technology* 54(7):3823–3830. https://doi.org/10.1021/acs.est.9b06443.

McMeeking, G. R., S. M. Kreidenweis, S. Baker, C. M. Carrico, J. C. Chow, J. L. Collett Jr, W. M. Hao, A. S. Holden, T. W. Kirchstetter, and W. C. Malm. 2009. "Emissions of Trace Gases and Aerosols during the Open Combustion of Biomass in the Laboratory." *Journal of Geophysical Research: Atmospheres* 114(D19). https://doi.org/10.1029/2009JD011836.

McNeill, V. F. 2015. "Aqueous Organic Chemistry in the Atmosphere: Sources and Chemical Processing of Organic Aerosols." *Environmental Science & Technology* 49(3):1237–1244. https://doi.org/10.1021/es5043707.

Ng, N. L., S. S. Brown, A. T. Archibald, E. Atlas, R. C. Cohen, J. N. Crowley, D. A. Day, N. M. Donahue, J. L. Fry, H. Fuchs, R. J. Griffin, M. I. Guzman, H. Herrmann, A. Hodzic, Y. Iinuma, J. L. Jimenez, A. Kiendler-Scharr, B. H. Lee, D. J. Luecken, J. Mao, R. McLaren, A. Mutzel, H. D. Osthoff, B. Ouyang, B. Picquet-Varrault, U. Platt, H. O. T. Pye, Y. Rudich, R. H. Schwantes, M. Shiraiwa, J. Stutz, J. A. Thornton, A. Tilgner, B. J. Williams, and R. A. Zaveri. 2017. "Nitrate Radicals and Biogenic Volatile Organic Compounds: Oxidation, Mechanisms, and Organic Aerosol." *Atmospheric Chemistry and Physics* 17(3):2103–2162. https://doi.org/10.5194/acp-17-2103-2017.

NOAA ARL (National Oceanic and Atmospheric Administration Air Resources Laboratory). 2012. *The HYSPLIT.* https://www.arl.noaa.gov/hysplit/ (accessed December 23, 2021).

NOAA ARL. 2021. *HYSPLIT-based Smoke Forecasts.* https://www.arl.noaa.gov/hysplit/smoke-forecasting/ (accessed October 28, 2021).

O'Dell, K., B. Ford, E. V. Fischer, and J. R. Pierce. 2019. "Contribution of Wildland-Fire Smoke to US $PM_{2.5}$ and Its Influence on Recent Trends." *Environmental Science & Technology* 53(4):1797–1804. https://doi.org/10.1021/acs.est.8b05430.

O'Dell, K., R. S. Hornbrook, W. Permar, E. J. T. Levin, L. A. Garofalo, E. C. Apel, N. J. Blake, A. Jarnot, M. A. Pothier, D. K. Farmer, L. Hu, T. Campos, B. Ford, J. R. Pierce, and E. V. Fischer. 2020. "Hazardous Air Pollutants in Fresh and Aged Western US Wildfire Smoke and Implications for Long-Term Exposure." *Environmental Science & Technology* 54(19):11838–11847. https://doi.org/10.1021/acs.est.0c04497.

Ohura, T., A. Kitazawa, T. Amagai, and M. Makino. 2005. "Occurrence, Profiles, and Photostabilities of Chlorinated Polycyclic Aromatic Hydrocarbons Associated with Particulates in Urban Air." *Environmental Science & Technology* 39(1):85–91. https://doi.org/10.1021/es040433s.

Ortiz-Montalvo, D. L., S. A. Häkkinen, A. N. Schwier, Y. B. Lim, V. F. McNeill, and B. J. Turpin. 2014. "Ammonium Addition (and Aerosol pH) Has a Dramatic Impact on the Volatility and Yield of Glyoxal Secondary Organic Aerosol." *Environmental Science & Technology* 48(1):255–262. https://doi.org/10.1021/es4035667.

Paglione, M., S. Gilardoni, M. Rinaldi, S. Decesari, N. Zanca, S. Sandrini, L. Giulianelli, D. Bacco, S. Ferrari, and V. Poluzzi. 2020. "The Impact of Biomass Burning and Aqueous-Phase Processing on Air Quality: A Multi-year Source Apportionment Study in the Po Valley, Italy." *Atmospheric Chemistry and Physics* 20(3):1233–1254. https://doi.org/10.5194/acp-20-1233-2020.

Palm, B. B., Q. Peng, C. D. Fredrickson, B. H. Lee, L. A. Garofalo, M. A. Pothier, S. M. Kreidenweis, D. K. Farmer, R. P. Pokhrel, Y. Shen, S. M. Murphy, W. Permar, L. Hu, T. L. Campos, S. R. Hall, K. Ullmann, X. Zhang, F. Flocke, E. V. Fischer, and J. A. Thornton. 2020. "Quantification of Organic Aerosol and Brown Carbon Evolution in Fresh Wildfire Plumes." *Proceedings of the National Academy of Sciences* 117(47):29469–29477. https://doi.org/10.1073/pnas.2012218117.

Pang, H., Q. Zhang, X. Lu, K. Li, H. Chen, J. Chen, X. Yang, Y. Ma, J. Ma, and C. Huang. 2019. "Nitrite-Mediated Photo-oxidation of Vanillin in the Atmospheric Aqueous Phase." *Environmental Science & Technology* 53(24):14253–14263. https://doi.org/10.1021/acs.est.9b03649.

Patoulias, D., E. Kallitsis, L. Posner, and S. N. Pandis. 2021. "Modeling Biomass Burning Organic Aerosol Atmospheric Evolution and Chemical Aging." *Atmosphere* 12(12):1638. https://doi.org/10.3390/atmos12121638.

Paugam, R., M. Wooster, S. Freitas, and M. Val Martin. 2016. "A Review of Approaches to Estimate Wildfire Plume Injection Height within Large-Scale Atmospheric Chemical Transport Models." *Atmospheric Chemistry and Physics* 16(2):907–925. https://doi.org/10.5194/acp-16-907-2016.

Paulson, S. E., P. J. Gallimore, X. M. Kuang, J. R. Chen, M. Kalberer, and D. H. Gonzalez. 2019. "A Light-Driven Burst of Hydroxyl Radicals Dominates Oxidation Chemistry in Newly Activated Cloud Droplets." *Science Advances* 5(5):eaav7689. https://doi.org/10.1126/sciadv.aav7689.

Peng, Q., B. B. Palm, C. D. Fredrickson, B. H. Lee, S. R. Hall, K. Ullmann, . . . and J. Thornton. 2021. "Observations and Modeling of NO_x Photochemistry and Fate in Fresh Wildfire Plumes." *ACS Earth and Space Chemistry* 5(10):2652–2667.

Permar, W., Q. Wang, V. Selimovic, C. Wielgasz, R. J. Yokelson, R. S. Hornbrook, A. J. Hills, E. C. Apel, I.-T. Ku, Y. Zhou, B. C. Sive, A. P. Sullivan, J. L. Collett Jr, T. L. Campos, B. B. Palm, Q. Peng, J. A. Thornton, L. A. Garofalo, D. K. Farmer, S. M. Kreidenweis, E. J. T. Levin, P. J. DeMott, F. Flocke, E. V. Fischer, and L. Hu. 2021. "Emissions of Trace Organic Gases from Western U.S. Wildfires Based on WE-CAN Aircraft Measurements." *Journal of Geophysical Research: Atmospheres* 126(11):e2020JD033838. https://doi.org/10.1029/2020JD033838.

Perri, M. J., S. Seitzinger, and B. J. Turpin. 2009. "Secondary Organic Aerosol Production from Aqueous Photooxidation of Glycolaldehyde: Laboratory Experiments." *Atmospheric Environment* 43(8):1487–1497. https://doi.org/10.1016/j.atmosenv.2008.11.037.

Pike, K. A., P. L. Edmiston, J. J. Morrison, and J. A. Faust. 2021. "Correlation Analysis of Perfluoroalkyl Substances in Regional US Precipitation Events." *Water Research* 190:116685. https://doi.org/10.1016/j.watres.2020.116685.

Posner, L. N., G. Theodoritsi, A. Robinson, G. Yarwood, B. Koo, R. Morris, M. Mavko, T. Moore, and S. N. Pandis. 2019. "Simulation of Fresh and Chemically-Aged Biomass Burning Organic Aerosol." *Atmospheric Environment* 196:27–37. https://doi.org/10.1016/j.atmosenv.2018.09.055.

Real, E., K. S. Law, B. Weinzierl, M. Fiebig, A. Petzold, O. Wild, J. Methven, S. Arnold, A. Stohl, and H. Huntrieser. 2007. "Processes Influencing Ozone Levels in Alaskan Forest Fire Plumes during Long-Range Transport over the North Atlantic." *Journal of Geophysical Research: Atmospheres* 112(D10). https://doi.org/10.1029/2006JD007576.

Reisch, M. S. 2018. "The Price of Fire Safety." *Chemical & Engineering News* 97(2):16–19. https://www.firefightingfoam.com/assets/Uploads/ARTICLES-/jan14-19-Chemical-and-Engineering-News.pdf.

Riemer, D. D., E. C. Apel, J. J. Orlando, G. S. Tyndall, W. H. Brune, E. J. Williams, W. A. Lonneman, and J. D. Neece. 2008. "Unique Isoprene Oxidation Products Demonstrate Chlorine Atom Chemistry Occurs in the Houston, Texas Urban Area." *Journal of Atmospheric Chemistry* 61(3):227–242. https://doi.org/10.1007/s10874-009-9134-5.

Riva, M., R. M. Healy, P. M. Flaud, E. Perraudin, J. C. Wenger, and E. Villenave. 2015. "Gas- and Particle-Phase Products from the Chlorine-Initiated Oxidation of Polycyclic Aromatic Hydrocarbons." *Journal of Physical Chemistry A* 119(45): 11170–11181. https://doi.org/10.1021/acs.jpca.5b04610.

Roberts, J. M., and Y. Liu. 2019. "Solubility and Solution-Phase Chemistry of Isocyanic Acid, Methyl Isocyanate, and Cyanogen Halides." *Atmospheric Chemistry and Physics* 19(7):4419–4437. https://doi.org/10.5194/acp-19-4419-2019.

Roberts, J. M., P. R. Veres, A. K. Cochran, C. Warneke, I. R. Burling, R. J. Yokelson, B. Lerner, J. B. Gilman, W. C. Kuster, R. Fall, and J. de Gouw. 2011. "Isocyanic Acid in the Atmosphere and Its Possible Link to Smoke-Related Health Effects." *Proceedings of the National Academy of Sciences* 108(22):8966–8971. https://doi.org/10.1073/pnas.1103352108.

Robinson, A. L., N. M. Donahue, M. K. Shrivastava, E. A. Weitkamp, A. M. Sage, A. P. Grieshop, T. E. Lane, J. R. Pierce, and S. N. Pandis. 2007. "Rethinking Organic Aerosols: Semivolatile Emissions and Photochemical Aging." *Science* 315(5816):1259–1262. https://doi.org/10.1126/science.1133061.

Rooney, B., Y. Wang, J. H. Jiang, B. Zhao, Z.-C. Zeng, and J. H. Seinfeld. 2020. "Air Quality Impact of the Northern California Camp Fire of November 2018." *Atmospheric Chemistry and Physics* 20(23):14597–14616. https://doi.org/10.5194/acp-20-14597-2020.

Rossignol, S., K. Z. Aregahegn, L. Tinel, L. Fine, B. Noziere, and C. George. 2014. "Glyoxal Induced Atmospheric Photosensitized Chemistry Leading to Organic Aerosol Growth." *Environmental Science & Technology* 48(6):3218–3227. https://doi.org/10.1021/es405581g.

Ruokojärvi, P., M. Aatamila, and J. Ruuskanen. 2000. "Toxic Chlorinated and Polyaromatic Hydrocarbons in Simulated House Fires." *Chemosphere* 41(6):825–828. https://doi.org/10.1016/s0045-6535(99)00549-4.

Schauer, J. J., M. J. Kleeman, G. R. Cass, and B. R. Simoneit. 2001. "Measurement of Emissions from Air Pollution Sources. 3. C_1–C_{29} Organic Compounds from Fireplace Combustion of Wood." *Environmental Science & Technology* 35(9): 1716–1728. https://doi.org/10.1021/es001331e.

Seinfeld, J. H., and S. N. Pandis. 2016. *Atmospheric Chemistry and Physics: From Air Pollution to Climate Change*. Hoboken, NJ: John Wiley & Sons.

Selimovic, V., R. J. Yokelson, C. Warneke, J. M. Roberts, J. d. Gouw, J. Reardon, and D. W. Griffith. 2018. "Aerosol Optical Properties and Trace Gas Emissions by PAX and OP-FTIR for Laboratory-Simulated Western US Wildfires during FIREX." *Atmospheric Chemistry and Physics* 18(4):2929–2948. https://doi.org/10.5194/acp-18-2929-2018.

Sha, B., A. K. Dahlberg, K. Wiberg, and L. Ahrens. 2018. "Fluorotelomer Alcohols (FTOHs), Brominated Flame Retardants (BFRs), Organophosphorus Flame Retardants (OPFRs) and Cyclic Volatile Methylsiloxanes (cVMSs) in Indoor Air from Occupational and Home Environments." *Environmental Pollution* 241:319–330. https://doi.org/10.1016/j.envpol.2018.04.032.

Shimizu, M. S., R. Mott, A. Potter, J. Zhou, K. Baumann, J. D. Surratt, B. Turpin, G. B. Avery, J. Harfmann, and R. J. Kieber. 2021. "Atmospheric Deposition and Annual Flux of Legacy Perfluoroalkyl Substances and Replacement Perfluoroalkyl Ether Carboxylic Acids in Wilmington, NC, USA." *Environmental Science & Technology Letters* 8(5):366–372. https://doi.org/10.1021/acs.estlett.1c00251.

Sokolik, I. N., A. J. Soja, P. J. DeMott, and D. Winker. 2019. "Progress and Challenges in Quantifying Wildfire Smoke Emissions, Their Properties, Transport, and Atmospheric Impacts." *Journal of Geophysical Research: Atmospheres* 124(23):13005–13025.

Spada, N. J., and N. P. Hyslop. 2018. "Comparison of Elemental and Organic Carbon Measurements between IMPROVE and CSN before and after Method Transitions." *Atmospheric Environment* 178:1783-180. https://doi.org/10.1016/j.atmosenv.2018.01.043.

Stein, A., R. R. Draxler, G. D. Rolph, B. J. Stunder, M. Cohen, and F. Ngan. 2015. "NOAA's HYSPLIT Atmospheric Transport and Dispersion Modeling System." *Bulletin of the American Meteorological Society* 96(12):2059–2077. https://doi.org/10.1175/BAMS-D-14-00110.1.

Sun, Y. L., Q. Zhang, C. Anastasio, and J. Sun. 2010. "Insights into Secondary Organic Aerosol Formed via Aqueous-Phase Reactions of Phenolic Compounds Based on High Resolution Mass Spectrometry." *Atmospheric Chemistry and Physics* 10(10):4809–4822. https://doi.org/10.5194/acp-10-4809-2010.

Surratt, J. D., A. W. H. Chan, N. C. Eddingsaas, M. Chan, C. L. Loza, A. J. Kwan, S. P. Hersey, R. C. Flagan, P. O. Wennberg, and J. H. Seinfeld. 2010. "Reactive Intermediates Revealed in Secondary Organic Aerosol Formation from Isoprene." *Proceedings of the National Academy of Sciences* 107(15):6640–6645. https://doi.org/10.1073/pnas.0911114107.

Tabazadeh, A., R. J. Yokelson, H. Singh, P. V. Hobbs, J. Crawford, and L. Iraci. 2004. "Heterogeneous Chemistry Involving Methanol in Tropospheric Clouds." *Geophysical Research Letters* 31(6). https://doi.org/10.1029/2003GL018775.

Tan, Y., Y. Lim, K. Altieri, S. Seitzinger, and B. Turpin. 2012. "Mechanisms Leading to Oligomers and SOA through Aqueous Photooxidation: Insights from OH Radical Oxidation of Acetic Acid and Methylglyoxal." *Atmospheric Chemistry and Physics* 12(2):801–813. https://doi.org/10.5194/acp-12-801-2012.

Tanaka, P. L., D. D. Riemer, S. Chang, G. Yarwood, E. C. McDonald-Buller, E. C. Apel, J. J. Orlando, P. J. Silva, J. L. Jimenez, M. R. Canagaratna, J. D. Neece, C. B. Mullins, and D. T. Allen. 2003. "Direct Evidence for Chlorine-Enhanced Urban Ozone Formation in Houston, Texas." *Atmospheric Environment* 37(9):1393–1400. https://doi.org/10.1016/S1352-2310(02)01007-5.

Theodoritsi, G. N., G. Ciarelli, and S. N. Pandis. 2021. "Simulation of the Evolution of Biomass Burning Organic Aerosol with Different Volatility Basis Set Schemes in PMCAMx-SRv1.0." *Geoscientific Model Development* 14(4):2041–2055. https://doi.org/10.5194/gmd-14-2041-2021.

Theys, N., R. Volkamer, J. F. Müller, K. J. Zarzana, N. Kille, L. Clarisse, I. De Smedt, C. Lerot, H. Finkenzeller, F. Hendrick, T. K. Koenig, C. F. Lee, C. Knote, H. Yu, and M. Van Roozendael. 2020. "Global Nitrous Acid Emissions and Levels of Regional Oxidants Enhanced by Wildfires." *Nature Geoscience* 13(10):681–686. https://doi.org/10.1038/s41561-020-0637-7.

Thornton, J. A., J. P. Kercher, T. P. Riedel, N. L. Wagner, J. Cozic, J. S. Holloway, W. P. Dubé, G. M. Wolfe, P. K. Quinn, A. M. Middlebrook, B. Alexander, and S. S. Brown. 2010. "A Large Atomic Chlorine Source Inferred from Mid-continental Reactive Nitrogen Chemistry." *Nature* 464(7286):271–274. https://doi.org/10.1038/nature08905.

Tilgner, A., P. Bräuer, R. Wolke, and H. Herrmann. 2013. "Modelling Multiphase Chemistry in Deliquescent Aerosols and Clouds Using CAPRAM3.0i." *Journal of Atmospheric Chemistry* 70(3):221–256. https://doi.org/10.1007/s10874-013-9267-4.

Tilgner, A., T. Schaefer, B. Alexander, M. Barth, J. L. Collett, Jr., K. M. Fahey, A. Nenes, H. O. T. Pye, H. Herrmann, and V. F. McNeill. 2021. "Acidity and the Multiphase Chemistry of Atmospheric Aqueous Particles and Clouds." *Atmospheric Chemistry and Physics* 21(17):13483–13536. https://doi.org/10.5194/acp-21-13483-2021.

Tkacik, D. S., A. A. Presto, N. M. Donahue, and A. L. Robinson. 2012. "Secondary Organic Aerosol Formation from Inter-mediate-Volatility Organic Compounds: Cyclic, Linear, and Branched Alkanes." *Environmental Science & Technology* 46(16):8773–8781. https://doi.org/10.1021/es301112c.

Tomaz, S., T. Cui, Y. Chen, K. G. Sexton, J. M. Roberts, C. Warneke, R. J. Yokelson, J. D. Surratt, and B. J. Turpin. 2018. "Photochemical Cloud Processing of Primary Wildfire Emissions as a Potential Source of Secondary Organic Aerosol." *Environmental Science & Technology* 52(19):11027–11037. https://doi.org/10.1021/acs.est.8b03293.

Tuet, W. Y., Y. Chen, S. Fok, D. Gao, R. J. Weber, J. A. Champion, and N. L. Ng. 2017. "Chemical and Cellular Oxidant Production Induced by Naphthalene Secondary Organic Aerosol (SOA): Effect of Redox-Active Metals and Photochemical Aging." *Scientific Reports* 7(1):15157. https://doi.org/10.1038/s41598-017-15071-8.

US Environmental Protection Agency. 2021. *EPI Suite™ -Estimation Program Interface*. https://www.epa.gov/tsca-screening-tools/epi-suitetm-estimation-program-interface (accessed December 22, 2021).

Veres, P., J. M. Roberts, I. R. Burling, C. Warneke, J. de Gouw, and R. J. Yokelson. 2010. "Measurements of Gas-Phase Inorganic and Organic Acids from Biomass Fires by Negative-Ion Proton-Transfer Chemical-Ionization Mass Spectrometry." *Journal of Geophysical Research: Atmospheres* 115(D23). https://doi.org/10.1029/2010JD014033.

Vidović, K., A. Kroflič, M. Šala, and I. Grgić. 2020. "Aqueous-Phase Brown Carbon Formation from Aromatic Precursors under Sunlight Conditions." *Atmosphere* 11(2):131. https://doi.org/10.3390/atmos11020131.

Volkamer, R., P. Ziemann, and M. Molina. 2009. "Secondary Organic Aerosol Formation from Acetylene (C_2H_2): Seed Effect on SOA Yields Due to Organic Photochemistry in the Aerosol Aqueous Phase." *Atmospheric Chemistry and Physics* 9(6):1907–1928. https://doi.org/10.5194/acp-9-1907-2009.

Wallington, T. J., M. D. Hurley, J. Xia, D. J. Wuebbles, S. Sillman, A. Ito, J. E. Penner, D. A. Ellis, J. Martin, S. A. Mabury, O. J. Nielsen, and M. P. Sulbaek Andersen. 2006. "Formation of $C_7F_{15}COOH$ (PFOA) and Other Perfluorocarboxylic Acids during the Atmospheric Oxidation of 8:2 Fluorotelomer Alcohol." *Environmental Science & Technology* 40(3):924–930. https://doi.org/10.1021/es051858x.

Wang, Z., J. C. DeWitt, C. P. Higgins, and I. T. Cousins. 2017. "A Never-Ending Story of Per- and Polyfluoroalkyl Substances (PFASs)?" *Environmental Science & Technology* 51(5):2508–2518.

Wei, Y., X. Chen, H. Chen, Y. Sun, W. Yang, H. Du, … and Z. Wang. 2021. "Investigating the Importance of Sub-grid Particle Formation in Point Source Plumes over Eastern China using IAP-AACM v1.0 with a Sub-grid Parameterization." *Geoscientific Model Development* 14(7):4411–4428.

Whitten, G. Z., G. Heo, Y. Kimura, E. McDonald-Buller, D. T. Allen, W. P. L. Carter, and G. Yarwood. 2010. "A New Condensed Toluene Mechanism for Carbon Bond: CB05-TU." *Atmospheric Environment* 44(40):5346–5355. https://doi.org/10.1016/j.atmosenv.2009.12.029.

Wilkins, J. L., G. Pouliot, K. Foley, W. Appel, and T. Pierce. 2018. "The Impact of US Wildland Fires on Ozone and Particulate Matter: A Comparison of Measurements and CMAQ Model Predictions from 2008 to 2012." *International Journal of Wildland Fire* 27(10):684–698. https://doi.org/10.1071/wf18053.

Wong, J. P. S., M. Tsagkaraki, I. Tsiodra, N. Mihalopoulos, K. Violaki, M. Kanakidou, J. Sciare, A. Nenes, and R. J. Weber. 2019. "Effects of Atmospheric Processing on the Oxidative Potential of Biomass Burning Organic Aerosols." *Environmental Science & Technology* 53(12):6747–6756. https://doi.org/10.1021/acs.est.9b01034.

Xu, L., J. D. Crounse, K. T. Vasquez, H. Allen, P. O. Wennberg, I. Bourgeois, ... and R. J. Yokelson. 2021. "Ozone Chemistry in Western US Wildfire Plumes." *Science Advances* 7(50):eabl3648.

Ye, X., P. Arab, and R. Ahmadov. 2021. "Evaluation and Intercomparison of Wildfire Smoke Forecasts from Multiple Modeling Systems for the 2019 Williams Flats Fire." *Atmospheric Chemistry and Physics* 21(18):14427–14469. https://doi.org/10.5194/acp-21-14427-2021.

Yee, L., K. Kautzman, C. Loza, K. Schilling, M. Coggon, P. Chhabra, M. Chan, A. Chan, S. Hersey, and J. Crounse. 2013. "Secondary Organic Aerosol Formation from Biomass Burning Intermediates: Phenol and Methoxyphenols." *Atmospheric Chemistry and Physics* 13(16):8019–8043. https://doi.org/10.5194/acp-13-8019-2013.

Ying, Q., J. Li, and S. H. Kota. 2015. "Significant Contributions of Isoprene to Summertime Secondary Organic Aerosol in Eastern United States." *Environmental Science & Technology* 49(13):7834–7842. https://doi.org/10.1021/acs.est.5b02514.

Yokelson, R. J., I. T. Bertschi, T. J. Christian, P. V. Hobbs, D. E. Ward, and W. M. Hao. 2003. "Trace Gas Measurements in Nascent, Aged, and Cloud-Processed Smoke from African Savanna Fires by Airborne Fourier Transform Infrared Spectroscopy (AFTIR)." *Journal of Geophysical Research: Atmospheres* 108(D13). https://doi.org/10.1029/2002JD002322.

Yokelson, R. J., J. Crounse, P. DeCarlo, T. Karl, S. Urbanski, E. Atlas, T. Campos, Y. Shinozuka, V. Kapustin, and A. Clarke. 2009. "Emissions from Biomass Burning in the Yucatan." *Atmospheric Chemistry and Physics* 9(15):5785–5812. https://doi.org/10.5194/acp-9-5785-2009.

Young, C. J., V. I. Furdui, J. Franklin, R. M. Koerner, D. C. Muir, and S. A. Mabury. 2007. "Perfluorinated Acids in Arctic Snow: New Evidence for Atmospheric Formation." *Environmental Science & Technology* 41(10):3455–3461.

Young, C., R. Washenfelder, P. Edwards, D. Parrish, J. Gilman, W. Kuster, L. Mielke, H. Osthoff, C. Tsai, and O. Pikelnaya. 2014. "Chlorine as a Primary Radical: Evaluation of Methods to Understand Its Role in Initiation of Oxidative Cycles." *Atmospheric Chemistry and Physics* 14(7):3427–3440. https://doi.org/10.5194/acp-14-3427-2014.

Yu, L., J. Smith, A. Laskin, C. Anastasio, J. Laskin, and Q. Zhang. 2014. "Chemical Characterization of SOA Formed from Aqueous-Phase Reactions of Phenols with the Triplet Excited State of Carbonyl and Hydroxyl Radical." *Atmospheric Chemistry and Physics* 14(24):13801–13816. https://doi.org/10.5194/acp-14-13801-2014.

Yu, L., J. Smith, A. Laskin, K. M. George, C. Anastasio, J. Laskin, A. M. Dillner, and Q. Zhang. 2016. "Molecular Transformations of Phenolic SOA during Photochemical Aging in the Aqueous Phase: Competition among Oligomerization, Functionalization, and Fragmentation." *Atmospheric Chemistry and Physics* 16(7):4511–4527. https://doi.org/10.5194/acp-16-4511-2016.

Yu, X., D. Li, D. Li, G. Zhang, H. Zhou, S. Li, W. Song, Y. Zhang, X. Bi, and J. Yu. 2020. "Enhanced Wet Deposition of Water-Soluble Organic Nitrogen During the Harvest Season: Influence of Biomass Burning and In-Cloud Scavenging." *Journal of Geophysical Research: Atmospheres* 125(18):e2020JD032699. https://doi.org/10.1029/2020JD032699.

Zhang, Y., H. Forrister, J. Liu, J. Dibb, B. Anderson, J. P. Schwarz, A. E. Perring, J. L. Jimenez, P. Campuzano-Jost, and Y. Wang. 2017. "Top-of-Atmosphere Radiative Forcing Affected by Brown Carbon in the Upper Troposphere." *Nature Geoscience* 10(7):486–489. https://doi.org/10.1038/ngeo2960.

Zhao, R., E. L. Mungall, A. K. Lee, D. Aljawhary, and J. P. Abbatt. 2014a. "Aqueous-Phase Photooxidation of Levoglucosan–A Mechanistic Study Using Aerosol Time-of-Flight Chemical Ionization Mass Spectrometry (Aerosol ToF-CIMS)." *Atmospheric Chemistry and Physics* 14(18):9695–9706. https://doi.org/10.5194/acp-14-9695-2014.

Zhao, Y., C. J. Hennigan, A. A. May, D. S. Tkacik, J. A. de Gouw, J. B. Gilman, W. C. Kuster, A. Borbon, and A. L. Robinson. 2014b. "Intermediate-Volatility Organic Compounds: A Large Source of Secondary Organic Aerosol." *Environmental Science & Technology* 48(23):13743–13750. https://doi.org/10.1021/es5035188.

Zhou, S., S. Collier, D. A. Jaffe, N. L. Briggs, J. Hee, A. J. Sedlacek III, L. Kleinman, T. B. Onasch, and Q. Zhang. 2017. "Regional Influence of Wildfires on Aerosol Chemistry in the Western US and Insights into Atmospheric Aging of Biomass Burning Organic Aerosol." *Atmospheric Chemistry and Physics* 17(3):2477–2493. https://doi.org/10.5194/acp-17-2477-2017.

Zuend, A., and J. H. Seinfeld. 2012. "Modeling the Gas-Particle Partitioning of Secondary Organic Aerosol: The Importance of Liquid-Liquid Phase Separation." *Atmospheric Chemistry and Physics* 12(9):3857–3882. https://doi.org/10.5194/acp-12-3857-2012.

5

Water and Soil Contamination

As earlier chapters in this report have discussed, combustion reactions for materials at the wildland-urban interface (WUI; e.g., household components such as siding and plastic, as well as biomass; see Chapter 3) result in emissions of chemical species to the surrounding environment. Although part of these emissions will remain in the air and be further transformed in the downwind plume, as described in Chapter 4, pathways also exist for the mobilization (or partitioning) of some of these chemical species into nearby soils and water streams, potentially impacting ecosystem health or even public health. The possibility also exists of pollutants in the plume being deposited downwind of the fire area, as long-term studies on the impacts of wildfires have observed. Finally, the active process of firefighting could add pollutants to the immediate area, in the form of flame retardants and other compounds (McGee et al., 2003).

All these processes can impact the immediate conditions after the fire, which are critical as first responders and community residents arrive back in the area to assess damage. There is a dearth of peer-reviewed data documenting the specific impacts of WUI fires on water and soil contamination, with only a few studies providing critical information on potential impacts (as will be discussed below). In addition, a number of reports commissioned by the State of California after major fires over the last few years document the potential for contamination from pollutants in ashes.

Perhaps the most widely known case for water contamination observed in the recent past is the case of the contamination of water distribution systems after two significant fires in California, the Tubbs Fire in 2017 and the Camp Fire in 2018 (see Chapter 2 for more details about the Camp Fire). As has been well summarized elsewhere, in these cases, benzene and other small-molecular-weight toxicants (e.g., volatile organic compounds, or VOCs) were measured in the potable water distribution systems. Although limited samples were taken across areas where communities had been burned, levels of benzene above 5 ppb (the US Environmental Protection Agency's (EPA's) maximum contaminant level recommendation) were measured in several samples collected from water mains, hydrants, and service connections (Proctor et al., 2020; Solomon et al., 2021). Subsequent investigations indicated that the source of contamination was more than likely the combustion of household materials, as opposed to contamination via other sources (e.g., hydrocarbon storage tanks). Researchers hypothesized that ingress into the distribution system occurred through negative pressures, in piping systems relative to external pressures (Proctor et al., 2020). In these cases, contamination persisted for months until remediation of the situation, which required extensive efforts by the water providers. Additionally, reports of data from WUI fires in California document levels of metals measured in ash samples collected in impacted neighborhoods, with elevated levels of antimony and other species (TetraTech Inc., 2019).

BOX 5-1
Key Terminology

Community water system (CWS): A water system serving populations greater than 25 people, year round

Dissolved organic matter (DOM): A mixture of organic compounds found ubiquitously in surface and groundwaters; derived from terrestrial and aquatic sources

Emissions (or effluents): Species emitted into the air, water, soil, or other media, from a process; these are sometimes called releases

Partitioning: The distribution of a species *i* between two media, such as air and particles, or air and water

Pollutant: A chemical or biological substance that harms water, air, or land quality

Toxicant: Any chemical that can injure or kill humans, animals, or plants (depending on the magnitude and duration of exposure); a poison

Volatile organic compounds (VOCs): Organic compounds with vapor pressures high enough to exist in the atmosphere primarily in the gas phase, typically excluding methane; VOCs can easily become airborne for inhalation exposure

---

Although the recent cases of water contamination after a WUI fire have become widely known and received significant press coverage, it has been known for decades that wildland fires can impact water and soil quality (Bladon et al., 2014; Hohner et al., 2019a; Olivella et al., 2006; Rhoades et al., 2019b; Smith et al., 2011). These issues will continue to be a concern as the frequency of WUI fires continues to increase.

The focus of this chapter is to provide a description of some of the concerns regarding water and soil contamination after WUI fires. These two media environments are discussed separately. One common theme in the subsequent sections is the lack of detailed information on the specific contamination of water and soils under conditions where the area burned included the built environment, for example, structures (see Chapter 2 for a more extensive definition of WUI fires). Therefore, the discussion cites studies that provide information originally collected for wildland fires, which can be extrapolated to WUI fires. As with other chapters, the last section provides a series of research needs.

Box 5-1 defines key terms that are used throughout this chapter.

IMPACTS TO WATER QUALITY RELATED TO COMMUNITY WATER SYSTEMS

This section discusses impacts to centralized community water systems (CWSs; Figure 5-1), generally in the vicinity near a fire, followed by potential long-range impacts. A CWS is defined as a system serving populations greater than 25 people, year round (EPA, 2021b). The committee does recognize that smaller communities can also be impacted by fires, and the concerns raised will also apply to those communities' systems, including groundwater systems. The contaminants considered will include largely organic materials and metals, but will not consider what it is typically referred to as black carbon or pyrogenic organic carbon (Jaffé et al., 2013). Also not explicitly covered is fire-impacted or fire-produced carbon in streams. This ill-defined mixture on its own is not considered toxic, even if some of its components (e.g., polycyclic aromatic hydrocarbons, or PAHs) have health considerations.

CWSs rely on stable and predictable water sources. Treatment operations are designed with an expectation that their systems will not be subject to abrupt changes in conditions. CWSs are regulated by federal statutes and therefore are required to meet stringent water quality standards, for example related to disinfection by-product formation levels (EPA, 2021a).

FIGURE 5-1 WUI fires can impact soil and water, ultimately also impacting CWSs. SOURCE: Hohner et al. (2019a).

A WUI fire can potentially impact the operations of a CWS in both the short and long term. Initially, operational concerns arise to address the immediate aftermath of the fire, focusing on returning potable water service to the impacted communities and possibly issuing notices to communities in and near the incident area (e.g., boil water advisories). In the aftermath of the fire, impacts related to water quality occur; different types of chemical species can contaminate the water, with different entry points.

Contamination to a CWS can occur via source water contamination (i.e., streams or reservoirs, some of which are in urban settings, and groundwater) or through critical failures in distribution systems, for example, negative pressure events that may draw in contamination (potentially gas phase or in the form of ashes). Source water issues relate directly to the runoff of contamination during or after the fire. For example, the use of firefighting chemicals can result in mobilization that ultimately impacts source waters, although the use of these firefighting chemicals is restricted near surface waters (Nolen et al., 2022). Mobilization of ashes and runoff from incident areas can impact surface waters as well.

Ashes and Sediments

The types of contamination considered important for CWSs include mobilization of ashes and sediments. These substances increase the turbidity in the water and can also act as a source of other pollutants, as chemicals partition between solid particles and the aqueous environment. The turbidity in the final water produced by a CWS is regulated, and for the most part, high turbidity (>100 NTUs [nephelometric turbidity units, the units from a calibrated nephelometer]) represents a challenge for CWSs (Becker et al., 2018). The particles leading to the turbidity can be mobilized by rain events in the aftermath of a fire and can impact CWS source waters.

Although a challenge to CWSs, turbidity is not directly related to potential health effects. However, it is used as a surrogate for pathogenic contamination (Fewtrell and Bartram, 2001; Sinclair et al., 2012). Microbiological contamination due to failures in drinking water–treatment systems (e.g., reduced disinfectant exposure) can cause immediate health concerns for communities. Treatment operations need to be able to remove the particles causing turbidity, most commonly using the process of coagulation. Work on the efficacy of coagulation at removing these particles suggests that higher coagulant doses are needed to achieve a target turbidity level in finished water (Hohner et al., 2016, 2019b). The physicochemical properties of the particles, including particle size distribution and zeta potential (charge at the surface of a particle), impact the performance of coagulation. Limited work exists on the characterization of these particles after wildland fires, and to the best of the committee's knowledge, no work has been done on any particles emanating from urban fires or WUI fires.

Ultimately, the particles that mobilize in water serve as a substrate from which the water can leach off other potential pollutants. Wildland fires impact the levels of dissolved organic matter (DOM) in surface waters (Hohner et al., 2019a). For example, recent studies in several burned watersheds provided evidence of changes in the concentration of DOM (Hohner et al., 2019a; Santos et al., 2016). The overall impact in the mobilization of DOM is expected to be a function of combustion temperature. The same studies indicate that mid-level combustion temperatures (~250°C–350°C) produce the highest levels of DOM in water after wildland fires. Once again, the committee found a dearth of information on DOM from particles (from sediment and ashes) after WUI fires. Given the different materials that combust in WUI fires, it is expected that the DOM mobilization from ashes in urban settings will be greater compared to that of wildland settings, although this hypothesis needs to be tested.

DOM is not directly toxic, but it is a substrate for the formation of disinfection by-products (DBPs) upon water disinfection (e.g., chlorination), which is why CWSs pay attention to this parameter. DBPs include a collection of halogenated compounds that are formed via the reaction between electron-rich components of DOM and the hypochlorous acid (or chloramines) used in water treatment (Richardson, 2003). DBPs include trihalomethanes and haloacetic acids, the two most studied classes of compounds in water disinfection, since they are regulated by the EPA at 80 and 60 ppb, respectively (EPA, 2006).

Changes in the concentration of DOM, as well as in the inherent reactivity of its components, can cause issues related to DBP formation. Post-fire, DBP levels can continue to change. Recent reports indicate that DOM levels impacted by fires can either increase or decrease the formation potential of different DBPs (Hohner et al., 2016; Wilkerson and Rosario-Ortiz, 2021). Work has focused mostly on wildland fires, and it is unclear what the impact would be under WUI fire conditions. Ultimately, the impact will relate to the kinds of compounds found in the overall DOM matrix, which is mostly unknown for WUI fires.

Organic Chemicals and Nutrients

The next level of contamination of concern for CWSs relates directly to specific chemicals impacting both surface waters and distribution systems. As one example, research on the impact of wildland fires on water quality has established that nutrient levels are impacted and can be elevated for years after the event. For example, Rhoades and coworkers examined the long-term mobilization of nutrients (e.g., nitrate) after fires, with data suggesting effects even beyond 10 years (Rhoades et al., 2019a). They attributed this observation to changes in the dynamics of nitrogen within the watershed, with changes in the amount of nitrogen retained after deposition. Other studies have also summarized the mobilization of nitrogen (and phosphorus); these studies predicted overall levels below that of any potential health concerns, but high enough to promote eutrophication in reservoirs (Smith et al., 2011). Eutrophication can result in issues with algal blooms, which can lead to the release of algal toxins and impacts to water quality.

In the context of WUI fires, limited data exist on the potential mobilization of nutrients from the combustion of urban materials; measurements are needed in this area. However, concerns related to eutrophication in reservoirs and surface waters are not an immediate concern in the aftermath of a fire; nitrate, for example, has a maximum contaminant level of 10 ppm, and it is unclear whether such levels can be reached in the immediate aftermath of a fire, even near the incident area.

As discussed above, recent reports demonstrated the potential for VOC contamination after WUI fires. Proctor et al. and others reported on the contamination of water distribution systems and households after recent fires in California (Proctor et al., 2020; Solomon et al., 2021). Table 5-1, taken from Proctor et al. (2020), shows the level of contamination measured after the Tubbs and Camp Fires. They provide concentrations for numerous VOCs measured, with benzene the most likely species to be found. Solomon and coworkers also reported benzene detection in samples collected after the 2018 Camp Fire (Solomon et al., 2021). They also detected methylene chloride, although the data suggest that this compound may be formed via other pathways not necessarily related to the wildfires. Work by Schulze and Fischer (2021) posits that local burn severity drives VOC contamination. The main

TABLE 5-1 VOCs Detected in Drinking Water Samples from the Tubbs and Camp Fires in California

	Exposure and Public Notification Limits				Tubbs Fire (21 months post-fire)		Camp Fire (8 months post-fire)				
	Long-term Limits		Short-term Limits		City of Santa Rosa		PID		SWR CB in PID	DOWC (three systems)[a]	
Chemical	US[b]	California[c]	HA[d]	NL[e]	n	Max	n	Max	n = 1	n	Max
Benzene	5	1	200, 26[i]	—	8,387	40,000	1,699[f]	923	>2,217	200[g] 40/20/140	530 8.1/5.3/530
Dichloromethane	5	200	10,000	—	6,254	41	NA[h]	28	—	NA	—
Naphthalene	—	100	500	17	661	6,800	NA	278	693	NA	—
Styrene	100	100	20,000	—	6,227	460	NA	6,800	378	NA	—
Tert-Butyl alcohol	—	—	—	12	339	29	NA	600	—	NA	—
Toluene	1,000	—	20,000	—	8,387	1,130	NA	1,400	676	NA	—
Vinyl chloride	2	—	3,000	—	6,227	16	NA	0.8	—	NA	—

NOTE: All units are μg/L. DOWC, Del Oro Water Company; PID, Paradise Irrigation District; SWRCB, State Water Resources Control Board.
[a]The three systems shown are Lime Saddle/Magalia/Paradise Pines.
[b]Federal maximum contaminant level (MCL) (USEPA, 2018)
[c]California MCL of the Office of Environmental Health Hazard Assessment (OEHHA) (SWRCB, 2019d).
[d]USEPA 1-day health advisory (HA) for a 10-kg child drinking 1 L/day (USEPA, 2018).
[e]California notification level (NL) (SWRCB, 2018a).
[f]This count does not include 98 blanks the were processed.
[g]This count does not include samples that were reported but later omitted due to long holding times.
[h]NA indicates that it is unknown how many samples were analyzed for this analyte because the utility did not record this information in their database (i.e., nondetects look the same as not-measured).
[i]California OEHHA (2019) reported that a benzene concentration of greater than or equal to 26 ppb may cause acute adverse health effects.
SOURCE: Table reproduced from Proctor et al. (2020, Table 2).

concern regarding VOC contamination is ingress through distribution systems, as recent work has rejected other sources (such as liquid fuel contamination).

These recent publications raise several concerns regarding water quality in the immediate aftermath of a WUI fire, as well as long-term (up to a year) concerns with the exposure of communities to unhealthy levels of VOCs. It is not clear what the formation mechanisms for these compounds are. Chapter 3 describes in general terms the complexity that is expected from combustion during WUI fires. These compounds may form in the gas phase and then partition to water or into ashes. If ashes are the main source, negative pressures in the distribution system could draw these materials into pipes, then mobilize them and raise concerns regarding long-term issues.

Given the potential contamination due to the species discussed in this section, CWSs need to have plans in place to isolate areas that may burn, with additional plans to limit the exposure to these compounds and to introduce higher sampling frequency after WUI fire incidents.

The VOC contamination after the Tubbs and Camp Fires raises the concern of what other chemicals, beyond those already reported, can be found in water after fires. Based on the complexities of the combustion reactions that are occurring, one might expect many other compounds to form. Recent work found numerous compounds in surface runoff and mobilized from ashes after wildland fires (Ferrer et al., 2021; Thurman et al., 2020). For example, researchers measured benzene polycarboxylic acids in both simulations of soil combustion and surface runoff samples (Thurman et al., 2020). The combustion reactions described in Chapter 3, and the emission factors measured for a wide range of organic compounds in gas-phase emissions, suggest that some of the compounds formed can partition to water or deposit downstream. Recent work also measured other compounds, in addition to benzene polycarboxylic acids, from ashes and runoff samples collected at different fires (Ferrer et al., 2021). Table 5-2 shows some of the chemical structures recently measured in runoff from wildland fires.

TABLE 5-2 Identifications of Compound Classes Found in Ash Leachates and Water Samples from Wildland Fires

Compound	Chemical Structure
Quinoline monocarboxylic acids	
Quinoline dicarboxylic acids	
Naphthoic acid	
Naphthalene dicarboxylic acids	
Naphthalene tricarboxylic acids	
Benzofuran monocarboxylates	
Benzofuran dicarboxylates	
Benzofuran tricarboxylates	

SOURCE: Adapted from Ferrer et al. (2021).

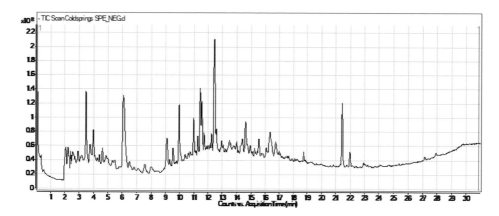

FIGURE 5-2 Total ion chromatogram of aqueous leachate of an ash sample from Cold Springs, collected in Nederland, Colorado. Although over 200 molecular species were detected, only 14 compounds were identified with standards using accurate mass fragmentation with liquid-chromatography quadrupole time-of-flight tandem mass spectrometry. This unpublished data illustrates the complexity of these types of samples and the need for non-targeted analyses to fully characterize them. SOURCE: Figure courtesy of Thurman and Ferrer, University of Colorado Boulder.

Although not every chemical structure is potentially toxic at the levels expected in water (at or below ppb levels), these results indicate that more work is needed on organic compounds in surface waters after fires, with an emphasis on fires at the WUI. Many other compounds will probably be found in surface streams near fires. Figure 5-2 is a total ion chromatogram showing the analysis of one aqueous sample collected in Colorado after a wildland fire. More than 200 molecular structures are observed that have expected concentrations above 0.3 ppb. Additional work to better understand the extent of potential contamination is needed; to better assess potential health impacts, researchers need information on the presence of compounds, their concentrations, and their dynamics.

Finding: VOCs have been found in distribution systems in two incidents related to WUI fires in California, and other chemicals have been identified in water systems after wildland fires.

Research need: Research is needed to further characterize potential chemical contamination to water resources (both surface waters and distribution systems) from WUI fires, and to better understand the formation pathways.

Metals and Other Organic Contaminants

Inorganic species can also be mobilized after fires. For example, many studies show the impact of fires on the mobilization of mercury and other metals after wildland fires (Caldwell et al., 2000; Kelly et al., 2006). In these cases, mercury mobilization can yield elevated levels of methylmercury, which presents a concern in water bodies. Studies have also measured different levels of metals and PAHs in runoff from wildland fires. For example, Stein and coworkers measured elevated concentrations of copper, lead, zinc, and PAHs in runoff after a wildland fire, compared to control sites (Stein et al., 2012). Other reports, as summarized in the review by Smith and coworkers, also indicate substantial mobilization of metals after wildland fires (Smith et al., 2011). Again, most of the published work regards wildland fires.

One source of important information regarding potential contamination from the combustion of urban materials comes from several reports commissioned by the State of California after fires. Although these reports are not peer reviewed, they were conducted following established methods, and the analysis of the samples was conducted using EPA methodologies in certified laboratories.

One of these reports characterized ash samples collected after the 2018 Camp Fire. It provides evidence of the potential for WUI fires to contaminate surface waters (TetraTech Inc., 2019). This study sampled 41 residential sites for a total of 150 individual samples of ash. Of these 41 sites, 37 sites had at least one metal above the screening level. The study identified a total of eight metals over the appropriate screening level, including antimony, arsenic, cobalt, copper, lead, nickel, and zinc. This report did not assess potential contamination in nearby source waters, although it is expected that these metals will partition into water. Given the content of modern homes and vehicles, it is reasonable to expect that ash will have metal contents that could result in contamination to both source waters and distribution systems.

Other similar reports, including those from fires in 2007 and 2015 in California, also confirm the presence of metal and PAHs in ash collected from urban settings. For example, ash samples collected from fires that impacted Southern California in 2007 (the Slide Fire in San Bernardino County and the Witch Creek Fire in San Diego County) showed elevated levels of arsenic, cadmium, copper, lead, and other metals, as well as PAHs including benzo[a] anthracene and naphthalene (Geosyntec Consultants, 2007). Dioxins have also been detected in runoff from burned wildland areas, although at low concentrations (no data are available from WUI fires; Gallaher and Koch, 2004).

Finding: Significant levels of metals were measured in ash samples collected from WUI fire incidents in California.

Research need: Research is needed to further characterize the potential for chemical contamination stemming from ash mobilization into both soil and water after WUI fires.

Although numerous studies characterize gas-phase emissions from the combustion of both urban materials and biomass (as summarized in Chapter 3), limited information exists on the potential mobilization of contaminants to water. Current work is largely observational, and conclusions are hard to generalize. Even though benzene was measured in both distribution systems and combustion studies with pipes, the main contamination mechanism is not clear. It is also not clear whether other compounds are present, even though studies from wildland fires suggest that many types of compounds could be present.

Fire Retardants

Potential contamination of surface waters and groundwaters by firefighting chemicals is also a concern. Fire suppressants are widely used in areas of active wildfires. The US Forest Service evaluates chemical products used in wildland firefighting activities for toxicological impacts and risks to human health (USFS, n.d.). Two major types of chemicals are commonly used: long-term retardants and foams (including Class A foam suppressants and water enhancers). Long-term fire retardants are used to decrease fire intensity and slow fire spread, and can be used preemptively ahead of an approaching fire.

Typical fire suppressants used for wildland and WUI fires are the Class A suppressants. Class A fire suppressants contain foaming and wetting agents. Perhaps the most common fire suppressant is Phos-Chek, which has been deployed by planes combating wildland fires. These Class A foams contain mostly inorganic salts (including ammonium, phosphates, and sulfates), in addition to other ingredients. Furthermore, as part of the application protocols, these suppressants are applied away from streams, although the substances could potentially reach streams. As such, limited impacts of these compounds are expected to surface waters. Mobilization of organic salts has the potential to result in eutrophication and changes to trophic levels (Angeler and Moreno, 2006; Angeler et al., 2005). But limited impacts to human health are expected since water enhancers or gels, composed of polymers and hydrocarbons, are used (Carratt et al., 2017).

Class A fire suppressants do not contain perfluorinated compounds, but Class B suppressants do. These substances, commonly known as PFASs, can cause adverse health effects. A significant number of cases of groundwater contamination with PFASs have been reported in different locations, including near sites where firefighting training occurs and the use of aqueous film–forming foams is documented. Class B fire suppressants could raise serious issues for CWSs in the aftermath of their use. For example, an incident in Galveston Bay in 2019 at a petrochemical plant resulted in significant contamination of water around the site (Nolen et al., 2022). Class B foams, which may contain fluorinated surfactants, have been used for liquid fuel fires but are not included in the

US Forest Service Wildland Fire Chemical Systems program and are unlikely to be used in a wildland fire situation, or for residential structures in a WUI fire.

Long-term fire suppressants are composed of some form of ammonium phosphate (monoammonium phosphate, diammonium phosphate, or ammonium polyphosphate), often mixed with dyes, wetting agents, anticorrosive agents, or other performance additives. Recently, durable, long-term ammonium polyphosphate fire retardants have been marketed for preventative application in areas prone to wildland fires, including WUI areas (Perimeter Solutions, 2022). Widespread use of preventative fire retardants may occur in areas not imminently threatened by wildland fire, which might change the chemicals of concern involved in a WUI fire, as well as the fire behavior in high-risk WUI areas. Furthermore, given that the chemical composition of these products is largely proprietary, it is difficult to determine what, if any, chemicals of concern these fire-suppression products contain.

IMPACTS TO GROUNDWATER AND SOIL CONTAMINATION

The impact of WUI fires on groundwater supplies needs to be considered. Mansilha and coworkers reported PAH contamination in groundwaters after a wildland fire (Mansilha et al., 2014). In their study, the total sum of PAHs in burned areas ranged from 23.1 to 95.1 ppt, about six times higher than for control samples. Once again, a dearth of information exists regarding groundwater contamination after a WUI fire, and environmental measurements need to be conducted to guide the research.

Soil contamination is also possible after a wildfire. As with water contamination, published data on soil contamination directly related to WUI fires are scarce. The literature on wildland fires and soil contamination includes examples of contamination due to different organic and inorganic compounds, such as PAHs and metals. As an example, García-Falcón et al. (2006) showed elevated levels of PAHs after wildland fires. More recently, Chen et al. (2018) investigated the levels of PAHs in soils and ash after fires. Recently, a metadata analysis of published PAH data in soils after wildland fires presented several conclusions, including how wildland fires increase PAH concentrations by approximately 205 percent compared to unburned soils (Yang et al., 2022). The same report demonstrated how the significant increase in PAHs overall in soils impacted by wildland fires corresponded to only a 73 percent increase in toxicity equivalence, mostly due to the production of low-molecular-weight PAHs. Finally, the study showed that the increase in PAH content was mostly observed in ash and topsoils, and that mild-intensity fires created higher PAH concentrations compared to both moderate and high-intensity events.

Once again, limited information exists regarding the expected levels of soil contamination after a WUI fire, though it is expected that a wide range of these organic compounds can be found.

Regarding fluorinated compounds, substantial evidence indicates widespread PFAS contamination in soils near sites where these compounds are used. These sites tend to be near firefighting training facilities, where aqueous film–forming foams are used for practice; or in cases where Class B foams were used for liquid fuel fires, such as at the petrochemical plant in Galveston Bay in 2019. Aqueous film–forming foam contamination results in mobilization to soils (Maizel et al., 2021).

As mentioned in the previous section on water quality, the State of California sponsored testing of ash in soils after the Camp Fire in 2018 (TetraTech Inc., 2019) and found numerous contaminants in soil. In addition, other chemicals are expected to be found in soils, although information is scarce regarding WUI fires. The studies from large-scale fire incidents also apply to soil contamination.

IMPACTS OF ATMOSPHERIC WET AND DRY DEPOSITION

Atmospheric deposition is a major contributor of nutrients (e.g., reactive nitrogen), acidic species (e.g., acidic sulfate, nitric acid, hydrochloric acid, acetic acid, formic acid), and toxicants (e.g., PAHs, mercury, chlorinated organics, PFASs) to soils, lakes, estuaries, and the remote ocean (Clark et al., 2018; Degrendele et al., 2020; Leister and Baker, 1994; McVeety and Hites, 1988; Shimizu et al., 2021; Valiela et al., 2018; Wiener et al., 2006; Wright et al., 2018; Young et al., 2007). The largest fluxes to terrestrial and aquatic bodies occur close to the source. However, atmospheric deposition can substantially impact ecological systems in remote locations as well, including pristine systems (e.g., alpine lakes and the arctic; Wright et al., 2018).

Deposition mechanisms (and chemical mechanisms) are different for gases and airborne particles, and thus the partitioning of a species between the gas and particle phases has a substantial impact on the species's atmospheric lifetime and the geographic distribution of the deposited species (Wania et al., 1998). Scavenging by precipitation (wet deposition) and gravitational settling (dry deposition) generally limits the atmospheric transport of particle-phase species to several thousands of miles (1–2 weeks of transport; NASA EO, 2017), but when particles are lofted into the upper troposphere, as often happens with wildland and WUI fires, interhemispheric transport can occur (NASA EO, 2017; Stohl et al., 2002). Wet deposition of gas-phase species occurs via precipitation after in-cloud or below-cloud scavenging, and can be an effective deposition mechanism for water-soluble gas-phase species (e.g., acid gases). Dry deposition via diffusion of vapors to terrestrial and aquatic surfaces (Wesley and Hicks, 2000) is also an important mechanism for many volatile species (Pan et al., 2012; Totten et al., 2006). VOCs and semi-VOCs with low atmospheric reactivities can be transported globally.

Wildland fires have been identified as a substantial contributor to the atmospheric deposition of some species. They are the largest contributor to the dry deposition of reactive nitrogen (ammonia and nitrogen dioxide), which is a nutrient for plant growth and a contributor to eutrophication. For example, Kharol et al. (2018) estimated deposition fluxes of 1.35 Tg of reactive nitrogen over the United States during the 2013 fire season, two to three times larger than deposition from other sources. Tang et al. (2021) argue that the anomalously widespread phytoplankton blooms from December 2019 to March 2020 in the iron-limited Southern Ocean resulted from atmospheric iron deposition from Australian wildland fires. Wildland fires are one of many contributors to mercury deposition, contributing about 10 percent (15 Mg/year) of the annual atmospheric flux of mercury to the Arctic (Kumar and Wu, 2019). Although wildland fires clearly contribute to PAH deposition, an increase in the PAH flux from wildland fires can be hard to discern, given the abundance of other anthropogenic sources (e.g., downwind of the 2016 Fort McMurray Fire in the oil sands region of Canada; Zhang et al., 2022). In cases like this where multiple sources exist, receptor modeling techniques that make use of differences between sources in the relative concentrations of many chemical tracers (source profiles; e.g., Positive Matrix Factorization) can be quite valuable in estimating source contributions.

While there are studies examining the impacts of wildland fires on atmospheric deposition, a paucity of data exists quantifying water and soil contamination caused by wildland fire–associated atmospheric deposition in the near field. One example is a study that concluded atmospheric deposition led to a 5-fold to 60-fold increase in stream phosphorus and nitrogen levels above the background in the vicinity of the 1988 Red Bench Fire in Glacier National Park (Spencer et al., 2003). Atmospheric deposition of WUI fire toxicants, nutrients, and acidic species (e.g., metals, reactive nitrogen, HCl, formaldehyde, isocyanic acid, chlorinated organics, phenols) could be a significant contributor to downwind soil and water contamination; however, deposition fluxes from WUI fires are largely unknown. Identification and measurement of unique atmospheric tracers of WUI fires could help address this gap.

Finding: Data that accurately characterize the (chemical) composition and concentration of WUI fire emissions present in water runoff and soil are missing.

Lessons Learned from Other Types of Disasters

Although significant data gaps exist regarding WUI fires and water contamination, reports do exist on contamination in water and soil from runoff of other types of large-scale fire disasters (see Table 5-3). These reports include the collapse of the World Trade Center and the Grenfell Tower fire in the United Kingdom. In these studies, as described by Guillaume (2020), the impact that any runoff has on the water and soil depends on a wide variety of factors, including the volume of runoff produced, the time of travel from the site of the fire to the receptor, and the dilution in the receiving water body. The chemical composition of the runoff, influenced to a great extent by the source of the fire, is probably the most important factor, as it may include soot, ash, and other suspended solids; the decomposition products of combustion of the building, storage vessels, and substances stored on site; the stored chemicals and their thermal-decomposition products, washed off the site by the runoff; and, if used as a firefighting agent, the firefighting foam.

TABLE 5-3 Published Results on Large-Scale Fire Incidents

Fire Incident	Details (Source)	Fire Emissions
World Trade Center, September 2001	The fire in and subsequent collapse of the World Trade Center released a wide range of eco-toxicants into the environment (Nordgrén et al., 2002)	Analysis of the dust revealed the presence of PAHs, PCBs, PCDDs/PCDFs, pesticides, phthalate esters, heavy metals, brominated diphenyl ethers, synthetic vitreous fibers, and asbestos
Buncefield, December 2005	A fire occurred in an oil storage depot and burned for several days (Lönnermark, 2005)	The groundwater under the site and up to 2 km to the north, east, and southeast of the site was heavily contaminated with hydrocarbons and firefighting foams
Grenfell Tower fire, June 2017	A high-rise fire broke out in the 24-story Grenfell Tower block of flats in London (Stec et al., 2019)	Analysis of soil revealed the presence of char fragments from building materials, PAHs, PCDDs/PCDFs, flame retardants, and synthetic vitreous fibers

NOTE: PCB = polychlorinated biphenyl; PCDD = polychlorinated dibenzo-p-dioxin; PCDF = polychlorinated dibenzofuran.

REFERENCES

Angeler, D. G., and J. M. Moreno. 2006. "Impact-Recovery Patterns of Water Quality in Temporary Wetlands After Fire Retardant Pollution." *Canadian Journal of Fisheries and Aquatic Sciences* 63(7):1617–1626. https://doi.org/10.1139/f06-062.

Angeler, D. G., S. Martín, and J. M. Moreno. 2005. "Daphnia Emergence: A Sensitive Indicator of Fire-Retardant Stress in Temporary Wetlands." *Environment International* 31(4):615–620. https://doi.org/10.1016/j.envint.2004.10.015.

Becker, W. C., A. Hohner, F. Rosario-Ortiz, and J. DeWolfe. 2018. "Preparing for Wildfires and Extreme Weather: Plant Design and Operation Recommendations." *Journal AWWA* 110(7):32–40. https://doi.org/10.1002/awwa.1113.

Bladon, K. D., M. B. Emelko, U. Silins, and M. Stone. 2014. "Wildfire and the Future of Water Supply." *Environmental Science & Technology* 48(16):8936–8943. https://doi.org/10.1021/es500130g.

Caldwell, C. A., C. M. Canavan, and N. S. Bloom. 2000. "Potential Effects of Forest Fire and Storm Flow on Total Mercury and Methylmercury in Sediments of an Arid-Lands Reservoir." *Science of the Total Environment* 260(1):125–133. https://doi.org/10.1016/S0048-9697(00)00554-4.

Carratt, S. A., C. H. Flayer, M. E. Kossack, and J. A. Last. 2017. "Pesticides, Wildfire Suppression Chemicals, and California Wildfires: A Human Health Perspective." *Current Topics in Toxicology* 13:1–12.

Chen, H., A. T. Chow, X.-W. Li, H.-G. Ni, R. A. Dahlgren, H. Zeng, and J.-J. Wang. 2018. "Wildfire Burn Intensity Affects the Quantity and Speciation of Polycyclic Aromatic Hydrocarbons in Soils." *ACS Earth and Space Chemistry* 2(12): 1262–1270. https://doi.org/10.1021/acsearthspacechem.8b00101.

Clark, C. M., J. Phelan, P. Doraiswamy, J. Buckley, J. C. Cajka, R. L. Dennis, . . . and T. L. Spero. 2018. "Atmospheric Deposition and Exceedances of Critical Loads from 1800–2025 for the Conterminous United States." *Ecological Applications* 28(4):978–1002.

Degrendele, C., H. Fiedler, A. Kočan, P. Kukučka, P. Přibylová, R. Prokeš, . . . and G. Lammel. 2020. "Multiyear Levels of PCDD/Fs, dl-PCBs and PAHs in Background Air in Central Europe and Implications for Deposition." *Chemosphere* 240:124852.

EPA (US Environmental Protection Agency). 2006. *Stage 2 Disinfectants and Disinfection Byproducts Rule: A Quick Reference Guide for Schedule 1 Systems.* https://www.epa.gov/dwreginfo/stage-1-and-stage-2-disinfectants-and-disinfection-byproducts-rules.

EPA. 2021a. *Drinking Water Regulations.* https://www.epa.gov/dwreginfo/drinking-water-regulations (accessed January 27, 2022).

EPA. 2021b. *Information about Public Water Systems.* https://www.epa.gov/dwreginfo/information-about-public-water-systems (accessed January 27, 2022).

Ferrer, I., E. M. Thurman, J. A. Zweigenbaum, S. F. Murphy, J. P. Webster, and F. L. Rosario-Ortiz. 2021. "Wildfires: Identification of a New Suite of Aromatic Polycarboxylic Acids in Ash and Surface Water." *Science of the Total Environment* 770:144661. https://doi.org/10.1016/j.scitotenv.2020.144661.

Fewtrell, L., and J. Bartram. 2001. *Water Quality: Guidelines, Standards & Health.* London, UK: IWA Publishing.

Gallaher, B. M., and R. J. Koch. 2004. *Cerro Grande Fire Impacts to Water Quality and Stream Flow near Los Alamos National Laboratory: Results of Four Years of Monitoring.* Los Alamos, NM: Los Alamos National Laboratory.

García-Falcón, M. S., B. Soto-González, and J. Simal-Gándara. 2006. "Evolution of the Concentrations of Polycyclic Aromatic Hydrocarbons in Burnt Woodland Soils." *Environmental Science & Technology* 40(3):759–763. https://doi.org/10.1021/es051803v.

Geosyntec Consultants. 2007. *Assessment of Burn Debris – 2007 Wildfires San Bernardino and San Diego Counties, California.*

Guillaume, E. 2020. "Emissions Measurements." In *Encyclopedia of Wildfires and Wildland-Urban Interface (WUI) Fires.* Cham: Springer International Publishing. 253–272.

Hohner, A. K., K. Cawley, J. Oropeza, R. S. Summers, and F. L. Rosario-Ortiz. 2016. "Drinking Water Treatment Response Following a Colorado Wildfire." *Water Research* 105:187–198. https://doi.org/10.1016/j.watres.2016.08.034.

Hohner, A. K., C. C. Rhoades, P. Wilkerson, and F. L. Rosario-Ortiz. 2019a. "Wildfires Alter Forest Watersheds and Threaten Drinking Water Quality." *Accounts of Chemical Research* 52(5):1234–1244. https://doi.org/10.1021/acs.accounts.8b00670.

Hohner, A. K., R. S. Summers, and F. L. Rosario-Ortiz. 2019b. "Laboratory Simulation of Postfire Effects on Conventional Drinking Water Treatment and Disinfection Byproduct Formation." *AWWA Water Science* 1(5):e1155. https://doi.org/10.1002/aws2.1155.

Jaffé, R., Y. Ding, J. Niggemann, A. V. Vähätalo, A. Stubbins, R. G. M. Spencer, J. Campbell, and T. Dittmar. 2013. "Global Charcoal Mobilization from Soils via Dissolution and Riverine Transport to the Oceans." *Science* 340(6130):345–347. https://doi.org/doi:10.1126/science.1231476.

Kelly, E. N., D. W. Schindler, V. L. St. Louis, D. B. Donald, and K. E. Vladicka. 2006. "Forest Fire Increases Mercury Accumulation by Fishes via Food Web Restructuring and Increased Mercury Inputs." *Proceedings of the National Academy of Sciences* 103(51):19380–19385. https://doi.org/10.1073/pnas.0609798104.

Kharol, S. K., M. W. Shephard, C. A. McLinden, L. Zhang, C. E. Sioris, J. M. O'Brien, . . . and N. A. Krotkov. 2018. "Dry Deposition of Reactive Nitrogen from Satellite Observations of Ammonia and Nitrogen Dioxide over North America." *Geophysical Research Letters* 45(2):1157–1166.

Kumar, A., and S. Wu. 2019. "Mercury Pollution in the Arctic from Wildfires: Source Attribution for the 2000s." *Environmental Science & Technology* 53(19):11269–11275.

Leister, D. L., and J. E. Baker. 1994. "Atmospheric Deposition of Organic Contaminants to the Chesapeake Bay." *Atmospheric Environment* 28(8):1499–1520.

Lönnermark, A. 2005. *Analyses of Fire Debris after Tyre Fires and Fires in Electrical and Electronics Waste.* SP Report 2005:44. Borås, Sweden: SP Swedish National Testing and Research Institute. https://www.diva-portal.org/smash/get/diva2:962335/FULLTEXT01.pdf.

Maizel, A. C., S. Shea, A. Nickerson, C. Schaefer, and C. P. Higgins. 2021. "Release of Per- and Polyfluoroalkyl Substances from Aqueous Film-Forming Foam Impacted Soils." *Environmental Science & Technology* 55(21):14617–14627. https://doi.org/10.1021/acs.est.1c02871.

Mansilha, C., A. Carvalho, P. Guimarães, and J. Espinha Marques. 2014. "Water Quality Concerns Due to Forest Fires: Polycyclic Aromatic Hydrocarbons (PAH) Contamination of Groundwater from Mountain Areas." *Journal of Toxicology and Environmental Health, Part A* 77(14–16):806–815. https://doi.org/10.1080/15287394.2014.909301.

McGee, J. K., L. C. Chen, M. D. Cohen, G. R. Chee, C. M. Prophete, N. Haykal-Coates, S. J. Wasson, T. L. Conner, D. L. Costa, and S. H. Gavett. 2003. "Chemical Analysis of World Trade Center Fine Particulate Matter for Use in Toxicologic Assessment." *Environmental Health Perspectives* 111(7):972–980. https://doi.org/doi:10.1289/ehp.5930.

McVeety, B. D., and R. A. Hites. 1988. "Atmospheric Deposition of Polycyclic Aromatic Hydrocarbons to Water Surfaces: A Mass Balance Approach." *Atmospheric Environment (1967)* 22(3):511–536. https://doi.org/10.1016/0004-6981(88)90196-5.

NASA EO (National Aeronautics and Space Administration Earth Observatory). 2017. "Record-Breaking Smoke over Canada." https://earthobservatory.nasa.gov/images/90759/record-breaking-smoke-over-canada.

Nolen, R. M., P. Faulkner, A. D. Ross, K. Kaiser, A. Quigg, and D. Hala. 2022. PFASs pollution in Galveston Bay surface waters and biota (shellfish and fish) following AFFFs use during the ITC fire at Deer Park (March 17th–20th 2019), Houston, TX. *Science of the Total Environment* 805:150361. https://doi.org/10.1016/j.scitotenv.2021.150361.

Nordgrén, M. D., E. A. Goldstein, and M. A. Izeman. 2002. *The Environmental Impacts of the World Trade Center Attacks: A Preliminary Assessment.* New York, NY: Natural Resources Defense Council.

Olivella, M. A., T. G. Ribalta, A. R. de Febrer, J. M. Mollet, and F. X. C. de las Heras. 2006. "Distribution of Polycyclic Aromatic Hydrocarbons in Riverine Waters after Mediterranean Forest Fires." *Science of the Total Environment* 355(1):156–166. https://doi.org/10.1016/j.scitotenv.2005.02.033.

Pan, Y. P., Y. S. Wang, G. Q. Tang, and D. Wu. 2012. "Wet and Dry Deposition of Atmospheric Nitrogen at Ten Sites in Northern China." *Atmospheric Chemistry and Physics* 12(14):6515–6535. https://doi.org/10.5194/acp-12-6515-2012.

Perimeter Solutions. 2022. *Season-Long Preventive Protection Against Wildfires.* https://www.perimeter-solutions.com/en/fire-safety-fire-retardants/phos-chek-fortify/.

Proctor, C. R., J. Lee, D. Yu, A. D. Shah, and A. J. Whelton. 2020. "Wildfire Caused Widespread Drinking Water Distribution Network Contamination." *AWWA Water Science* 2(4):e1183. https://doi.org/10.1002/aws2.1183.

Rhoades, C. C., A. T. Chow, T. P. Covino, T. S. Fegel, D. N. Pierson, and A. E. Rhea. 2019a. "The Legacy of a Severe Wildfire on Stream Nitrogen and Carbon in Headwater Catchments." *Ecosystems* 22(3):643–657. https://doi.org/10.1007/s10021-018-0293-6.

Rhoades, C. C., J. P. Nunes, U. Silins, and S. H. Doerr. 2019b. "The Influence of Wildfire on Water Quality and Watershed Processes: New Insights and Remaining Challenges." *International Journal of Wildland Fire* 28(10):721–725. https://doi.org/10.1071/WFv28n10_FO.

Richardson, S. D. 2003. "Disinfection By-products and Other Emerging Contaminants in Drinking Water." *TrAC Trends in Analytical Chemistry* 22(10):666–684. https://doi.org/10.1016/S0165-9936(03)01003-3.

Santos, F., D. Russell, and A. A. Berhe. 2016. "Thermal Alteration of Water Extractable Organic Matter in Climosequence Soils from the Sierra Nevada, California." *Journal of Geophysical Research: Biogeosciences* 121(11):2877–2885. https://doi.org/10.1002/2016JG003597.

Schulze, S. S., and E. C. Fischer. 2021. "Prediction of Water Distribution System Contamination Based on Wildfire Burn Severity in Wildland Urban Interface Communities." *ACS ES&T Water* 1(2):291–299. https://doi.org/10.1021/acsestwater.0c00073.

Shimizu, M. S., R. Mott, A. Potter, J. Zhou, K. Baumann, J. D. Surratt, . . . and J. D. Willey. 2021. "Atmospheric Deposition and Annual Flux of Legacy Perfluoroalkyl Substances and Replacement Perfluoroalkyl Ether Carboxylic Acids in Wilmington, NC, USA." *Environmental Science & Technology Letters* 8(5):366–372.

Sinclair, R. G., J. B. Rose, S. A. Hashsham, C. P. Gerba, and C. N. Haas. 2012. "Criteria for Selection of Surrogates Used to Study the Fate and Control of Pathogens in the Environment." *Applied and Environmental Microbiology* 78(6). https://doi.org/10.1128/AEM.06582-11.

Smith, H. G., G. J. Sheridan, P. N. J. Lane, P. Nyman, and S. Haydon. 2011. "Wildfire Effects on Water Quality in Forest Catchments: A Review with Implications for Water Supply." *Journal of Hydrology* 396(1):170–192. https://doi.org/10.1016/j.jhydrol.2010.10.043.

Solomon, G. M., S. Hurley, C. Carpenter, T. M. Young, P. English, and P. Reynolds. 2021. "Fire and Water: Assessing Drinking Water Contamination after a Major Wildfire." *ACS ES&T Water* 1(8):1878–1886. https://doi.org/10.1021/acsestwater.1c00129.

Spencer, C. N., K. O. Gabel, and F. R. Hauer. 2003. "Wildfire Effects on Stream Food Webs and Nutrient Dynamics in Glacier National Park, USA." *Forest Ecology and Management* 178(1–2):141–153.

Stec, A. A., K. Dickens, J. L. J. Barnes, and C. Bedford. 2019. "Environmental Contamination Following the Grenfell Tower Fire." *Chemosphere* 226:576–586. https://doi.org/10.1016/j.chemosphere.2019.03.153.

Stein, E. D., J. S. Brown, T. S. Hogue, M. P. Burke, and A. Kinoshita. 2012. "Stormwater contaminant Loading Following Southern California Wildfires." *Environmental Toxicology and Chemistry* 31(11):2625–2638. https://doi.org/10.1002/etc.1994.

Stohl, A., S. Eckhardt, C. Forster, P. James, and N. Spichtinger. 2002. "On the Pathways and Timescales of Intercontinental Air Pollution Transport." *Journal of Geophysical Research: Atmospheres* 107(D23):ACH 6-1–ACH 6-17. https://doi.org/10.1029/2001JD001396.

Tang, W., J. Llort, J. Weis, M. M. Perron, S. Basart, Z. Li, S. Sathyendranath, T. Jackson, E. S. Rodriguez, B. C. Proemse, A. R. Bowie, C. Schallenberg, P.G. Strutton, R. Matear, and N. Cassar. 2021. "Widespread Phytoplankton Blooms Triggered by 2019–2020 Australian Wildfires." *Nature* 597(7876):370–375. https://doi.org/10.1038/s41586-021-03805-8.

TetraTech Inc. 2019. *Final Assessment of Ash Sampling – Camp Fire Incident.*

Thurman, E. M., Y. Yu, I. Ferrer, K. A. Thorn, and F. L. Rosario-Ortiz. 2020. "Molecular Identification of Water-Extractable Organic Carbon from Thermally Heated Soils: C-13 NMR and Accurate Mass Analyses Find Benzene and Pyridine Carboxylic Acids." *Environmental Science & Technology* 54(5):2994–3001. https://doi.org/10.1021/acs.est.9b05230.

Totten, L. A., M. Panangadan, S. J. Eisenreich, G. J. Cavallo, and T. J. Fikslin. 2006. "Direct and Indirect Atmospheric Deposition of PCBs to the Delaware River Watershed." *Environmental Science & Technology* 40(7):2171–2176. https://doi.org/10.1021/es052149m.

USFS (US Forest Service National Technology and Development Program). n.d. *Wildland Fire Chemical Systems and Aerial Delivery Systems.* https://www.fs.fed.us/rm/fire/wfcs/wildland-fire-chemicals.php (accessed February 18, 2022).

Valiela, I., D. Liu, J. Lloret, K. Chenoweth, and D. Hanacek. 2018. "Stable Isotopic Evidence of Nitrogen Sources and C4 Metabolism Driving the World's Largest Macroalgal Green Tides in the Yellow Sea." *Scientific Reports* 8(1):17437.

Wania, F., J. Axelman, and D. Broman. 1998. "A Review of Processes Involved in the Exchange of Persistent Organic Pollutants across the Air–Sea Interface." *Environmental Pollution* 102(1):3–23.

Wesely, M. L., and B. B. Hicks. 2000. "A Review of the Current Status of Knowledge on Dry Deposition." *Atmospheric Environment* 34(12–14):2261–2282. https://doi.org/10.1016/S1352-2310(99)00467-7.

Wiener, J. G., B. C. Knights, M. B. Sandheinrich, J. D. Jeremiason, M. E. Brigham, D. R. Engstrom, L. G. Woodruff, W. F. Cannon, and S. J. Balogh. 2006. "Mercury in Soils, Lakes, and Fish in Voyageurs National Park (Minnesota): Importance of Atmospheric Deposition and Ecosystem Factors." *Environmental Science & Technology* 40(20): 6261–6268.

Wilkerson, P. J., and F. L. Rosario-Ortiz. 2021. "Impact of Simulated Wildfire on Disinfection Byproduct Formation Potential." *AWWA Water Science* 3(1):e1217. https://doi.org/10.1002/aws2.1217.

Wright, L. P., L. Zhang, I. Cheng, J. Aherne, and G. R. Wentworth. 2018. "Impacts and Effects Indicators of Atmospheric Deposition of Major Pollutants to Various Ecosystems—A Review." *Aerosol and Air Quality Research* 18(8):1953–1992.

Yang, B., Y. Shi, S. Xu, Y. Wang, S. Kong, Z. Cai, and J. Wang. 2022. "Polycyclic Aromatic Hydrocarbon Occurrence in Forest Soils in Response to Fires: A Summary Across Sites." *Environmental Science: Processes & Impacts* 24(1):32–41. https://doi.org/10.1039/D1EM00377A.

Young, C. J., V. I. Furdui, J. Franklin, R. M. Koerner, D. C. Muir, and S. A. Mabury. 2007. "Perfluorinated Acids in Arctic Snow: New Evidence for Atmospheric Formation." *Environmental Science & Technology* 41(10):3455–3461.

Zhang, Y., R. Pelletier, T. Noernberg, M. W. Donner, I. Grant-Weaver, J. W. Martin, and W. Shotyk. 2022. "Impact of the 2016 Fort McMurray Wildfires on Atmospheric Deposition of Polycyclic Aromatic Hydrocarbons and Trace Elements to Surrounding Ombrotrophic Bogs." *Environment International* 158:106910.

6

Human Exposures, Health Impacts, and Mitigation

This chapter explores chemicals and other pollutants of concern for human exposure and health, as well as ways to reduce exposure and mitigate the impacts of wildland-urban interface (WUI) fires. It addresses all relevant distance scales, from the immediate fire zone, where emergency responders such as firefighters experience direct exposure to heat and high concentrations of fire emissions, to the regional and continental levels, where WUI fire smoke can have an extended impact. The committee has noted where WUI-specific information is available; however, such information is scarce. Therefore, much of the discussion presented in this chapter draws on studies from wildland fires. It is important to note, however, that this chapter is not a comprehensive review of the overall health effects of wildland fire smoke.

This chapter begins by identifying chemicals and other pollutants of concern for human exposure and describing populations who may be more vulnerable to the impacts of WUI fires, and those who experience environmental injustice and health inequities. The next section describes what is known about routes of exposure, followed by a discussion of the acute and chronic health impacts for all near- and far-field populations, with a dedicated section on exposures and impacts for firefighters. The last section explores ways to reduce exposure to WUI fire pollutants, as well as specific interventions for firefighters and affected communities.

Box 6-1 defines key terms that are used throughout this chapter.

BOX 6-1
Key Terminology

Spatial Scales

Near-field scale: From 1 to 10 km downwind of the fire, where the plume remains quite concentrated, dilution has a major effect on the gas-particle partitioning, and chemistry is driven by fast processes that occur on a timescale of minutes

Local scale: From 10 to 100 km, where chemistry is driven by processes that occur on a timescale of minutes to hours

---

BOX 6-1 Continued

Regional scale: From 100 to 1,000 km, where chemistry is driven by processes that occur on a timescale of hours to days

Continental scale: Greater than 1,000 km, where chemistry is driven by processes that occur over days

Terms and Definitions

Acute exposure: Contact with a substance that occurs once or for only a short time (acute exposure can occur for up to 14 days) compared with intermediate duration exposure and chronic exposure

Acute health effect: A health effect that develops immediately or within minutes, hours, or even days after an exposure

Bioavailability: The potential for uptake (ability to be absorbed and used by the body) of a substance by a living organism; it is usually expressed as the fraction that can be taken up by the organism in relation to the total amount available

Chronic exposure: Continuous or repeated contact with a toxic substance over a long period of time (months or years)

Chronic health effect: An adverse health effect resulting from long-term exposure to a substance; examples could include diabetes, bronchitis, cancer, or any other long-term medical condition

Coarse particulate matter (PM_{10}): Inhalable particles with diameters equal to or less than 10 micrometers in diameter

Environmental justice: A social movement developed in response to environmental racism; environmental justice research and practice involves identifying the disproportionate health burdens that populations experience from environmental exposures and social vulnerabilities, and focusing on solutions to alleviate those burdens in partnership with affected communities

Fine particulate matter ($PM_{2.5}$): Airborne particles with diameters of 2.5 micrometers or less, small enough to enter the lungs and bloodstream, posing risks to human health

Health equity: Identifying health disparities between populations that are driven largely by social, economic, and environmental factors, and focusing on solutions to eliminate those disparities

Particulate matter (PM): A complex mixture of solid particles and liquid droplets found in the air

Pollutant: A chemical or biological substance that harms water, air, or land quality

Semi-volatile organic compounds (SVOCs): Organic compounds that, based on their vapor pressure, tend to evaporate from the particle phase with near-field dilution of plumes; SVOCs are of concern because of their abundance in the indoor environment and their ability to accumulate and persist in the human body, the infrastructure of buildings, and environmental dust

Structural racism: The totality of ways in which societies foster racial discrimination through mutually reinforcing systems of housing, education, employment, earnings, benefits, credit, media, health care, and criminal justice; these patterns and practices in turn reinforce discriminatory beliefs, values, and the distribution of resources[a]

Toxicant: Any chemical that can injure or kill humans, animals, or plants (depending on the magnitude and duration of exposure); a poison

Toxic: Related to harmful effects on the body by either inhalation (breathing), ingestion (eating), or dermal absorption of the chemical

Ultrafine particles: Particles with a diameter less than or equal to 0.1 micrometers

Volatile organic compounds (VOCs): Organic compounds with vapor pressures high enough to exist in the atmosphere primarily in the gas phase, typically excluding methane; VOCs can easily become airborne for inhalation exposure

Vulnerable populations: Individuals or communities at higher risk of adverse health effects from exposures, such as from greater pollutant exposure concentrations, higher health response to a given level of exposure, or reduced capacity to adapt

---

[a]Bailey et al., 2017. "Structural Racism and Health Inequities in the USA: Evidence and Interventions." *Lancet* 389 (10077): 1453–1463. https://dol.org/10.1016/S0140-6736(17)30569-X.

CHEMICALS OF CONCERN FOR HUMAN EXPOSURE

Wildland fires are major sources of organic and inorganic gases and aerosols containing ultrafine particles (0.1 μm or less), fine particulate matter ($PM_{2.5}$), and coarse particulate matter (PM_{10}; EPA, 2021e). Specific composition is dependent on the fire fuel sources, direct emissions, secondary chemical and physical processes, ventilation, and meteorological conditions. A complex mixture of emission gases is present and can include greenhouse gases such as carbon dioxide (CO_2) and methane (CH_4); photochemically reactive substances such as as nitrogen oxides (NO_x), carbon monoxide (CO), and volatile organic compounds (VOCs); and PM of various diameters. These emissions all contribute to available air pollution that can be detrimental to human health and ecosystems. Specific studies of fire-generated smoke have identified numerous hazardous substances including VOCs and semi-VOCs (SVOCs) like benzene, dioxins, furans, flame retardants, plasticizers, polycyclic aromatic hydrocarbons (PAHs), and polychlorinated biphenyls (PCBs), as well as carbon monoxide (CO), hydrogen cyanide (HCN), hydrogen chloride (HCl), nitrogen oxides (NO_x), sulfur oxides (SO_x), and heavy metals (CARB, 2021). Many of these substances represent known or probable carcinogens, irritants, respiratory sensitizers, or reproductive and developmental toxicants with linkages to cardiovascular impacts and neurological impairments (IARC, 2010). Smoke from WUI and wildland fires consists of complex mixtures of the aforementioned toxicants, which have the potential to interact with each other to modify the toxicity of single compounds (Zeliger, 2003). However, analysis of mixture toxicity and health effects is complicated. Thus, assessing the toxicity of individual constituents is most often used to predict the toxicity of the mixture. Table 6-1 summarizes some examples of key pollutants recognized in fire events along with general health impacts and common routes of exposure.

Research need: A need exists to pioritize WUI-related chemicals of concern, their exposure potentials, and their human health risks, with consideration of vulnerable populations, so that mitigation strategies can be developed.

TABLE 6-1 Examples of Chemical Pollutants Related to WUI Fire Events and Human Exposures

Group of Pollutants	Common Examples	Routes of Exposure	Potential Health Outcomes	Selected References
Asbestos	Fibrous asbestos, chrysotile	Inhalation Ingestion	Cancer; asbestosis; respiratory irritation; pleural disease	EPA, 2021a; ATSDR, 2016
Asphyxiant gases	Carbon monoxide (CO), carbon dioxide (CO_2), hydrogen cyanide (HCN)	Inhalation	Depression of central nervous system and hypoxia; acute respiratory effects	NIOSH, 2011; Gold and Perera, 2021
Dioxins and furans (polychlorinated dibenzo-p-dioxins [PCDDs] and polychlorinated dibenzofurans [PCDFs])	2,3,7,8-Tetrachloro-dibenzo-p-dioxin/furan	Inhalation Ingestion Dermal (low penetration into skin by itself, but can cause skin lesions)	Cancer or predisposition to cancer; reproductive and developmental effects; immune suppression; dermal toxicity; endocrine disruption	CDC, 2017; EPA, 2021b
Flame retardants	Tris(1-chloro-2-propyl) phosphate (TCPP), tris(2-chloroethyl) phosphate (TCEP), tris isobutylated triphenyl phosphate, methyl phenyl phosphate	Inhalation Ingestion	Neurotoxicity or neurodevelopmental damage; reproduction and fetal development effects; endocrine and thyroid disruption	ATSDR, 2015; NIEHS, 2021a
Inorganic acid gases	Hydrogen chloride (HCl), hydrogen fluoride (HF), phosphoric acid, SO_x, NO_x	Inhalation	Chemical burns; increased risk of laryngeal and lung cancer	ATSDR, 2002; NCI, 2019
Inorganic and organic metals	Lead, lithium, iron, mercury, methylmercury, nickel, cadmium, palladium chloride	Inhalation Ingestion Dermal	Neurotoxicity; reproductive and developmental effects, dermal irritation or allergen; respiratory irritation	Goyer, 2004; NIOSH, 2019a
Isocyanates	Methyl isocyanate, methylene diphenyl diisocyanate, toluene diisocyanate	Inhalation	Irritation and pulmonary sensitivity	ATSDR, 2014a; NIOSH, 2014
Organic and other gases	Phosgene ($COCl_2$), ammonia (NH_3)	Inhalation	Acute effects; pulmonary edema and irritation	EPA, 2021c; CDC, 2018
Ozone (O_3)		Inhalation	Acute and chronic respiratory symptoms including coughing and exacerbation of chronic diseases such as bronchitis and asthma; increased risk of pulmonary infections	NEPHT, 2020
Particulate matter (PM)	PM is often classified by size, where the size is based on the aerodynamic diameter in micrometers (e.g., $PM_{2.5}$); smaller particles penetrate deeper into the respiratory system	Inhalation Ingestion Dermal	Cancer; cardiopulmonary toxicant; immunosuppressant; neurotoxicant; reproductive and developmental toxicity	EPA, 2021d; CDC, 2019

Group	Examples	Exposure Routes	Health Effects	References
Plasticizers	Ortho phthalates including dibutyl phthalate, terephthalates, adipates, benzoates	Inhalation Ingestion	Endocrine disruptors; reproductive and developmental toxicity	EPA, 2022a; CDC, 2021
Polycyclic aromatic hydrocarbons (PAHs)	Benzo[a]pyrene, benzo[a]anthracene, benzo[b]fluoranthene, chrysene, pyrene, fluoranthene, naphthalene, anthracene	Inhalation Ingestion Dermal	Cancer; reproductive and developmental (teratogenic) toxicity; kidney and liver damage	CDC, 2022; ATSDR, 2014b
Polychlorinated biphenyls (PCBs)	2-Chlorobiphenyl, 2,2-dichlorobiphenyl, 2,4,5-trichlorobiphenyl; PCBs typically occur as a mixture of PCB congeners (i.e., aroclors)	Inhalation Ingestion Dermal	Cancer; neurotoxicity; immune suppression; endocrine disruption; reproductive and developmental toxicity; respiratory toxicity	ATSDR, 2000; Ahlborg et al., 1992
Volatile organic compounds (VOCs)	Formaldehyde, acetaldehyde, acrolein, benzene, toluene, ethylbenzene, para-xylene, ortho-xylene, meta-xylene, styrene, naphthalene; complex mixtures of VOCs are classified as total VOCs, and some are not toxic	Inhalation Ingestion	Cancer; reproductive and developmental toxicity; neurotoxicity; respiratory irritation; odorants	EPA, 2021c
Other emission and transformation products that are currently unidentified	Per- or polyfluoroalkyl substances (PFASs), including perfluorooctane sulfate and perfluorooctanoic acid	Inhalation Ingestion	Cancer; respiratory and developmental toxicity	NTP, 2016
	Reactive oxygen species, including peroxides (R-O-O-R) and superoxides (O_2^-)			

NOTE: Table 6-1 focuses on groups of chemicals with sound available data. The lists are meant as common examples and not meant to be exhaustive. A wide range of VOCs and SVOCs may be present including amines, amides, nitrosamines, trihalomethanes, glycols, and ethers.

VULNERABLE POPULATIONS, HEALTH EQUITY, AND ENVIRONMENTAL JUSTICE

Health is a product of multiple determinants. Social, economic, environmental, and structural factors and their unequal distribution matter more than health care in shaping health disparities (NASEM, 2017). To characterize what is known about human exposure, health impacts, and mitigation strategies for WUI fires, the committee examined individual and structural factors that may affect human health. For the purpose of this report, vulnerable populations are defined as individuals or communities at higher risk of adverse health effects from exposures, such as from greater pollutant exposure concentrations, higher health response to a given level of exposure, or reduced capacity to adapt, including lack of information or resources for personal mitigative strategies to reduce exposure. Another vulnerability is inequity in access to information, such as inequity based on the location of air pollution monitoring networks (Sun et al., 2022), which could result in a differential exposure misclassification that is propagated through human health studies.

A growing number of studies have shown that human health vulnerability to wildland fires can be influenced by several factors such as those related to life stage, location (e.g., at the WUI), socioeconomic status, race/ethnicity, occupation (e.g., outdoor workers), and underlying health conditions (Burke et al., 2021; Chan et al., 2013; Davies et al., 2018; Rudolph et al., 2018). The studies differ in their approaches, especially the methods to assess where and when fire exposures occur, such as for air pollution, and results are not perfectly consistent. For example, a study of wildland fire smoke in Alaska that examined the effects of wildland fire–associated $PM_{2.5}$ on the general population and different subpopulations to assess vulnerability found that exposure to wildland fire–associated $PM_{2.5}$ varied between specific Alaskan Native tribes, with the greatest effect levels seen for the Alaskan Athabascan tribe (Woo et al., 2020).

In another study, however, $PM_{2.5}$ from wildland fire smoke was found to be at higher levels for non-Hispanic whites (Burke et al., 2021). The authors noted that actual differential health impacts were affected by social determinants other than exposure, such as those factors briefly described later in this section. Collectively, studies indicate that some populations may suffer a higher risk of related health outcomes from wildland fires, which is consistent with the broader research on environmental justice and health equity.

At a population level, health disparities (or health inequities) arise from root causes that could be organized in two clusters: (1) intrapersonal, interpersonal, institutional, and systemic mechanisms (sometimes called structural inequities) that organize the distribution of power and resources differentially across lines of race, gender, class, sexual orientation, gender expression, and other dimensions of individual and group identity; and (2) the unequal allocation of power and resources—including goods, services, and societal attention—which manifests itself in unequal social, economic, and environmental conditions, also called the determinants of health.

Health inequities are largely a result of poverty, structural racism, and discrimination (NASEM, 2017). Structural racism refers to the "totality of ways in which societies foster racial discrimination through mutually reinforcing systems of housing, education, employment, earnings, benefits, credit, media, health care, and criminal justice. These patterns and practices in turn reinforce discriminatory beliefs, values, and the distribution of resources" (Bailey et al., 2017, p. 1453). Additionally, structural intersectionality highlights that individual life chances are shaped not by a single status hierarchy but by multiple overlapping systems of oppression such as racism, sexism, and classism (Hardeman et al., 2022; Homan et al., 2021).

The reasons for greater vulnerability for some conditions are multifaceted and based on both historical and current social, cultural, and political factors, including structural racism. As a few examples, baseline health status, nutrition and diet, access to quality health care including language-appropriate providers, occupational opportunities, and ability to adapt (e.g., move residence) can impact one's health response to environmental conditions. This health response includes not only the more commonly studied physical responses such as respiratory symptoms, but also mental health and well-being impacts that come from such stressors. For instance, building envelope airtightness affects how well ambient air, including smoke, can penetrate indoors, which can result in differential impacts for those in lower-income housing (Chan et al., 2013). Additionally, low-income housing units are less likely to have adequate cooling systems to allow them to keep windows closed during the summer heat when wildland fires and WUI fires are typically occurring (HHS, 2018). They also may not be able to afford to purchase home air cleaners for particle and chemical filtration. The nonprofit Resources for the Future reported that projects to reduce risk to wildland fires are disproportionately located in communities of higher socioeconomic status, of higher education, and with a higher proportion of white residents (Anderson et al., 2020).

Table 6-2 describes vulnerable populations affected by WUI fires. Note that adults may face additional exposure based on occupation and may experience multiple aspects of vulnerability. Environmental injustice in the workplace

TABLE 6-2 Vulnerable Populations at the WUI

Vulnerable Population	Example Pathways to Vulnerability (vulnerability includes increased exposure, inability to adapt, and health response)	Selected References
Community- or Life Stage–Based Vulnerability		
Children	• Higher respiratory rate and more water ingested per body weight than adults (e.g., babies take a breath 40 times per minute compared to 12–20 times per minute for adults) • More time spent outdoors • Developing organs and immature immune response • Inability to discern risks	Sacks et al., 2011; Vanos, 2015; Perera, 2008; Revi et al., 2014
Older Adults	• Preexisting conditions • Limited mobility • Compromised immune response • Social isolation	Rudolph et al., 2018; EPA, 2019; Liu et al., 2017
Pregnant People	• Physiological changes (e.g., higher respiratory rates, increase in blood and lung volumes)	EPA, 2009
People with Respiratory and Cardiovascular Disease	• Fine particle pollution further exacerbates disease and triggers symptoms	Wettstein et al., 2018; DeFlorio-Barker et al., 2019;
Tribal Communities	• Structural racism • Disproportionate health burdens from environmental conditions • Existing health disparities • Less access to resources by tribal governments	Rudolph et al., 2018; Woo et al., 2020
Communities of Color, and Immigrant, Migrant, and Refugee Communities	• Structural racism • Disproportionate health burdens from environmental conditions • Existing health disparities (e.g., higher burden of asthma and cardiovascular disease) • Less access to resources (e.g., quality health care) • Barriers to receiving language- and culturally appropriate care	Rudolph et al., 2018; Brim et al., 2008; Fussell et al., 2018; Davies et al., 2018; Liu et al., 2017
Low-Income Communities	• Fewer resources and means to evacuate • Less access to indoor spaces with air cleaners and air cooling • Higher burden of asthma and cardiovascular disease • Existing health disparities • Lack of safety nets for missing work	Rudolph et al., 2018; Reid et al., 2016a; Brim et al., 2008
Rural Communities	• Less municipal infrastructure, including access to drinking water and safe spaces • Lack of extensive evacuation routes • Fewer environmental monitoring units available • Less access to indoor spaces with air cleaners and cooling • Existing health disparities	Rudolph et al., 2018
Unhoused/Homeless Communities	• Lack of access to multiple basic resources • Existing health disparities	Rudolph et al., 2018
People with One or More Disabilities	• Fewer resources and less means to evacuate • Inadequate community infrastructure • Less mobility and ability to assess risks	Rudolph et al., 2018
Occupation-Based Vulnerability		
Wildland Firefighters and Emergency Responders (e.g., Emergency Health Care Personnel)	• Lack of personal protective equipment during firefighting • Live and/or work in active fire zone • Some states use incarcerated populations for firefighting (already considered vulnerable) • Increased duration of exposure and elevated concentrations of smoke-associated chemicals and particles • Post-traumatic stress	Rothman et al., 1991; Fent et al., 2017; Navarro et al., 2019
Outdoor Workers (e.g., Farmworkers, Construction Workers, Utility Workers)	• Increased exposure to outdoor wildland and WUI fire smoke and other potential occupational exposures (e.g., pesticides) • Outdoor physical exertion increases respiratory exchange	Austin et al., 2021

continued

TABLE 6-2 Continued

Environmental Remediation Workers (e.g., Hazardous and Solid Waste Removal)	• First to reenter fire zone • Direct contact with chemical hazards • Increased exposure to lingering wildland and WUI fire smoke	EPA, 2019; Fussell et al., 2018
Domestic Workers and Day Laborers	• May participate in ash and debris cleanup as an informal labor workforce with little or no personal protective equipment	IDEPSCA, 2020

Referenced in developing this table:
EPA. 2021g. "Which Populations Experience Greater Risks of Adverse Health Effects Resulting from Wildfire Smoke Exposure?" https://www.epa.gov/wildfire-smoke-course/which-populations-experience-greater-risks-adverse-health-effects-resulting.
American Public Health Association. n.d. "Climate Changes Health: Vulnerable Populations." https://www.apha.org/topics-and-issues/climate-change/vulnerable-populations.
EPA. 2019. *Wildfire Smoke: A Guide for Public Health Officials.* EPA-452/R-19-901. Washington, DC: EPA.

occurs as disproportionate exposure to occupational hazards and as vulnerable populations disproportionately working at high-risk jobs, such as wildland firefighting, emergency response, and environmental remediation.

Finding: Vulnerable populations may exhibit a higher risk of adverse health response and exacerbation of ongoing disease, particularly asthma and pulmonary disorders, as a result of WUI smoke exposure.

Research need: More studies are needed on exposure and potential health effects of WUI fires on children and other vulnerable groups.

ROUTES OF EXPOSURE

Human exposure to smoke emissions is possible at all spatial scales away from WUI fires (Figure 6-1). As depicted in the figure, the subpopulations that are exposed and the most important pathways and routes of exposure vary across the different WUI zones.

The environmental matrices of immediate concern for health effects of exposure to emissions from a WUI fire to persons in the immediate (<1 km) and near-field (1–10 km) regions include the following:

- Outdoor air contaminated directly from smoke emissions associated with combusted fuels
- Indoor air contaminated via infiltration from outdoors
- Indoor surfaces contaminated by deposition of airborne particulates
- Soil contaminated via ash residue from combusted fuel or deposition from airborne emissions
- Surface water contaminated from ash and aerosol deposition, runoff from contaminated soils, and increased runoff due to reduced land cover
- Drinking water

Air that is polluted by plume transport is the main environmental matrix of concern for populations in the local (10–100 km), regional (100–1,000 km), and continental (>1,000 km) zones. There is also potential for exposure in the local and regional zones via contact with settled dust and ash and contaminated surface water. However, a major difficulty in understanding the extent and the impact of wildland fire is the limitation of ambient monitoring networks that were designed primarily to support environmental regulation. The insufficiency of ambient monitoring networks is amplified for WUI fires, where additional chemical composition information is needed to understand exposure. As expected, the potential interactions of different population subgroups with the contaminated matrices and the patterns of exposure (i.e., intensity, frequency, and duration) will vary and be time dependent relative to the fire event (during vs post fire). These interactions and the potential exposure levels for possible exposure scenarios, when data are available, are discussed in the following subsections.

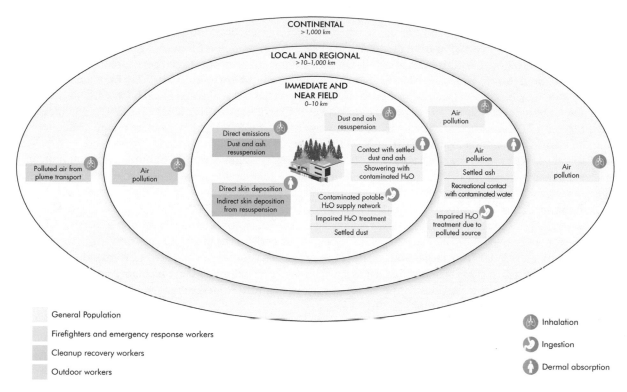

FIGURE 6-1 Potential exposure pathways to WUI fire pollutants.

Inhalation Exposures

This section explores inhalation exposures in the various zones of a WUI fire.

Immediate and Near-Field Inhalation Exposure

Ideally, the general public is evacuated from the immediate and near-field zones during WUI fires; however, complete and timely evacuation is not always possible. In cases of full evacuation, wildland firefighters and other emergency response workers are the most likely population subgroups that will be exposed to air pollution from WUI fires in these zones, while the fire is burning. These firefighters will experience the highest exposure concentration of pollutants because of firefighting duties that involve protecting structures and other properties in urban areas and the consequent proximity to the emission source. A few individuals who do not comply with evacuation orders could also experience elevated exposure levels of air pollutants.

However, data on the personal inhalation exposures of wildland firefighters and other emergency response workers in the immediate and near-field zones during WUI fires have yet to be reported. While such data could be extrapolated from that of wildland fires, burning structures could contribute to increased exposure concentrations and smoke components that are mostly absent in purely vegetative wildland fires (Fent et al., 2018; Jaffe et al., 2020). Also, wildland firefighters, unlike structural firefighters, routinely do not wear any respiratory protection against inhalation exposures, as there is no approved respirator for wildland firefighting. Moreover, they may experience repeated exposures annually and across their careers.

Personal exposure monitoring of wildland firefighters suggests that the exposure concentration of $PM_{2.5}$ at WUI fires could average hundreds to thousands of micrograms per cubic meter ($\mu g/m^3$; Cherry et al., 2019). Currently, no regulated occupational standard exists for wildland firefighters that is relevant to such PM exposure, since wildland fire–generated particles have multiple reactive and toxic components. Relatedly, wildland firefighters are specifically excluded from the California Occupational Safety and Health Administration's standard on protection of workers from wildland fire smoke when the US Environmental Protection Agency (EPA) Air Quality Index (AQI) for $PM_{2.5}$ is equal to or higher than 151 ($PM_{2.5}$ concentration $\geq$ 55 $\mu g/m^3$; State of California, n.d.). Nonetheless, the National Wildfire Coordinating Group recommends an occupational exposure limit of 750 $\mu g/m^3$ for wildland firefighting, which is often exceeded. Moreover, smoke particles from WUI fires are likely to be enriched with metals, asbestos, and other fibrous particles, as well as PAHs, halogenated hydrocarbons, and pesticides, due to burning structures and depending on land use (Carratt et al., 2017; Kohl et al., 2019; Plumlee et al., 2007; Stec, 2017). Radionuclide enrichment of airborne particles is also possible if nuclear facilities are involved, as in the Cerro Grande, New Mexico, fire in 2000 (Eberhart, 2010).

Personal exposure and area concentration data at wildland fires are available only for a limited number of gaseous contaminants, with results indicating that fireline exposure could occasionally exceed the Occupational Safety and Health Administration (OSHA) 8-hour permissible exposure limits of 100 ppb, 1 ppm, and 50 ppm for acrolein, benzene, and carbon monoxide, respectively (Barboni et al., 2010; Miranda et al., 2010; Reinhardt and Ottmar, 2004; Reisen and Brown, 2009). Personal exposure of firefighters at the fireline to benzene, a blood carcinogen, may be even more elevated at WUI fires due to its enrichment in smoke emissions from the combustion of non-biomass WUI fuels (structural materials, vehicles, and furniture), compared to wildland fires (Jaffe et al., 2020). Emissions of hydrogen cyanide, an asphyxiant, and respiratory irritants including hydrogen chloride, nitrogen dioxide, and sulfur dioxide are also enriched in smoke emitted from the combustion of non-biomass WUI fuels versus wildland fires (Jaffe et al., 2020).

Direct measurement of the exposures of other emergency response workers within the immediate and near-field zones during fires has yet to be reported in the literature (Navarro, 2020). However, their inhalation exposures are expected to be less than those experienced by wildland firefighters because they spend less time in proximity to the fireline. The amount inhaled and effective dose of smoke constituents in wildland firefighters are also increased relative to other emergency responders and the public due to the increased breathing (ventilation) rates required during the physically exerting tasks of firefighting (Navarro et al., 2019).

Area measurements of air pollutants within the near-field zone reveal possible exposures of individuals who do not or cannot comply with evacuation orders. $PM_{2.5}$ concentrations up to an order of magnitude above the National Ambient Air Quality Standard of 35 $\mu g/m^3$ per day, and considered unhealthy for the general public on the EPA AQI, were measured in this zone during the 2016 Horse River Fire in Alberta, Canada, and the 2018 Camp Fire in California (CARB, 2021; Landis et al., 2018; Tam and Adams, 2019; Wentworth et al., 2018). Elevated concentrations of lead, but below the National Ambient Air Quality Standard of 0.15 $\mu g/m^3$, were also measured during the Camp Fire. Exceedances of the National Ambient Air Quality Standard for the major gases (nitrogen dioxide, carbon monoxide, sulfur dioxide, and ozone) were not observed in area measurements in either of the aforementioned fires.

Ultimately, the effective exposure of those who do not or cannot evacuate will depend on their activity indoors or outdoors and/or the building ventilation. Since the size of particles in combustion emissions is typically below 1 micrometer (Kleinman et al., 2020; Sparks and Wagner, 2021; Zauscher et al., 2013), the penetrability of particulates together with gaseous contaminants in smoke from the outdoors to the indoor airspace will be high (Xiang et al., 2021). A large portion of wildland fire smoke is PM with a higher proportion of ultrafine particles than typical ambient air pollution. The average $PM_{2.5}$ infiltration factor (the fraction of outdoor $PM_{2.5}$ that enters and remains suspended in indoor air) during a period of wildland fire impact ranged between 0.33 and 0.76 for seven homes that were located hundreds of miles away from the fire (Xiang et al., 2021). Infiltration in the near-field zone may be even higher since particle size distribution tends to be smaller in freshly emitted smoke (Kleinman et al., 2020; Zauscher et al., 2013).

Since reentry decisions after an evacuation are made based on monitoring, ambient air concentrations of fire-related pollutants are expected to be of less concern for residents returning to evacuated communities, as observed for the Horse River Fire (Tam and Adams, 2019). In addition, careful debris removal will mitigate the contribution of this removal to air pollution and the exposure of returning residents (DuTeaux, 2019). Nevertheless, inhalation exposure of residents and recovery (cleanup) workers to fire emissions can still occur due to the resuspension of fire ash and dust from surfaces both indoors and outdoors, and offgassing of VOCs adsorbed onto surfaces indoors. Also, inhalation exposure to VOCs from contaminated potable water supply networks is a possibility; elevated concentrations of benzene, toluene, and styrene were measured in the water supply system after the 2017 Tubbs Fire and 2018 Camp Fire (Proctor et al., 2020). Although these are *possible* routes to inhalation exposure, the committee chose to focus on the most *likely* routes to inhalation exposure.

Regional and Continental Inhalation Exposure

Farther away from the fire, the plume dilutes and ages, resulting in different inhalation exposure considerations from those of the near-field region. Concentrations of air pollutants can drop off rapidly away from the fire (O'Neill et al., 2021), but as the plume spreads, it may impact large population centers and thus affect a far greater number of people than in the near-field region (Koman et al., 2019). Numerous examples from the 2020 and 2021 fire seasons show large parts of the western United States experiencing sustained unhealthy air pollutant concentrations, with several days or longer of daily averaged $PM_{2.5}$ concentrations near 150 $\mu g/m^3$ (Filonchyk et al., 2022; Zhou et al., 2021). Some of these plumes degraded air quality thousands of kilometers downwind, as was observed on the US East Coast during July of 2021 (NASA EO, 2021). This long-range smoke transport makes wildland fires more than just a regional concern, but a national or international air quality issue (Dreessen et al., 2016). Additionally, local topography can lead to severe smoke episodes outside of the near-field range. Smoke can settle into valleys and remain trapped by smoke-induced inversions, leading to extremely high PM concentrations (Kochanski et al., 2019). An example of this was seen in 1999 in Northern California, where daily PM_{10} concentrations exceeded 500 $\mu g/m^3$ (Mott et al., 2002).

Measurements in aging plumes have shown that gas-phase compounds have varying atmospheric lifetimes, and that the species that are important drivers of health effects in the near field may not be as important for far-field exposures. For example, reactive compounds like acrolein rapidly decrease in concentration with aging, while nonreactive compounds like hydrogen cyanide can remain elevated after 3 days of aging (O'Dell et al., 2020). Particle composition also changes, with the rapid formation of tar balls (Adachi et al., 2019) and the oxidation of organic PM constituents (Hodshire et al., 2019) downwind. The fly ash in the plume will also fall out downwind of the fire, but the lifetime of ash in the atmosphere is not well known (Bodi et al., 2014). The extent of ash dispersion may be particularly important for WUI fires since toxic metals can concentrate in the ash (Alexakis, 2020). Many unknowns remain in the aging of smoke from WUI fires (as discussed in Chapter 4) and how those factors could impact the potential health effects in downwind communities.

In addition to the changing chemistry of the plume, mitigation measures taken farther afield may differ from those closer to the fire. Public health messaging efforts may be limited or nonexistent in communities hundreds of kilometers downwind, and therefore fewer people might take mitigating measures to reduce potential exposure, such as staying indoors, operating heating, ventilating, and air-conditioning (HVAC) filtration systems, or using portable air cleaners. The awareness of smoke, either through decreased visibility or by odor, can prompt people to take mitigating actions (Kolbe and Gilchrist, 2009), and the lower concentrations far away from the fire may not meet the threshold for public awareness. Magzamen et al. (2021) observed differential health effects between local and long-range transported smoke with lower incidence of adverse health outcomes being associated with local smoke. Although engagement in mitigation actions and exposure misclassification could have biased the result, the authors hypothesized that aging led to additional oxidant production in the long-range transported smoke that could have increased inflammatory response (Magzamen et al., 2021). The authors' results highlight the importance of a better understanding of exposures in the far-field range. Identification of compounds that can serve as surrogates for WUI fire exposures (i.e., source tracers) is one of the tools that could support future epidemiological analyses.

> **Finding:** Based on information from wildland fires, smoke inhalation is an exposure route of concern for WUI fires, including at longer distance scales (i.e., regional and continental), with the best available information on particulate matter exposure.

> **Research need:** There is a need to better understand the composition of gas and particulate exposures in WUI fires, and how they differ from wildland fires.

> **Research need:** Assessments of smoke exposure typically rely on outdoor concentrations. A continued understanding of indoor penetration and persistence of smoke during and after fires, and an accounting for the chemical emissions profile of WUI fires, is needed.

Dermal Exposures

Although likely for residents in the near-field zone, dermal exposure is of most concern for wildland firefighters during responses to WUI fires. Evidence from non-WUI fires indicates that dermal absorption contributes significantly to the exposures of wildland firefighters working at fires (Banks et al., 2021; Fent et al., 2017, 2019a,b; Pleil et al., 2014; Stec et al., 2018). Such exposure could occur either through direct deposition on exposed skin, by cross-contamination between the turnout gear and skin, and/or via permeation and penetration through firefighter protective clothing (Kirk and Logan, 2015). Biomonitoring studies have measured increased concentrations of PAHs on the skin, their metabolites in urine, and VOCs (including benzene, toluene, ethylbenzene, xylenes, styrene, and naphthalene) in exhaled breath following training or controlled structure fires. This was true even among structural firefighters who, in addition to fireproof clothing, consistently wore self-contained breathing apparatuses (Banks et al., 2021; Fent et al., 2017, 2019a,b; Pleil et al., 2014; Stec et al., 2018), which essentially eliminate inhalation as an exposure route.

Increased concentrations of PAHs on skin and their metabolites in urine have also been measured among wildland firefighters following work at prescribed fires (Adetona et al., 2017a; Cherry et al., 2021). Furthermore, the potential for dermal exposure is likely higher during wildland firefighting since looser fitting protective clothing with fewer layers is typically worn at wildland fires compared to structure fires. The significance of dermal absorption during wildland firefighting is demonstrated by a more pronounced reduction in urinary concentration of the PAH metabolite, hydroxypyrene, from immediately after a firefighting shift to the following morning among firefighters who followed an enhanced skin hygiene intervention (change to new clothes, washing of firefighting clothes, and showering) immediately post-shift compared to those who did not. Similarly, urinary hydroxypyrene concentration was lower among wildland firefighters who reported having more opportunities for skin hygiene (e.g., wash, shower, and change of clothes) while working at the 2016 Horse River Fire (Cherry et al., 2019).

In addition to PAHs and VOCs, firefighters may also be dermally exposed to higher concentrations of halogenated organics (e.g., PCDD/Fs, PCBs, and flame retardants) at WUI fires. However, no study of dermal exposures of firefighters to these compounds during WUI fires has been reported. Likewise, the committee found no study of dermal exposure of recovery workers to contaminants in WUI fire ash; the committee expects that such exposure can be prevented by the use and cleaning of appropriate protective clothing and personal washing procedures.

While dermal absorption of contaminants in settled indoor dust is a potential exposure pathway for residents returning after WUI fires (Meng, 2018; Shi, 2021), trace metal and PAH concentrations were similar or lower in dust vacuumed from home floors (carpet or hardwood) in the city of Fort McMurray, which was affected by the 2016 Horse River Fire, compared to the typical home in Canada (Kohl et al., 2019). Similar concentrations of PAHs and metals in vacuumed dust were also measured in homes in Fort McMurray neighborhoods irrespective of the severity of fire damage based on the number of burned buildings (Kohl et al., 2019). However, dust samples were collected only once, 14 months after the wildland fire event, rendering comparisons with pre-wildland fire dust concentrations in the same homes impossible.

Appropriate post-fire cleanup procedures and regular home cleaning are expected to reduce concentrations of fire-related smoke contaminants in homes and potential exposures over time. However, it is uncertain that cleanup guidance is consistently issued to returning residents, and the potential increased exposure risk that can result from noncompliance with issued guidance is a source of concern. Moreover, the concentrations of arsenic, chromium, copper, tin, vanadium, and zinc in ash samples collected from burned buildings in Fort McMurray during the fire event were one to two orders of magnitude higher than those measured in the house dust 14 months later. The building ash samples were also enriched in trace metals, but not PAHs, relative to the ash samples collected in burned forests outside of the city during the fire event.

The committee identified no other studies of actual or potential dermal exposures due to indoor dust. Although enrichment of nutrients and metals in fire-impacted surface water and increased concentrations of VOCs and heavy metals in the public water supply have been documented (Burton et al., 2016; Emmerton et al., 2020; Proctor et al., 2020; USGS, 2012), actual dermal exposure via these pathways has yet to be studied. Finally, dermal exposures of outdoor workers and the general population beyond the immediate and near-field zones around fires may be of concern: researchers observed an association between WUI fire–impacted air pollution and clinical visits due to atopic dermatitis and itch in San Francisco, California, which is 280 km away from the areas burned by the 2018 Camp Fire (Fadadu et al., 2021).

Finding: The potential for dermal exposure to smoke constituents among firefighters at WUI fires is demonstrated by dermal exposures of firefighters to PAHs and VOCs at structural and wildland fires.

Research need: Studies are needed to characterize the dermal exposures of emergency and cleanup workers, and residents upon reentry.

Research need: There is a need for research to develop standard procedures for the cleanup of contaminated homes post-fire.

Ingestion

Ingestion of contaminants emitted in fire smoke most likely results from exposure to contaminated water post-fire in the immediate, near-field, local, and regional zones (see Chapter 5). Ingestion can also occur from hand-to-mouth transfer when hands are contaminated with settled dust or surfaces are contaminated with fire by-products. Consistent with wildland fires in general (Rust et al., 2018), studies of WUI fires have observed an increased loading of sediments, carbon, nutrients, and metals into surface waters, because of exposed soil and ash, in storm flows following WUI fires (Burke et al., 2013; Burton et al., 2016; Emmerton et al., 2020; Hohner et al., 2016; Stein et al., 2012; USGS, 2012; Writer et al., 2014). Studies have reported storm-associated elevation of the concentration of dissolved organic carbon in surface or source water above the threshold (5 mg/L; USGS, 2012) set for ideal water treatment; levels below this threshold are more conducive to water treatment and disadvantage the formation of unwanted disinfection by-products (Burton et al., 2016; Hohner et al., 2016; USGS, 2012; Writer et al., 2014). Ash from fires could also reduce the effectiveness of dissolved organic carbon removal by alum coagulation for drinking water treatment (Chen et al., 2020).

Hohner et al. (2016) reported the formation of the disinfection by-products trihalomethanes and haloacetic acids above the EPA maximum concentration limits of 80 and 60 μg/L, respectively, in post-storm water samples at the intake of the Fort Collins, Colorado, drinking water treatment plant downstream of the burned areas of the 2012 Hewlett Gulch and High Park WUI fires. However, the committee found no studies of disinfection by-product concentration in drinking water supply following chlorine or chloramine disinfection in areas affected by fires. The more toxic, but unregulated, bromine-substituted trihalomethanes and haloacetic acids preferentially form in the presence of bromine (Chen et al., 2020; Uzun et al., 2020), which is likely to be more concentrated in WUI fire emissions due to its presence in plastics and household products. Nevertheless, no actual measurements of these compounds in drinking water supplies following WUI fires have been reported.

Evidence indicates that trace metals may be enriched in contaminated stormwater that is associated with fires that involve residences or other structures compared to strictly wildland fires. Burton et al. (2016) and Wolf et al. (2011) reported higher concentrations of trace metals in fire ash and its leachate (water that has percolated through a solid and leached out some of the constituents), respectively, in samples collected from residential areas compared to those from wildlands burned by fires. WUI fire studies measured post-fire concentrations of arsenic, cadmium, and lead in stormwater that were above the EPA primary drinking water standards (10, 5, and 15 μg/L, respectively; Burton et al., 2016; Burke et al., 2013; Murphy et al., 2020; Stein et al., 2012). Arsenic concentrations also exceeded the EPA ambient water quality criteria of 18 μg/L in some studies (Burton et al., 2016; Murphy et al., 2020; Stein et al., 2012). EPA secondary drinking water standards were exceeded in some instances for other metals: aluminum (200 μg/L), iron (300 μg/L), manganese (50 μg/L), and zinc (7.4 mg/L; Burton et al., 2016; Burke et al., 2013; Murphy et al., 2020). Nonetheless, evidence suggests that bioavailability of the metals may be low as they partition preferentially to the particulate phase and are adsorbed efficiently by ash (Cerrato et al., 2016).

Radionuclides may also be mobilized into surface water after WUI fires depending on existing land use (Gallaher and Koch, 2004; Igarashi et al., 2020). Persistent organic pollutants emitted in WUI fires, including PAHs, dioxins, and PCBs, have also been measured in stormwater runoff (Gallaher and Koch, 2004; Stein et al., 2012). Of these measurements, only PCBs following the Cerro Grande Fire were reported to exceed standards, including that for primary drinking water at 500 ng/L and ambient water criteria at 64 pg/L.

Proctor et al. (2020) recently reported potential post-WUI-fire exposures to elevated concentrations of VOCs in drinking water via the contamination of the plumbing network in fire-affected areas during the Tubbs and

Camp Fires (Proctor et al., 2020). They found that maximum concentrations of VOCs in the impacted area, measured in the public drinking water supply of the affected communities, were multiple times the EPA primary drinking water standard for benzene (40,000 vs 5 µg/L), dichloromethane (41 vs 5 µg/L), styrene (460 vs 100 µg/L), toluene (1.4 vs 1.0 mg/L), and vinyl chloride (16 vs 2 µg/L; Proctor et al., 2020). The study suggested that contamination of the drinking water supply could have been caused by the heating/burning of the components of the water supply plumbing network and/or depressurization of the system (Proctor et al., 2020). Although recent laboratory experiments have demonstrated the thermal degradation of plastic pipe products used for drinking water sources with the emission of VOCs and SVOCs (including benzene, toluene, styrene, and phthalates) leaching into the water (Chong et al., 2019; Isaacson et al., 2020), findings similar to those of Proctor et al. (2020) or other measurements of contamination of drinking water by WUI fires have not been reported. More research is needed to better understand potential sources of contamination into drinking water and how community water systems could minimize impacts on public health.

Finally, ingestion of deposited WUI-emitted dust- and ash-borne contaminants on indoor surfaces and crops is possible. However, ingestion of contacted contaminants from surfaces by residents upon reentry into their homes and/or community can be minimized through effective cleanup activities and personal hygiene measures such as frequent hand washing. As previously noted, indoor dust concentrations of PAHs and trace metals in homes in Fort McMurray, Alberta, Canada, were apparently unimpacted by the Horse River Fire when measured 14 months after the fire event (Kohl et al., 2019). Also, information is lacking about potential exposure due to the ingestion of contaminated crops.

Finding: Ingestion exposures can occur through three primary mechanisms: compromise of the water supply network, compromise of water treatment, and hand-to-mouth transfer of contaminated dust.

Research need: Studies that characterize the contamination of the indoor environment, pathways of pollutant exposure to returning residents, and effective mitigation strategies for pollutant removal are needed.

Research need: Research is needed to further characterize the chemicals and health impacts of WUI fire emissions via different routes of exposure, including inhalation, dermal exposure, and ingestion. Affected communities need to be fully engaged in this research to better inform approaches and implementation.

HEALTH IMPACTS

Wildland and WUI fires adversely impact public health as they represent a dramatic source of air, water, and soil pollution in neighboring and even more distant populations. Epidemiological studies estimate that the global mortality burden attributable to landscape fire smoke is approximately 339,000 deaths annually (Johnston et al., 2012). Adverse health effects from fires have been measured hundreds of kilometers away, such as the impacts of the 2018 Camp Fire on California Bay Area residents 240 km downwind (Rooney et al., 2020). They have also been measured well beyond 1,000 km away, such as with the 2016 Horse River Fire, which started near Fort McMurray, Alberta, Canada, and resulted in smoke transport and impacts on air quality more than 4,000 km away in New York City, New York, in the United States (Wu et al., 2018). This section first describes health effects associated with exposure to wildland and WUI fires, in particular WUI fires, as they impact communities and regional populations. It then explores health implications in near-field firefighters and emergency responders, which could include law enforcement, emergency medical technicians, and paramedics, all of whom may be at higher risk due to proximity. The discussion is divided into acute and chronic health conditions and further divided into epidemiological and toxicological findings related to wildland fires and WUI fires. Publications selected for inclusion in this section were based upon a literature search by the authors and the National Academies of Sciences, Engineering, and Medicine, and studies were selected that have clearly described health outcomes in communities and regional populations associated with smoke from wildland fires or WUI fires.

Health effect studies in the general population specific to WUI fires are extremely limited. However, researchers can learn much from the studies that identify health impacts associated with exposure of general populations to wildland fire smoke, as well as the decades of research on outdoor PM that can inform our understanding of the health effects associated with fire-associated smoke. Key health outcomes from these studies include mortality from heart disease, chronic obstructive pulmonary disease (COPD), and respiratory disease (Malig et al., 2021). Lung cancer has not yet been identified as a health outcome of wildland fires or WUI fires for smoke-exposed community and regional populations; however, the International Agency for Research on Cancer (IARC, Group 1) and the National Cancer Institute have concluded that exposure to PM in outdoor air is carcinogenic to humans (https://dceg.cancer.gov/research/what-we-study/ambient-outdoor-matter). Moreover, a recent IARC determination that there was sufficient evidence for increased mesothelioma and bladder cancer rates in firefighters, as well as limited evidence for an additional five cancers, demonstrates the need to study these health conditions in regional populations (Demers et al., 2022).

Available data demonstrated elevated risks of emergency room visits during the 2015 wildland fire season in Northern California with specific health outcomes more prevalent, including myocardial infarction, ischemic heart disease, dysrhythmia, heart failure, pulmonary embolism, ischemic stroke, and transient ischemic attack (Wettstein et al., 2018). Studies (reviewed in Reid et al., 2016a) consistently confirm the association between wildland fire smoke exposure and some respiratory health outcomes, with the clearest evidence related to exacerbation of asthma (Gan et al., 2017). In a study by Malig et al. (2021) examining $PM_{2.5}$ and morbidity during the 2017 northern San Francisco Bay wildland fires, increased PM levels were consistently linked with emergency department visits with asthma, chronic lower respiratory disease, and acute myocardial infarction; increases in acute respiratory infection and decreases in mental health/behavioral hospital visits were observed, but were sensitive to model specifications. In contrast to the preponderance of data demonstrating the association between wildland fires and asthma, the links between wildland fire smoke and both COPD and increased susceptibility to bacterial infections, including those associated with pneumonia, are not as clear (Malig et al., 2021; Reid and Maestas, 2019; Reid et al., 2016a). Studies also suggest that toxicity of $PM_{2.5}$ from wildland fires is more harmful than equal concentrations of non–wildland fire $PM_{2.5}$ (Aguilera et al., 2021).

Both current and past research concludes that more studies are needed to fill the gaps on population subgroups that are most vulnerable to the health implications of wildland fire smoke and the far-field respiratory health impacts of these events with high air pollution. As the pulmonary, cardiovascular, cerebrovascular, and reproductive systems appear to be most representative of the health effects associated with wildland fires, diseases associated with those target organs, as well as mental health, will be the focus of this section. In addition to acute and chronic health conditions associated with smoke exposure in the general population, delayed effects (i.e., effects that occur later, after the fire is over) have been observed. Because the committee identified only one study on delayed effects, it is grouped with chronic health studies in the section on "Chronic Conditions and Exacerbation of Disease." It is important to note that smoke-associated PM is the primary constituent considered among all of the health studies discussed in this report. Also worth noting is the fact that comparing health effects of wildland and WUI fire events across world regions is difficult due to variation in intensity, fuel type, population, PM sources and composition, and other factors (Tinling et al., 2016).

Finding: WUI-specific studies are rare, making it extremely difficult to understand related health effects for this specific, complex type of fire. Most information about acute and chronic health conditions is based on wildland fire studies, because of the greater amount of data available.

Research need: Studies that explicitly define the particular type of fire (i.e., wildland fires vs WUI fires) to which the general population is exposed are needed to clearly delineate and possibly mitigate the types of health conditions that could emerge, as pollutant emissions (as well as toxicant concentrations) between wildland fires and WUI fires can be quite different.

Acute Conditions and Exacerbation of Disease

Epidemiological Studies

Respiratory Effects

A growing body of evidence documents an association between an exacerbation of asthma and wildland fire smoke exposure. For example, a positive association has been observed between hospitalizations, emergency department (ED) visits, and outpatient visits for asthma exacerbation and wildland fire smoke exposure (Gan et al., 2017; Hutchinson et al., 2018; Malig et al., 2021; Reid et al., 2016a,b). Wildland fire smoke inhalation also increased the risk for asthma morbidity during the 2013 wildland fire season in Oregon (Gan et al., 2020). By contrast, however, Tinling et al. (2016) found only weak associations between wildland fire smoke exposure and asthma.

As discussed by Reid et al. (2016a,b), multiple studies have described a decrease in lung function associated with wildland fire smoke exposure among individuals without asthma or bronchial hyperreactivity. In the Wallow Fire, researchers observed an increased risk of ED visits for some respiratory and cardiovascular conditions during heavy smoke conditions, and risk varied by age and sex. The subpopulation of 65+ years of age was especially at risk for increased ED visits. The majority of epidemiology studies found that wildland fire smoke was associated with an increased risk of respiratory and cardiovascular diseases. In one study, a significantly increased risk of ED visits was identified among the 65+ population for asthma (relative rate [RR] = 1.73, 95% confidence interval [CI] = 1.03–2.93) and for diseases of the veins and lymphatic and circulatory systems (RR = 1.56, 95% CI = 1.00–2.43). For the age group of 20 to 64 years, there was an increase in ED visits for diseases of pulmonary circulation (RR = 2.64, 95% CI = 1.42–4.9) and for cerebrovascular disease (RR = 1.69, 95% CI = 1.03–2.77; Resnick et al., 2015). Additional findings suggest that $PM_{2.5}$ exposure during wildland fire smoke days is more likely to trigger strong, acute responses such as asthma, bronchitis, and wheezing, compared with non-smoke days, in the older population (DeFlorio-Barker et al., 2019).

Findings from studies examining infection risk associated with exposure to wildland fire smoke are inconsistent. In a study of the impacts of Indonesian wildland fires on air pollution and health in Singapore, clinic visits for acute respiratory infections increased during weeks with high fire levels (as estimated from satellite-derived values of fire radiative power) in 2010–2016 (Sheldon and Sankaran, 2017). Alman et al. (2016) reported a borderline association for combined hospitalizations and ED visits for upper respiratory infections and $PM_{2.5}$ during wildland fires in 2012 in Colorado. However, the committee found two recent studies with seven different analyses of the association between wildland fire smoke and pneumonia, of which all analyses revealed negative associations except two: (1) The analysis of outpatient presentations (but not hospitalizations or ED visits) by Hutchinson et al. (2018) found a weak relationship with infections. (2) Gan et al. (2017) reported a strong association between pneumonia hospitalizations and wildland fire smoke during the 2012 Washington state fires only when assessing exposure from kriging monitoring data, but not from an aerosol measurement or blended model. A recent study by Zhou et al. (2021) examined the relationship between increased ambient concentrations of $PM_{2.5}$ from recent California wildland fires and the incidence of COVID-19 infection. The authors concluded that high particulate pollution levels were temporally associated with an increase in the incidence and mortality rates of COVID-19 after adjusting for several time-varying confounding factors such as weather, seasonality, long-term trends, mobility, and population size.

Studies investigating the association between wildland fire smoke and acute bronchitis show mixed findings, with results from a single study using temporal comparisons demonstrating an association for ED visits and outpatient presentations, but not for hospitalizations, among Medi-Cal patients in San Diego (Reid and Maestas, 2019). By contrast, a study of the 2012 Washington state wildland fires found no association between acute bronchitis hospitalizations and wildland fire smoke using three different methods to estimate smoke. In addition, no association was found for combined hospitalizations and ED visits for bronchitis during the 2012 wildland fire season in Colorado.

The null findings associated with pneumonia and bronchitis contrast with previous papers that collectively hinted at an association between wildland fire smoke and pneumonia and bronchitis. Most of the previous studies grouped pneumonia and bronchitis together rather than separating them, as is the norm in the recent studies. One earlier study that did separate pneumonia and bronchitis found a strong association between $PM_{2.5}$ and pneumonia, but not acute bronchitis, during the 2003 wildland fires in Southern California (Delfino et al., 2008).

Several recent papers investigated the relationship between wildland fire smoke and all respiratory health outcomes grouped together. Studies consistently found strong associations between smoke exposure and hospitalizations, hospitalizations and ED visits combined, ED visits, and outpatient presentations (Reid and Maestas, 2019). However, findings from a few studies reported conflicting results. One of these authors noted that one of the investigations examined smoke transported long-range rather than fresh smoke, which could have a different chemical composition (Magzamen et al., 2021). For more information on smoke plume aging and composition, please see Chapter 3 of this report.

Cardiovascular Effects

A growing number of studies have implicated exposure to wildland fire smoke as a risk factor for cardiovascular disease. Of the 48 wildland fire epidemiological studies that currently exist (as of 2021; Chen et al., 2021), 38 are on cardiovascular morbidity, and most outcomes focus on ED visits and hospitalizations for cardiovascular symptoms and illness. Twenty-five of the 38 studies report a positive association between wildland fire smoke exposure and increased health care for cardiovascular disease (Chen et al., 2021). For example, Haikerwal et al. (2015) demonstrated that $PM_{2.5}$ exposure was associated with increased risk of out-of-hospital cardiac arrests and ischemic heart disease during the 2006–2007 wildland fires in Victoria, Canada. The evidence indicates that $PM_{2.5}$ may act as a triggering factor for acute coronary events during wildland fire episodes. In a study by Haikerwal et al. (2015), an interquartile range increase of 9.04 $\mu g/m^3$ in $PM_{2.5}$ over a 2-day moving average during the 2006–2007 wildland fire season in Victoria, Australia, was significantly associated with an increase in risk of out-of-hospital cardiac arrests, with the strongest associations observed among men and older adults, and was also associated with an increased risk for ischemic heart disease–related ED visits (2.07 percent) and hospitalizations (1.86 percent), particularly among women and the elderly.

While the preponderance of evidence suggests an association between cardiovascular disease and wildland fire–generated PM, a conclusive link between wildland fire smoke exposure and adverse cardiovascular effects still remains somewhat elusive. However, a recent review by Chen et al. (2021) assessing epidemiological data, controlled clinical exposure studies, and toxicological studies has concluded that wildland fire smoke–associated PM is a risk factor for adverse cardiovascular disease, especially among vulnerable populations. In addition, findings also concluded that young and healthy people may also develop biological responses including systemic inflammation and vascular activation, both associated with increased cardiovascular disease risk. Future epidemiological studies could benefit from improved exposure assessments and the implementation of more sensitive indicators of cardiovascular dysfunction. More data are also needed to accurately define the risk of cardiovascular health effects including common, life-threatening, disabling, and costly clinical outcomes that include myocardial infarction, stroke, heart failure, heart rhythm disturbances, and sudden death.

Reproductive Effects

In 2012, Kessler compared the average birth weight of infants carried in utero during wildland fires with that of infants who were either born before or conceived after the fires. The author also compared birth weights of infants exposed to wildland fire smoke during the first, second, and third trimesters of pregnancy. After adjusting for infant sex, gestational age at birth, and other factors known to influence birth weight, the researcher reported that smoke-exposed infants weighed an average of 6.1 g (0.21 oz) less at birth than unexposed infants. Infants exposed to smoke as fetuses during the second trimester of pregnancy demonstrated the largest average size reduction at 9.7 g (0.34 oz), while those exposed during the third trimester showed an average reduction of 7.0 g (0.24 oz). Alternatively, infants exposed during the first trimester showed an average reduction of 3.3 g (0.11 oz), which was no different from the control. The trends were corroborated by a sensitivity analysis that compared pregnant mothers according to whether the air-pollution monitor nearest their residence recorded high or low PM levels during the wildland fires.

Findings from O'Donnell and Behie (2015), who used data from the Australian Capital Territory and examined the influence of the 2003 Canberra wildland fires on the weight of babies born to mothers living in fire-affected regions, demonstrated heightened responsiveness in the male cohort. While this was not a direct effect of smoke exposure, the authors also showed that elevated maternal stress due to wildland fires acted to accelerate the growth of male fetuses, potentially through an elevation of maternal blood glucose levels. These studies concluded that effects of disaster exposure alter fetal growth patterns in response to maternal signals. However, the direction of the change in birth weight is opposite to that of many earlier studies. While this study has a number of limitations, the findings indicate a risk of macrosomia to exposed male infants that could pose immediate health risks to both mother and child, as well as longer-term health risks to the child.

The work of Abdo et al. (2019) supports these findings, demonstrating that exposure to wildland fire smoke, specifically smoke-associated $PM_{2.5}$, over the full gestation period and during the second trimester was positively associated with preterm birth, while exposure during the first trimester was associated with decreased birth weight. In a recent systematic review by Amjad et al. (2021), the authors concluded that while some evidence existed that maternal exposure to wildland fire smoke is associated with birth weight reduction and preterm birth, particularly when exposure occurs late in the pregnancy, based on all data, the association still remains inconclusive, despite strong evidence linking $PM_{2.5}$ exposure with reduced birth weight (Sun et al., 2016).

Chronic Conditions and Exacerbation of Disease

Epidemiological Studies

Respiratory Effects

Recent data on wildland fires and COPD outcomes have caused doubt about a relationship between them, which previously had been established. Researchers observed strong associations between wildland fire smoke and COPD ED visits, but saw null results for hospitalizations during the 2008 Northern California wildland fires (Reid et al., 2016b). An analysis of the 2012 Washington state fires demonstrated an association between hospitalizations for COPD when using monitoring data or $PM_{2.5}$ exposures from a model that blended monitoring, aerosol optical depth, and aerosol measurement data, but not from $PM_{2.5}$ estimates derived from aerosol measurements (Gan et al., 2017). Alman et al. (2016) also reported a strong association between combined hospitalizations and ED visits for COPD and smoke-associated $PM_{2.5}$ levels during the 2012 Colorado fire season. Analyses using temporal comparisons were null for outpatient visits, ED visits, and hospitalizations (Hutchinson et al., 2018), as were results from two analyses using aerosol measurement–derived PM exposures for ED visits (Haikerwal et al., 2016; Tinling et al., 2016). In a study by Hutchinson et al. (2018) that examined the health care utilization of Medi-Cal recipients during the fall 2007 San Diego wildland fires, which exposed millions of people to wildland fire smoke, young children appeared at highest risk for respiratory problems, especially asthma, which is a cause of particular concern because of the potential for long-term harm to children's lung development. Regarding delayed effects, a study by Reid et al. (2016a) 10 years after wildland fire smoke exposure found lung function to be reduced among individuals without asthma or bronchial hyperreactivity.

Cardiovascular and Cerebrovascular Effects

Wettstein et al. (2018) conducted population-based epidemiological analyses for daily cardiovascular and cerebrovascular ED visits and wildland fire smoke exposure in 2015 among adults in eight California air basins. Findings from this study revealed elevated risks for individual diagnoses of myocardial infarction, ischemic heart disease, heart failure, dysrhythmia, and pulmonary embolism. The same study also showed that smoke exposure was associated with cerebrovascular ED visits for all adults (ischemic stroke and transient ischemic attack), particularly for those 65+ years. The authors of the study concluded that individuals with underlying cardiovascular disease risk factors may be at greater risk for a cardiovascular or cerebrovascular event during a period of wildland fire smoke exposure.

Mental Health Effects

Mental health is included in this report as recent research has shown that small increases in particulate air pollution are linked to significant rises in depression and anxiety. For example, a recent study (Petrowski et al., 2021) examining nationally representative data of German populations revealed an association between mental health and well-being and particulate air pollution. Moreover, studies have also found an increased rate of post–wildland fire mental health disorders in both adult and pediatric populations, with a number of associated risk factors, the most significant being characteristics of the wildland fire trauma itself. A recent study by To et al. (2021) demonstrated that direct exposure to large-scale fires significantly increased the risk for mental health disorders, particularly for post-traumatic stress disorder, depression, and generalized anxiety, at several follow-up times after the wildland fire, from the subacute phase to years after.

Brown et al. (2019) examined mental health in children following the Fort McMurray fire in Alberta, Canada, and noted a high incidence of post-traumatic stress disorder, depression, anxiety, and alcohol- or substance-abuse disorders. The authors concluded that disasters have a negative impact on the mental health of youth, particularly those who directly experienced fires. However, they also noted the role of resilience on mental health among this group, with lower resilience associated with substantially lower mental health outcomes. A follow-up study by some of the same authors (Brown et al., 2021) following the Fort McMurray fire not only validated their first study, but showed that mental health issues worsened rather than improved over time. Moreover, the authors observed higher levels of mental health distress among older students, in females compared to male students, and in individuals with a minority gender identity, including transgender and gender-nonconforming individuals. These follow-up findings illustrate that deleterious mental health effects can persist in youth for years following a fire disaster and therefore are critical to discuss in this report.

Controlled Human Exposure Studies

This section discusses controlled human wood-smoke exposure studies, in concert with in vivo animal studies, to complement the wildland fire studies in this report. Controlled human exposure studies complement the cardiovascular mortality and morbidity observations in epidemiology as they offer a chance to examine the associated pathophysiological changes. While wood smoke used for clinical and experimental studies does not entirely reflect smoke from a wildland fire or WUI fire, chamber studies do allow for a better understanding of mechanisms underlying observed health effects. Moreover, low or moderate exercise can be introduced in the study protocols to increase the ventilatory rate to simulate physical activity in the vicinity of a wildland fire.

Most human wood-smoke exposure studies conducted to date have investigated pulmonary function, respiratory inflammation, systemic inflammatory responses, and cardiovascular endpoints or markers related to vascular pathophysiology. In one study (Andersen et al., 2017), 43 subjects participated in various training exercises that included activities to extinguish fires fueled by either wood alone or by wood and household materials (a mattress). They found that firefighting activity was associated with decreased heart rate variability and microvascular function measured as reactive hyperemia index pressure. While no definitive conclusions can be reached in these types of studies due to differences in such factors as wood types and duration of exposure, the overall outcomes support the positive cardiopulmonary findings observed in wildland fire epidemiological studies.

In Vivo Toxicology Studies

While humans best represent themselves for identifying contaminant-associated health effects, studies using animal models provide biological plausibility for the associations revealed in epidemiological studies, as well as provide a mechanistic understanding for such observed effects. Several studies have shown that exposure to wood smoke leads to increased cardiovascular risk in animals (reviewed in Chen et al., 2021). For example, Kim and colleagues (2014) established an in vivo model wherein female CD-1 mice were exposed through oropharyngeal aspiration to 100 µg peat smoke condensate from different combustion stages (smoldering vs near extinguished). Mice exposed to smoke produced during the smoldering stage revealed significantly decreased cardiac function and increased post-ischemic cardiac death in an experimentally induced ischemia model. Animal studies have shown that the toxicity of wood smoke depends on the wood type, as well as the combustion conditions, with wood smoke from smoldering wood producing fewer effects relative to smoke produced by the burning of specific wood types.

In a study by Black et al. (2017), 3-year-old macaque monkeys were housed in outdoor facilities at the California National Primate Research Center and exposed to California wildland fire smoke during pregnancy and in the first 3 years of life. Adolescent monkeys demonstrated reduced lung function and long-term immune alterations dependent on sex. The authors concluded that modulation of peripheral blood cytokine synthesis and altered later-in-life lung function was significantly associated with early-life wildland fire smoke exposure (Black et al., 2017). These results support the findings by Orr et al. (2020) which show that initial smoke exposure can lead to health effects in subsequent years. Overall, in vivo animal studies are challenging to interpret in a clinical context, as exposure dose and duration may be difficult to compare with real-world conditions impacting public health. However, as previously stated, these studies provide plausibility for epidemiological studies demonstrating wildland fire–induced respiratory dysfunction, as well as insight into underlying mechanisms.

Finding: Particulate matter is the primary metric related to health outcomes in almost all existing studies on wildland fire and WUI fire smoke.

Research need: Further studies are critical to understanding whether physiological and psychological chronic health disorders associated with regional exposures among the general public are similar for WUI and wildland fires.

Research need: Additional studies are needed to evaluate the health implications of wildland fire and WUI fire smoke constituents other than PM, including volatile organic compounds, semi-volatile organic compounds, and polycyclic aromatic compounds.

Research need: A need exists to better understand the persistent and delayed effects of smoke exposures, especially for far-field community and regional populations.

Research need: A need exists to better understand the effects of smoke exposures to WUI fire components, which may include biomonitoring studies of general populations and vulnerable subpopulations.

Firefighter Acute Exposures and Effects

Although ongoing research exists on WUI firefighter acute exposures and health effects, the committee found no published studies specific to WUI fires. A limited amount of information is available on acute exposures and health effects for wildland firefighters, with additional data available on structural firefighters from municipal fire departments. Measurement of urinary hydroxylated metabolites of PAHs has been used for the assessment of exposure to combustion products (Fent et al., 2020; Keir et al., 2017). Among PAHs, the known, probable, and possible carcinogens include benzo[a]pyrene, dibenz[a,h]anthracene, chrysene, benzo[a]anthracene, and naphthalene (IARC, 2010). Evaluation of PAH metabolites in urine also provides a biological measure of combined inhalation and dermal exposure.

Wildland Firefighters

Wildland firefighters are exposed to carcinogens including benzene, formaldehyde, and PAHs (Broyles, 2013; Reinhardt and Ottmar, 2004; Ward and Hardy, 1991). More research studies have been published on prescribed burns than on larger campaign fires. Researchers have found urinary PAH metabolites to be elevated in wildland firefighters after a prescribed burn (Adetona et al., 2017a). However, firefighters can minimize their exposures during certain types of prescribed burns by avoiding smoke plumes. For example, in a study of Bureau of Indian Affairs/Fort Apache Agency wildland firefighters before, during, and after conducting pile burns (prescribed fires), despite finding 20 of the 21 PAHs analyzed in *area* (smoke plume) samples, researchers found only 3 (naphthalene, phenanthrene, and fluorine) of the 16 PAHs analyzed in *personal* exposures, based on the ability of the firefighters to move away from the smoke, and there were no significant changes in urinary 1-hydroxypyrene concentrations (Robinson et al., 2008). Urinary methoxyphenols have been used to evaluate exposure to wood smoke (Dills et al., 2006). In US wildland firefighters, PAH DNA adducts were associated with diet, but not occupational exposure (Rothman et al., 1993).

Wildland firefighting is known to be associated with acute adverse health effects. Previous studies of wildland firefighters have recorded cross-seasonal changes in lung function (Gaughan et al., 2008; Liu et al., 1991; Rothman et al., 1991), and decreases in lung function have been associated with hours of firefighting activity (Rothman et al., 1991) and continuous exposure to wood smoke (Adetona et al., 2011). Additionally, several studies have measured elevated inflammation and oxidative stress among wildland firefighters after prescribed burns (Adetona et al., 2017b, 2019) and wildland fires (Main et al., 2020). Inflammation and oxidative stress are known risk factors for cancer (Coussens and Werb, 2002; Reuter et al., 2010).

Structural Firefighters

Among structural firefighters, exposure during the overhaul phase (the phase after the main fire has been extinguished, during which firefighters search a fire scene to detect hidden fires or smoldering areas that may rekindle) increased lung inflammation in firefighters wearing multipurpose (high-efficiency particulate, acid gas, and organic vapor) air-purifying respirators, which suggested inadequate protection offered by the air-purifying respirators

(Burgess et al., 2001). Air-purifying respirators can provide protection against many products of combustion at concentrations relevant to overhaul, although formaldehyde may break through under certain conditions, particularly at elevated ambient temperatures (Jones et al., 2015; Staack et al., 2021). Fent et al. (2014) evaluated exposures in structure fires during controlled burns when firefighters wore self-contained breathing apparatus for all phases including overhaul (Fent et al., 2014). Within this study, dermal swabs of skin surfaces that were collected before and after the fires showed that the neck was the most exposed part of the body. The results also showed an increase in urinary PAH metabolites 3 hours after the fire response. Kirk and Logan (2015) also analyzed firefighting ensembles for deposition and off-gassing of contaminants following controlled burns and found VOCs, carbonyl compounds, PAHs, and hydrogen cyanide off-gassing from the gear. They determined that deposition of contaminants on firefighting ensembles is cumulative, and that laundering reduced contamination to pre-exposure levels.

One potential mechanism of acute toxicity following exposures among structural firefighters is activation of the aryl hydrocarbon receptor (Beitel et al., 2020). This activation is reportedly the same toxic mechanism by which dioxin causes cancer, and the aryl hydrocarbon receptor is also a potential target for cancer chemotherapy (Kolluri et al., 2017).

Firefighter Chronic Effects

As with acute exposures and health effects, although there is ongoing research on WUI firefighter chronic health effects, the committee found no published studies specific to WUI fires. Limited information is available on chronic health effects from smoke exposure for wildland firefighters, with additional data available from municipal fire departments. Researchers have linked PAH exposures to a number of cancers, including skin, lung, bladder, and gastrointestinal cancers (Boffetta et al., 1997; Diggs et al., 2011; Rota et al., 2014). Research has identified elevated levels of persistent organic contaminants, such as PCDD/Fs, associated with cancer and other adverse chronic health effects in California firefighters (Shaw et al., 2013).

Wildland Firefighters

A survey study associated the employment duration of wildland firefighters with a history of being diagnosed with hypertension and arrhythmias (Semmens et al., 2016). Wildland firefighters have an estimated increased risk of lung cancer mortality and cardiovascular disease (Navarro et al., 2019). To our knowledge, no epidemiological studies of cancer in wildland firefighters have been published.

Structural Firefighters

Cancer is a leading cause of mortality and morbidity in the municipal fire service. The National Institute for Occupational Safety and Health (NIOSH) study of Philadelphia, Chicago, and San Francisco firefighters demonstrated a 14 percent increase in overall cancer mortality compared to the general population, and specific excesses in lung (10%), gastrointestinal (30–45%), kidney (29%), and mesothelioma (100%) cancer deaths, with similar increases in cancer incidence (Daniels et al., 2014). In Australian volunteer firefighters, prostate cancer was increased compared to the general population, and kidney cancer increased with the number of fire exposures (Glass et al., 2016). Among female firefighters, colorectal cancer increased with the increasing number of fire responses. A study of cancer among firefighters in five Nordic countries found a significant excess risk of prostate cancer and melanoma among those 30–49 years of age, as well as an increase in non-melanoma skin cancer, multiple myeloma, lung adenocarcinoma, and mesothelioma among older firefighters (Pukkala et al., 2014). A previous meta-analysis of 32 studies of cancer in the fire service found an elevated risk for non-Hodgkin's lymphoma, prostate cancer, and testicular cancer (LeMasters et al., 2006).

Cancer is a multistage process with a latency period between exposure and the onset of disease from as short as 5 to over 30 years (Nadler and Zurbenko, 2014). Previous firefighter studies have shown limited evidence of direct DNA damage (change in DNA sequence) as a result of occupational exposure. A study of structural firefighters in India found increased buccal cell micronuclei compared with controls (Ray et al., 2005). Beyond direct DNA damage, advances in technology have led to a greater appreciation of the role of epigenetic (not involving a change in DNA sequence) contributions to carcinogenesis: epigenetic markers are known to have important roles in carcinogenesis. Key among these are DNA methylation and miRNA expression, which researchers have shown to be different in incumbent and new-recruit structural firefighters (Jeong et al., 2018; Zhou et al., 2019) and which change over time in new recruits (Jung et al., 2021). Numerous studies have shown how DNA methylation (Dugue et al., 2021; Jaenisch and Bird, 2003; Jones and Laird, 1999; Onwuka et al., 2020; Wang et al., 2010) and miRNA expression (Calin and Croce, 2006; Carter et al., 2017) are associated with cancers.

> **Finding:** Insufficient research exists on acute exposures and acute and chronic health effects in WUI firefighters.

REDUCING EXPOSURE RISKS AND HEALTH EFFECTS

This section focuses primarily on reducing the risk of smoke (inhalation) exposure and its effects. However, dermal and ingestion exposures are also a concern, particularly in the near field, that can be reduced through actions discussed in this section. Mitigation opportunities at a structural level to reduce the incidence of WUI fires, such as adjusting land-use policy, also exist; these are not discussed because the scope of this report is focused on chemistry.

Reducing Near-Field Exposure Risks

This section discusses actions that reduce near-field exposure for the general public in the evacuation zone and for firefighters working in close proximity. While there is a range of individual and community actions (or interventions; e.g., use of portable air cleaners, wearing of N95 masks [see Box 6-2]) to mitigate fire smoke exposure (EPA, 2021f; Xu et al, 2020), there is limited information about the effectiveness of their use during wildland and WUI fires (EPA, 2021f). Information is also lacking about best practices by residents to reduce their exposures upon reentry post-fire. These limitations are underscored by the lack of comprehensive information about fire emissions specific to WUI fires, such as the chemical composition of PM and pollutants that may not be monitored in targeted analyses.

Evacuation

While no data sets systematically track the occurrence of emergency evacuations or other exposure reduction actions for wildland fires (CCST, 2020), data specific to state and fire incidents are available. For example, CCST (2020) noted one study estimated that 1.1 million people were ordered to evacuate during 11 major wildland fires between 2017 and 2019 in California (Wong et al., 2020) with non-compliance to mandatory evacuation orders for three wildland fires in California ranging from 3 to 13 percent of the population (Wong et al., 2020). Similar results

---

BOX 6-2
Reducing Smoke Exposure through Respirator Use

Respirators or filtering face masks, like NIOSH-certified N95 masks, can be used to reduce exposure to $PM_{2.5}$. Filtering face masks have been widely used due to the COVID-19 pandemic and may now have increased acceptance as a way to reduce exposure to wildland fire smoke (Holm et al., 2021). However, personal respiratory protection, which has long been viewed as an occupational exposure reduction method (Samet et al., 2022), may have limited effectiveness for the general public since it depends on the fit of the mask and the length of time it was used (Kodros et al., 2021). Additionally, masks or respirators may not be effective for individuals with preexisting conditions that make it difficult to breathe or for children since there are no certified N95 masks for their size range (Samet et al., 2022). Despite these challenges, masks have been found to provide a health benefit in polluted environments (Holm et al., 2021).

have been observed in other studies. In the western United States, 11–20 percent of survey respondents who resided in areas within or adjacent to perimeters of recent wildland fires and/or were under evacuation orders did not evacuate during the fires (Kuligowski, 2021; Kuligowski et al., 2022; McCaffrey et al., 2018), while a similar 9–10 percent of survey respondents reported an intention to stay and defend property regardless of authorities' evacuation orders in cases of future wildland fires (Edgeley and Paveglio, 2019; Mozumder et al., 2008; Paveglio et al., 2014; Stasiewicz and Paveglio, 2021).

In a study conducted in New Mexico, 89 percent of survey respondents reported that they would evacuate under mandatory orders compared to 57 percent under voluntary ones (Mozumder et al., 2008). Nonetheless, the implementation of mandatory evacuation orders and the enforcement of penalties for disobeying such orders, if any exist, are inconsistent across the country, and WUI fire–specific data are largely unavailable (Kuligowski, 2021; Mozumder et al., 2008). Even without research, WUI fire evacuation planning can benefit from the availability and use of contemporaneous real-time collection of data about evacuation decisions or progress, such as location data from cellular information (Melendez et al., 2021). But the inadequacy of communication and transport infrastructure or their degradation by wildland fire activities can impair the availability of such data and the evacuation performance (Cova et al., 2011; Kuligowski, 2021). In addition, evacuation is not always the best safety decision, and evacuation that is delayed or made under congested traffic conditions can result in catastrophic unintended consequences including death (Cova et al., 2011; Kuligowski, 2021; Li et al., 2019; McLennan et al., 2018).

Reentry Exposure Risk and Mitigation

Exposure to lingering smoke and ash from a fire after WUI residents are allowed to return to their homes can cause significant health effects. Guidance provided by NIOSH and other agencies provides general information on health and safety hazards for homeowners and environmental remediation workers after the area is declared safe for reentry. According to NIOSH, these activities may pose health and safety hazards that require necessary precautions, and in most cases, it may be more appropriate for professional cleanup and disaster restoration companies, rather than homeowners or volunteers, to conduct this work (NIOSH, 2019b).

Cleanup recovery workers undertake demanding and often dangerous work to remove debris and demolish damaged structures. Many immigrants make up this recovery workforce because of their substantial representation in the formal and informal construction sectors (Eckstein and Peri, 2018). Related research on immigrant experiences with cleanup and reconstruction work illustrates the many risk factors that make these workers vulnerable to hazardous work conditions and exploitation (Vinck et al., 2009). Despite this, specific research on vulnerable populations and workers in WUI fire–affected areas is sparse. One resource that extends education and training to those responding to disaster events is the National Institute of Environmental Health Sciences Worker Training Program, which funds a network of labor organizations, worker centers, and university-based programs (NIEHS, 2021b).

One survey conducted by a community-based organization in Southern California surveyed 195 blue collar workers in the area following the 2018 Woolsey Fire (house cleaners, childcare providers, gardeners, construction workers, and caregivers to the elderly) to learn how their work was shaped by the fire and to document the consequent health, safety, and economic impacts. Some domestic workers reported engaging in fire cleanup tasks, such as cleaning ash and debris that thickly coated furniture inside homes that were adjacent to fully burned structures, and respondents indicated that employers provided little more than simple face masks to carry out dangerous tasks (IDEPSCA, 2020).

Research need: Research is needed to better identify at-risk and vulnerable residents and workers affected by near-field WUI fires, and to identify culturally appropriate opportunities for interventions to reduce adverse health impacts.

Reducing Exposure Risk and Health Effects for Firefighters

Potential mitigation interventions for wildland firefighters are grouped into dermal, inhalation, mixed, and administrative categories (Table 6-3). Dermal interventions include directly removing soot from the skin and more

TABLE 6-3 Potential Interventions to Reduce Exposures and Health Effects in Wildland Firefighters

Dermal	Take shower after operational period, or use skin wipes when shower not available
	Carry clean change of clothes for after operational period if not returning to base camp
	Prevent contamination of backpack (14-day bag with personal items) contents using sealed, impervious bags
	Use improved wildland turnout gear to reduce dermal exposure
Inhalation	Use respiratory protection if you are a vehicle driver / pump operator
	Provide exposure monitoring equipment to firefighters, when appropriate, to help them identify and avoid inhalation hazards (implement with support of safety officers on incidents and risk management personnel at land agencies)
	Change cabin air filters in fire department apparatus
Mixed	Clean tents and sleeping bags after each assignment
	Implement clean cab protocol
Administrative	Use job rotations to reduce exposures (fire incident management personnel can rotate and reassign crews and resources after completing job tasks associated with higher smoke exposures to job tasks that have lower expected smoke exposures)
	Allow firefighters to patrol for spots rather than stand stationary on the fireline

frequently cleaning dirty gear. Wildland firefighting gear is different from municipal turnout gear, being constructed of lighter materials. Many wildland firefighters continue to wear dirty gear for multiple shifts in a row during a campaign fire, although recent changes in some departments encourage cleaning after each shift. The wildland firefighter interventions shown in Table 6-3 may also apply to WUI firefighters.

By early 2022, some US wildland departments expected to have new personal protective equipment with a built-in protective barrier to help reduce skin exposure. They are also working on decontamination protocols for wildland firefighters, focused on dermal post-fire decontamination, and a clean cab protocol for wildland firefighting. Researchers have shown that the use of air-purifying respirators reduces respiratory symptoms in wildland firefighters in Australia (De Vos et al., 2009).

> **Research need:** A need exists for research on the effectiveness of interventions to mitigate WUI firefighter smoke exposures and the associated adverse health effects.

A limited number of published studies are available evaluating exposure reduction interventions in municipal firefighters. A study (Hoppe-Jones et al., 2021) measured urinary PAH metabolites (naphthols, phenanthrols, fluorenols, and 1-hydroxypyrene) in over 140 municipal fire department personnel at a baseline and 2 hours after completing their fireground responses. Contrary to expectations, the exposure was similar in engineers and paramedics at the scene as compared to the fire attack teams, which may be explained by exposure to smoke from the fire and diesel exhaust from fire service vehicles coupled with the nonuse of respiratory protection. Furthermore, the study found that interior responses (as contrasted to exterior responses), smoke odor on the skin after showering, and a lack of recent (greater than 1 month) laundering or changing of hoods were all significantly associated with an increased level of urinary PAH metabolites.

Based on these study findings, the fire department chose to test interventions to reduce fireground exposure, including the use of self-contained breathing apparatuses by engineers, washing procedures for the entry team prior to doffing gear at the fire scene, the isolation of contaminated equipment, and personnel showering and the washing of gear upon return to the station (Burgess et al., 2020). Compared to exposures in the pre-intervention period, the interventions significantly reduced the mean total post-fire urinary PAH metabolite level in engineers (−40.4%, 95% CI = −63.9%, −2.3%) and firefighters (−36.2%, 95% CI = −56.7%, −6.0%).

The potential effectiveness of skin hygiene intervention for the reduction of exposure at wildland fires was also demonstrated by Cherry et al. (2021). They observed that the reduction in urinary 1-hydroxypyrene excretion from immediately post-shift to the following morning among firefighters with confirmed hand PAH contamination was more than six times higher among those who followed the intervention, including a change of clothes immediately after firefighting and returning to base, washing of firefighting clothes, and showering (Cherry et al., 2021). While reported opportunities for engaging in skin hygiene while deployed to the Fort McMurray WUI fire were associated with reduced urinary 1-hydroxypyrene, samples were collected at least 37 days after deployment, which is potentially longer than the half-life of the biomarker (Cherry et al., 2019).

Reducing Regional Exposures

Communities face a range of potential exposures from air, water, and ash or soil deposition. However, the health effects of smoke inhalation, and ways to mitigate risk, are the most widely studied exposures at this time. Smoke from large fires can spread over thousands of square kilometers, potentially affecting the air breathed by millions of people. While people can take numerous actions to reduce exposure to smoke, a large degree of variability exists in the efficacy of each (Laumbach, 2019; Xu et al., 2020). Table 6-4 lists some interventions.

TABLE 6-4 Potential Interventions to Reduce Exposures and Health Effects in Communities

Inhalation	Reduce activity level and time outdoors. Follow any state or municipality guidance levels or standards for acceptable exposure to smoke, such as the threshold of 151 µg/m³ or less of $PM_{2.5}$, as provided in California (by the California Division of Occupational Safety and Health) for outdoor workers.
	Consider wearing a respirator outdoors, or a properly fitted N95 or P100 mask. Standard dust or surgical masks do not keep out fire smoke. Make sure the mask fits as recommended by the manufacturer.
	If you are traveling in a vehicle, close the windows and doors and run the air system in recirculation mode.
	Stay indoors with windows and doors closed, and use an HVAC system in recirculation mode, not bringing in any outdoor air. Use the highest rated minimum efficiency reporting value (MERV) filter in the HVAC system, with a MERV 13 recommended.
	Use indoor air cleaners (if available) that contain a high-efficiency particulate air (HEPA) filter for fine particle removal. If the option is available, add a charcoal filter for chemical collection. Place air cleaners in primary living and sleeping areas and ensure that the air cleaner has Association of Home Appliance Manufacturers approval with an appropriately sized clean air delivery rate provided. Do not use ozone-generating air cleaners, since ozone is a strong pulmonary irritant. Other air cleaners may release unintentional ozone during operation; thus, select only products that meet California's Air Cleaner Regulation AB 2276 of not releasing more than 0.05 ppm ozone (CARB, 2008).
	If retail air cleaners are not available or are economically prohibitive, consider using a DIY air cleaner consisting of a simple box fan with an attached HVAC filter. Follow assembly and safe operating instructions (Underwriters Laboratories, 2021).
	Keep the indoor space clean by wiping all surfaces with an antistatic or damp cloth. Use a HEPA vacuum to clean surfaces and upholstered furniture.
	Consider using $PM_{2.5}$ personal devices or real-time air monitors for measuring indoor levels, so that mitigative strategies including air cleaning/filtration can be used for high PM measurement events.
	Reduce $PM_{2.5}$ generation indoors by limiting the use of chemical cleaners and aerosols, air fresheners, gas cooking, and processes like frying.
	Locate indoor "safe air spaces" provided by the community, especially for vulnerable populations. These may include specially prepared spaces like sports arenas, theaters, and shopping malls.
	When the outdoor air becomes safer, consider "airing out" your space by opening windows and doors and providing as much air circulation as possible. EPA's AirNow Fire and Smoke map (https://fire.airnow.gov/) is a source of real-time air quality and fire information that can be used to identify when outdoor air quality has improved.
Dermal/ Ingestion	Keep the indoor space clean and reduce indoor dust by using air cleaners and wiping all surfaces with a static-free or damp cloth. If available, use a HEPA vacuum to clean all surfaces including flooring, countertops, shelving, and upholstered furniture.
	Wash hands frequently, especially those of children who have frequent hand-to-mouth activity.
	Consider treatment/filtration of drinking water to remove particles and chemical contamination from fire emissions or thermally degraded water lines.

Reducing Indoor Exposure

EPA (2021f) describes means of reducing smoke, particularly $PM_{2.5}$ exposure, and the impacts of interventions; selected text follows:

> The primary focus for several actions is reducing indoor $PM_{2.5}$ concentrations while at home where people spend most of their time. Housing characteristics, such as age of the home and presence and type of HVAC system, influence the infiltration of particles indoors under normal conditions, and also influence the efficacy and availability of these actions for reducing smoke exposures in homes (Davison et al., 2021; EPA, 2020; Joseph et al., 2020). Therefore, housing characteristics of the geographic area impacted by smoke is another important factor, and if variability in the housing stock is not accounted for in some way then estimates of exposure reduction could be under- or overestimated. . . .
>
> EPA (2018) reviewed residential measurement studies that used portable air cleaners and central HVAC system filters to reduce indoor $PM_{2.5}$ exposures overall, not $PM_{2.5}$ specific to wildfire smoke. Portable air cleaners were found to substantially reduce indoor concentrations of PM of both indoor and outdoor origin, often reducing indoor $PM_{2.5}$ concentrations by around 50% on average. . . .
>
> EPA (2018) also noted a few residential measurement studies that showed higher efficiency central HVAC system filters such as those rated minimum efficiency reporting value (MERV) 13 or above can reduce indoor $PM_{2.5}$ concentrations. Singer et al. (2017) reported a 90% reduction in $PM_{2.5}$ using HVAC filtration with high efficiency MERV filters in a single test house in California during typical ambient $PM_{2.5}$ concentrations, which was comparable to running a portable air cleaner in the home. However, results from a recent study by Alavy and Siegel (2020) showed actual in-home effectiveness of HVAC filtration for $PM_{2.5}$ was much lower (average ~40%) and varied widely across homes even for filters with the same MERV rating depending on the home. Filter performance was strongly linked to home- and system-specific parameters including ventilation rate and system run time.
>
> Of the studies evaluated, Reisen et al. (2019) is the only available residential measurement study that examined the effectiveness of closing windows and doors during a smoke event. However, the study only included four homes in Australia that experienced smoke due to a prescribed fire. Simple infiltration modeling of the measurements showed that remaining indoors with windows and doors closed reduced exposure to peak $PM_{2.5}$ concentrations by 29 to 76% across the homes and that a tighter house, in terms of reduced ventilation, provided greater protection against particle infiltration. (EPA, 2021f)

Most of the previous research has focused on achieving PM reductions indoors; however, WUI fires may emit a wide variety of chemical compounds, and very little is known about the infiltration of these compounds indoors and effective air cleaning approaches. Recent work by Ye et al. (2021) focused on evaluating VOC air cleaning technologies for a range of compounds. They observed a wide variety of reduction rates across appliances and for different compounds. They also observed that some air cleaners themselves were a source of VOC emissions and in some cases their usage led to the formation of oxidized by-products. Further research is needed to identify air cleaning technology that can be used to mitigate exposure to VOCs and other WUI fire smoke compounds.

Research need: A need exists to identify the air cleaning and exposure reduction approaches that are effective at reducing the concentrations of WUI fire compounds other than particulate matter.

Finally, making spaces with cleaner air available during the day can provide an alternative for people unable to reduce smoke levels in their homes, especially vulnerable populations. Public health official guidance describes opportunities to provide such public "safe air spaces" with cleaner air than outdoors (EPA, 2019). These spaces could include schools, senior centers, libraries, and shopping malls, as long as adequate filtration is used, such as a MERV 13 or higher filter in the HVAC system or an adequately sized portable air cleaner with HEPA filtration. However, information about the efficacy of this type of intervention is not available.

Standards/Guidance for Outdoor Workers and Communities

In addition to the workers who are directly involved with wildland fire management and suppression, workers are engaged in supporting the fire response (e.g., at base camp or evacuation centers) or in cleanup efforts

TABLE 6-5 Occupational Wildland Fire Smoke Standards

Jurisdiction	Threshold Value or PM$_{2.5}$ Concentration	Action
State of California	AQI ≥ 151 (55.5 μg/m^3)	Implement engineering and administrative controls; provide respirators for voluntary use
	AQI ≥ 500 (500.4 μg/m^3)	Require respirator use
State of Oregon	AQI ≥ 101 (35.5 μg/m^3)	Develop training and communication programs; provide respirators for voluntary use
	AQI ≥ 201 (150.5 μg/m^3)	Implement engineering and administrative controls; require respirator use
	AQI ≥ 501 (500.4 μg/m^3)	Require respirator use; implement a respiratory protection program
State of Washington	20.5 μg/m^3	Develop information and hazard communication plan; encourage use of exposure controls
	55.5 μg/m^3	Implement engineering and administrative controls; provide respirators for voluntary use at no cost

SOURCE: Table adapted from EPA, 2022b.

(e.g., demolition crews). Many other workers continue to do their usual non-fire-related outdoor jobs, such as agricultural workers, landscapers, utility workers, and park personnel, during an incident (EPA, 2021e). OSHA is the regulatory entity for employee health and safety, but in about half of US states, an OSHA-approved state program regulates nonfederal workplaces.

Currently, three occupational standards specifically for wildland fire smoke exist in the United States (Table 6-5). On July 18, 2019, the California Safety and Health Standards Board adopted an emergency regulation for a California Division of Occupational Safety and Health standard to protect workers from hazards associated with wildland fire smoke (Navarro, 2020), which was revised and readopted in 2020 (Cal/OSHA, 2020). The State of Washington adopted a temporary rule in July 2021, and Oregon adopted a temporary rule in August 2021 (Oregon DCBS, 2021; State of Washington, 2021). Common features of these three rules include use of threshold AQI or PM$_{2.5}$ values for smoke exposure reduction actions; exemption of some workplaces, such as workplaces involved in emergency response; use of direct-read PM$_{2.5}$ monitors for ambient measurements; and stipulations for instrument accuracy and operation. To date, recommended standards and guidelines address particulate emissions (PM$_{2.5}$) only, and not emissions of chemicals or combustion gases.

Although no specific federal occupational standards for wildland fire smoke exist, federal occupational standards do exist for total dust and specific chemicals like benzene. These standards for toxic and hazardous substances are included in OSHA standard 19CFR1910 and include levels for long- and short-term exposure periods for work/industrial environments. These occupational standards are not applicable in nonindustrial environments or among the general public, for example, in schools, homes, or recreational spaces, where vulnerable populations exist. In these unregulated environments, guidance is often provided by organizations related to healthy buildings and healthy indoor environments. Key sources of such guidance are voluntary ASHRAE Standards (ASHRAE, 2020, n.d.).

Reducing Other Routes of Exposure

While interventions to mitigate dermal exposure to firefighters working at wildland fires have been studied, the committee found no published studies related to mitigation to reduce dermal exposure to contaminants in other outdoor workers, fire recovery (cleanup) workers, and the general population. While there are engineering, administrative, or personal protective equipment controls that are potentially applicable to prevent and mitigate dermal exposure, there is a lack of published empirical evidence in this area. Individual fire incident reports, work-site monitoring data, and information from occupational safety agencies may provide opportunities for future epidemiological research and recommended mitigations to prevent potential exposure and protect worker health.

No published studies are available on specific interventions designed to mitigate ingestion of contaminants related to WUI fires. Ingestion of contaminants emitted in fire smoke is most likely to result from post-fire exposure to contaminated water, either through fire emissions or leached chemicals from thermally degraded components of water systems. Therefore, increased monitoring, reporting, and potential treatment of drinking water for such contaminants has the potential to mitigate this exposure route. Ingestion may also be possible through dust- and ash-borne contaminants at homes or other indoor settings, depending on the scope of the post-fire remediation activities performed. Reducing indoor dust may mitigate potential ingestion of WUI fire–related contaminants.

Health Risk Communication

Risk communication is critical for minimizing harm to residents affected by any disaster. Studies on reducing human smoke exposure for all types of major fires have mostly been conducted through retrospective surveys of communities. Surveys have focused on determining awareness of smoke through health risk communications such as public service announcements and tying those health communication methods to any resulting action by a community or individual (Kolbe and Gilchrist, 2009; Mott et al., 2002; Sugerman et al., 2012). However, risk communication can help only to some extent. For example, while staying indoors is often advised by public health authorities to reduce exposure to regional smoke, staying indoors is not possible for people who have to work or travel outdoors during fire events, and people experiencing homelessness.

Reviews by Xu et al. (2020) and Laumbach (2019) have found that engineering controls such as closing windows and doors or indoor air filtration can be effective (20–80 percent exposure reduction), as can administrative controls such as staying indoors and avoiding outdoor activities (~50 percent exposure reduction). These reviews and studies are limited by not clearly differentiating WUI and non-WUI fires, and by having limited information on the accessibility of exposure reduction actions (e.g., portable air cleaners) for vulnerable populations, such as low-income communities.

Findings on the effectiveness of risk communication concerning fires also vary by study location, population, source of the communication, and levels of language- and culturally appropriate information. For example, if local radio and television broadcast disaster communications only in English, immigrants with limited English proficiency are language isolated. Studies show that Hispanic immigrants, especially those who do not speak or understand English well and who lack legal status, are less likely to be prepared for an emergency event (Burke et al., 2012; Eisenman et al., 2009). Trusted sources of information about health risks may come from community-based organizations and trusted social networks in some areas.

Health care providers also play a critical role in advising patients on actions they can take to reduce exposure related to fires. EPA provides a training endorsed by the Centers for Disease Control and Prevention (https://www.epa.gov/wildfire-smoke-course) for health care providers to better advise their patients on actions they can take before and during a fire to reduce exposure. The training includes information to help providers identify groups at greater risk from particle pollution in fire smoke and information on how to use the AQI to advise patients on how to reduce exposure and protect their health during periods of fire smoke.

CONCLUSION

This chapter on exposures, adverse health effects, and mitigation provides an overview of the human impacts of WUI fires at spatial scales from the immediate up to the continental level. Many chemicals of concern for human health are generated in WUI fires and, through a variety of exposure routes, can be introduced to different populations. Some populations are more vulnerable to the impacts of WUI fire exposure than others, for a variety of reasons discussed here. There is limited information available on health effects that is specific to WUI fire exposure; therefore, the chapter takes an expanded look at what is known about chronic and acute health effects in the general population and for firefighters for the more well-studied cases of wildland fires and structural fires. Finally, the chapter presents some ways to reduce exposures that lead to health effects.

An important common thread runs through the narrative: the need for more research specific to WUI fires. The committee hypothesizes that some conclusions can be drawn from what is known about wildland fires and structural fires.

Still, the complexity and heterogeneity that is characteristic of WUI fires make them difficult yet important to study directly. The committee's examination of the issues in this chapter led them to identify several findings as well as crucial research needs, which are highlighted throughout and reiterated in Chapter 8.

REFERENCES

Abdo, M., I. Ward, K. O'Dell, B. Ford, J. R. Pierce, E. V. Fischer, and J. L. Crooks. 2019. "Impact of Wildfire Smoke on Adverse Pregnancy Outcomes in Colorado, 2007–2015." *International Journal of Environmental Research and Public Health* 19. https://doi.org/10.3390%2Fijerph16193720.

Adachi, K., A. J. Sedlacek III, L. Kleinman, S. R. Springston, J. Wang, D. Chand, J. M. Hubbe, J. E. Shilling, T. B. Onasch, T. Kinase, K. Sakata, Y. Takahashi, and P. R. Buseck. 2019. "Spherical Tarball Particles Form through Rapid Chemical and Physical Changes of Organic Matter in Biomass-Burning Smoke. *Proceedings of the National Academy of Sciences* 116 (39): 19336–19341. https://doi.org/10.1073/pnas.1900129116.

Adetona, O., D. B. Hall, and L. P. Naeher. 2011. "Lung Function Changes in Wildland Firefighters Working at Prescribed Burns." *Inhalation Toxicology* 23 (13). https://doi.org/10.3109/08958378.2011.617790.

Adetona, O., C. D. Simpson, Z. Li, A. Sjodin, A. M. Calafat, and L. P. Naeher. 2017a. "Hydroxylated Polycyclic Aromatic Hydrocarbons as Biomarkers of Exposure to Wood Smoke in Wildland Firefighters." *Journal of Exposure Science and Environmental Epidemiology* 27 (1): 78–83. https://doi.org/10.1038/jes.2015.75.

Adetona, A. M., O. Adetona, R. M. Gogal Jr., D. Diaz-Sanchez, S. L. Rathbun, and L. P. Naeher. 2017b. "Impact of Work Task-Related Acute Occupational Smoke Exposures on Select Proinflammatory Immune Parameters in Wildland Firefighters." *Journal of Occupational and Environmental Medicine* 59 (7): 679–690. https://doi.org/10.1097/JOM.0000000000001053.

Adetona, A. M., W. K. Martin, S. H. Warren, N. M. Hanley, O. Adetona, J. J. Zhang, C. Simpson, M. Paulsen, S. Rathbun, J. S. Wang, D. M. DeMarini, and L. P. Naeher. 2019. "Urinary Mutagenicity and Other Biomarkers of Occupational Smoke Exposure of Wildland Firefighters and Oxidative Stress." *Inhalation Toxicology* 31 (2): 73–87. https://doi.org/10.1080/08958378.2019.1600079.

Aguilera, R., T. Corringham, A. Gershunov, and T. Benmarhnia. 2021. "Wildfire Smoke Impacts Respiratory Health More Than Fine Particles from Other Sources: Observational Evidence from Southern California." *Nature Communications* 12 (1): 1493. https://doi.org/10.1038/s41467-021-21708-0.

Ahlborg, U. G., A. Brouwer, M. A. Fingerhut, J. L. Jacobson, S. W. Jacobson, S. W. Kennedy, A. A. F. Kettrup, J. H. Koeman, H. Poiger, C. Rappe, S. H. Safe, R. F. Seegal, T. Jouko, and M. van den Berg. 1992. "Impact of Polychlorinated Dibenzo-p-dioxins, Dibenzofurans, and Biphenyls on Human and Environmental Health, with Special Emphasis on Application of the Toxic Equivalency Factor Concept." *European Journal of Pharmacology: Environmental Toxicology and Pharmacology* 228 (4): 179–199. https://doi.org/10.1016/0926-6917(92)90029-C.

Alavy, M., and J. A. Siegel. 2019. "In-situ Effectiveness of Residential HVAC Filters." *Indoor Air* 30 (1): 156–166. https://doi.org/10.1111/ina.12617.

Alexakis, D. E. 2020. "Suburban Areas in Flames: Dispersion of Potentially Toxic Elements from Burned Vegetation and Buildings. Estimation of the Associated Ecological and Human Health Risk." *Environmental Research* 183. https://doi.org/10.1016/j.envres.2020.109153.

Alman, B. L., G. Pfister, H. Hao, J. Stowell, X. Hu, Y. Liu, and M. J. Strickland. 2016. "The Association of Wildfire Smoke with Respiratory and Cardiovascular Emergency Department Visits in Colorado in 2012: A Case Crossover Study." *Environmental Health* 15. https://doi.org/10.1186/s12940-016-0146-8.

American Public Health Association. n.d. "Climate Changes Health: Vulnerable Populations." https://www.apha.org/topics-and-issues/climate-change/vulnerable-populations (accessed February 19, 2022).

Amjad, S., D. Chojecki, A. Osornio-Vargas, and M. B. Ospina. 2021. "Wildfire Exposure during Pregnancy and the Risk of Adverse Birth Outcomes: A Systematic Review." *Environment International* 156. https://doi.org/10.1016/j.envint.2021.106644.

Andersen, M. H. G., A. T. Saber, P. B. Pedersen, et al. 2017. "Cardiovascular Health Effects Following Exposure of Human Volunteers during Fire Extinction Exercises." *Environmental Health* 16 (1): 96. https://doi.org/10.1186/s12940-017-0303-8.

Anderson, S., A. J. Plantinga, and M. Wibbenmeyer. 2020. *Inequality in Agency Responsiveness: Evidence from Salient Wildfire Events*. Washington, DC: Resources for the Future.

ASHRAE. n.d. *Planning Framework for Protecting Commercial Building Occupants from Smoke During Wildfire Events*. Atlanta, GA: ASHRAE. https://www.ashrae.org/file%20library/technical%20resources/covid-19/planning-framework-for-protecting-commercial-building-occupants-from-smoke-during-wildfire-events.pdf.

ASHRAE. 2020. *ASHRAE 189.1 Standard for the Design of High-Performance Green Buildings Except Low-Rise Residential Buildings*. Atlanta, GA: ASHRAE. https://www.epa.gov/smartgrowth/ansiashraeusgbcies-standard-1891-2014-standard-design-high-performance-green-buildings.

ATSDR (US Agency for Toxic Substances and Disease Registry). 2000. *Public Health Statement Polychlorinated Biphenyls (PCBS)*: 1–9. https://www.atsdr.cdc.gov/toxprofiles/tp17.pdf.

ATSDR. 2002. *Hydrogen Chloride*. https://www.atsdr.cdc.gov/toxfaqs/tfacts173.pdf.

ATSDR. 2014a. *Methyl Isocyanate*. https://www.atsdr.cdc.gov/toxfaqs/tfacts182.pdf (accessed March 22, 2022).

ATSDR. 2014b. *Polycyclic Aromatic Hydrocarbons (PAHs)*. https://wwwn.cdc.gov/TSP/ToxFAQs/ToxFAQsDetails.aspx?faqid=121&toxid=25 (accessed March 24, 2022).

ATSDR. 2015. *Phosphate Ester Flame Retardants*. https://wwwn.cdc.gov/TSP/ToxFAQs/ToxFAQsDetails.aspx?faqid=1164&toxid=239 (accessed March 22, 2022).

ATSDR. 2016. *Health Effects of Asbestos*. https://www.atsdr.cdc.gov/asbestos/health_effects_asbestos.html (accessed March 22, 2022).

Austin, E., E. Kasner, E. Seto, and J. Spector. 2021. "Combined Burden of Heat and Particulate Matter Air Quality in WA Agriculture." *Journal of Agromedicine* 26 (1): 18–27. https://doi.org/10.1080/1059924x.2020.1795032.

Bailey, Z. D., N. Krieger, M. Agénor, J. Graves, N. Linos, and M. T. Bassett. 2017. "Structural Racism and Health Inequities in the USA: Evidence and Interventions." *Lancet* 389 (10077): 1453–1463. https://doi.org/10.1016/S0140-6736(17)30569-X.

Banks, A. P. W., P. Thai, M. Engelsman, X. Wang, A. F. Osorio, and J. F. Mueller. 2021. "Characterising the Exposure of Australian Firefighters to Polycyclic Aromatic Hydrocarbons Generated in Simulated Compartment Fires." *International Journal of Hygiene and Environmental Health* 231. https://doi.org/10.1016/j.ijheh.2020.113637.

Barboni, T., M. Cannac, V. Pasqualini, A. Simeoni, E. Leoni, and N. Chiaramonti. 2010. "Volatile and Semi-volatile Organic Compounds in Smoke Exposure of Firefighters during Prescribed Burning in the Mediterranean Region." *International Journal of Wildland Fire* 19 (5). https://doi.org/10.1071/wf08121.

Beitel, S. C., L. M. Flahr, C. Hoppe-Jones, J. L. Burgess, S. R. Littau, J. Gulotta, P. Moore, D. Wallentine, and S. A. Snyder. 2020. "Assessment of the Toxicity of Firefighter Exposures Using the PAH CALUX Bioassay." *Environment International* 135: 105207. https://doi.org/10.1016/j.envint.2019.105207.

Black, C., J. E. Gerriets, J. H. Fontaine, R. W. Harper, N. J. Kenyon, F. Tablin, ... and L. A. Miller. 2017. "Early Life Wildfire Smoke Exposure Is Associated with Immune Dysregulation and Lung Function Decrements in Adolescence." *American Journal of Respiratory Cell and Molecular Biology* 56 (5): 657–666.

Bodi, M. B., D. A. Martin, V. N. Balfour, C. Santin, S. H. Doerr, P. Pereira, A. Cerda, and J. Mataix-Solera. 2014. "Wildland Fire Ash: Production, Composition and Eco-hydro-geomorphic Effects." *Earth-Science Reviews* 130: 103–127. https://doi.org/10.1016/j.earscirev.2013.12.007.

Boffetta, P., N. Jourenkova, and P. Gustavsson. 1997. "Cancer Risk from Occupational and Environmental Exposure to Polycyclic Aromatic Hydrocarbons." *Cancer Causes and Control* 8: 444–472. https://doi.org/10.1023/a:1018465507029.

Brim, H., P. Mokarram, F. Naghibalhossaini, and M. Saberi-Firoozi. 2008. "Impact of BRAF, MLH1 on the Incidence of Microsatellite Instability High Colorectal Cancer in Populations Based Study." *Molecular Cancer* 7 (68). https://doi.org/10.1186/1476-4598-7-68.

Brown, M. R. G., V. I. O. Agyapong, A. J. Greenshaw, I. Cribben, P. Brett-MacLean, J. Drolet, C. B. McDonald-Harker, J. Omeje, M. Mankowski, S. Noble, D. T. Kitching, and P. H. Silverstone. 2019. "Significant PTSD and Other Mental Health Effects Present 18 Months After the Fort Mcmurray Wildfire: Findings from 3,070 Grades 7–12 Students." *Frontiers in Psychiatry* 10. https://doi.org/10.3389/fpsyt.2019.00623.

Brown, M. R. G., H. Pazderka, V. I. O. Agyapong, A. J. Greenshaw, I. Cribben, P. Brett-MacLean, J. Drolet, C. B. McDonald-Harker, J. Omeje, B. Lee, M. Mankowski, S. Noble, D. T. Kitching, and P. H. Silverstone. 2021. "Mental Health Symptoms Unexpectedly Increased in Students Aged 11–19 Years During the 3.5 Years After the 2016 Fort McMurray Wildfire: Findings from 9,376 Survey Responses." *Frontiers in Psychiatry* 12. https://doi.org/10.3389%2Ffpsyt.2021.676256.

Broyles, G. A. 2013. *Wildland Firefighter Smoke Expsoure Study*. Utah State, Logan, Utah. Masters of Natural Resources.

Burgess, J. L., C. Nanson, D. Bolstad-Johnson, R. Gerkin, T. Hysong, R. C. Lantz, D. Sherrill, C. Crutchfield, S. Quan, A. Bernard, and M. Witten. 2001. "Adverse Respiratory Effects Following Overhaul in Firefighters." *Journal of Occupational and Environmental Medicine* 43 (5): 467–473. https://doi.org/10.1097/00043764-200105000-00007.

Burgess, J. L., C. Hoppe-Jones, S. C. Griffin, J. J. Zhou, J. J. Gulotta, D. D. Wallentine, P. K. Moore, E. A. Valliere, S. R. Weller, S. C. Beitel, L. M. Flahr, S. R. Littau, D. Dearmon-Moore, J. Zhai, A. M. Jung, F. Garavito, and S. A. Snyder. 2020. "Evaluation of Interventions to Reduce Firefighter Exposures." *Journal of Occupational and Environmental Medicine* 62 (4): 279–288. https://doi.org/10.1097/JOM.0000000000001815.

Burke, M., A. Driscoll, S. Heft-Neal, J. Xue, J. Burney, and M. Wara. 2021. "The Changing Risk and Burden of Wildfire in the United States." *Proceedings of the National Academy of Sciences* 118 (2).

Burke, M. P., T. S. Hogue, A. M. Kinoshita, J. Barco, C. Wessel, and E. D. Stein. 2013. "Pre- and Post-Fire Pollutant Loads in an Urban Fringe Watershed in Southern California." *Environmental Monitoring and Assessment* 185 (12): 10131–10145. https://doi.org/10.1007/s10661-013-3318-9.

Burke, S., J. W. Bethel, and A. F. Britt. 2012. "Assessing Disaster Preparedness among Latino Migrant and Seasonal Farmworkers in Eastern North Carolina." *International Journal of Environmental Research and Public Health* 9 (9): 3115–3133. https://doi.org/10.3390/ijerph9093115.

Burton, C. A., T. M. Hoefen, G. S. Plumlee, K. L. Baumberger, A. R. Backlin, E. Gallegos, and R. N. Fisher. 2016. "Trace Elements in Stormflow, Ash, and Burned Soil Following the 2009 Station Fire in Southern California." *PLoS One* 11 (5): e0153372. https://doi.org/10.1371/journal.pone.0153372.

CARB (California Air Resources Board). 2008. *Evaluation of Ozone Emissions from Portable Indoor Air Cleaners: Electrostatic Precipitators and Ionizers.* Sacramento, CA: California Air Resources Board. https://www.arb.ca.gov/research/indoor/esp_report.pdf?_ga=2.88136991.567876424.1668176695-1290726036.1668176695.

CARB. 2021. *Camp Fire Air Quality Data Analysis.* Sacramento, CA: California Air Resources Board. https://ww2.arb.ca.gov/resources/documents/camp-fire-air-quality-data-analysis.

Cal/OSHA (California Division of Occupational Safety and Health). 2020. "Protection from Wildfire Smoke." https://www.dir.ca.gov/Title8/5141_1.html.

Calin, G. A., and C. M. Croce. 2006. "MicroRNA-Cancer Connection: The Beginning of a New Tale." *Cancer Research* 66 (15): 7390–7394. https://doi.org/10.1158/0008-5472.CAN-06-0800.

Carratt, S. A., C. Flayer, M. Kossack, and J. A. Last. 2017. "Pesticides, wildfire suppression chemicals, and California wildfires: A human health perspective." *Current Topics in Toxicology* 13.

Carter, J. V., N. J. Galbraith, D. Yang, J. F. Burton, S. P. Walker, and S. Galandiuk. 2017. "Blood-Based MicroRNAs as Biomarkers for the Diagnosis of Colorectal Cancer: A Systematic Review and Meta-Analysis." *British Journal of Cancer* 116 (6): 762–774. https://doi.org/10.1038/bjc.2017.12.

CCST (California Council on Science and Technology). 2020. *The Costs of Wildfire in California.*

CDC (Centers for Disease Control and Prevention). 2017. *Dioxins, Furans and Dioxin-Like Polychlorinated Biphenyls Factsheet.* https://www.cdc.gov/biomonitoring/DioxinLikeChemicals_FactSheet.html (accessed March 22, 2022).

CDC. 2018. *Facts About Phosgene.* https://emergency.cdc.gov/agent/phosgene/basics/facts.asp (accessed March 22, 2022).

CDC. 2019. *Particle Pollution.* https://www.cdc.gov/air/particulate_matter.html (accessed March 22, 2022).

CDC. 2021. *Phthalates Factsheet.* https://www.cdc.gov/biomonitoring/Phthalates_FactSheet.html (accessed March 23, 2022).

CDC. 2022. *Polycyclic Aromatic Hydrocarbons (PAHs) Factsheet.* https://www.cdc.gov/biomonitoring/PAHs_FactSheet.html (accessed March 24, 2022).

Cerrato, J. M., J. M. Blake, C. Hirani, A. L. Clark, A. M. Ali, K. Artyushkova, E. Peterson, and R. J. Bixby. 2016. "Wildfires and Water Chemistry: Effect of Metals Associated with Wood Ash." *Environmental Science: Processes & Impacts* 18 (8): 1078–1089. https://doi.org/10.1039/c6em00123h.

Chan, W. R., J. Joh, and M. H. Sherman. 2013. "Analysis of Air Leakage Measurements of US Houses." *Energy and Buildings* 66: 616–625. https://doi.org/10.1016/j.enbuild.2013.07.047.

Chen, H., H. Uzun, A. T. Chow, and T. Karanfil. 2020. "Low Water Treatability Efficiency of Wildfire-Induced Dissolved Organic Matter and Disinfection By-product Precursors." *Water Research* 184. https://doi.org/10.1016/j.watres.2020.116111.

Chen, H., J. M. Samet, P. A. Bromberg, et al. 2021. "Cardiovascular Health Impacts of Wildfire Smoke Exposure." *Particle and Fibre Toxicology* 18: 2. https://doi.org/10.1186/s12989-020-00394-8.

Cherry, N., Y. A. Aklilu, J. Beach, P. Britz-McKibbin, R. Elbourne, J. M. Galarneau, B. Gill, D. Kinniburgh, and X. Zhang. 2019. "Urinary 1-Hydroxypyrene and Skin Contamination in Firefighters Deployed to the Fort McMurray Fire." *Annals of Work Exposures and Health* 63 (4): 448–458. https://doi.org/10.1093/annweh/wxz006.

Cherry, N., J. M. Galarneau, D. Kinniburgh, B. Quemerais, S. Tiu, and X. Zhang. 2021. "Exposure and Absorption of PAHs in Wildland Firefighters: A Field Study with Pilot Interventions." *Annals of Work Exposures and Health* 65 (2): 148–161. https://doi.org/10.1093/annweh/wxaa064.

Chong, N. S., S. Abdulramoni, D. Patterson, and H. Brown. 2019. "Releases of Fire-Derived Contaminants from Polymer Pipes Made of Polyvinyl Chloride." *Toxics* 7 (4). https://doi.org/10.3390/toxics7040057.

Coussens, L., and Z. Werb. 2002. "Inflammation and Cancer." *Nature* 420. https://doi.org/10.1038/nature01322.

Cova, T. J., D. M. Theobald, J. B. Norman, and L. K. Siebeneck. 2011. "Mapping Wildfire Evacuation Vulnerability in the Western US: The Limits of Infrastructure. *GeoJournal* 78 (2): 273–285. https://doi.org/10.1007/s10708-011-9419-5.

Daniels, R. D., T. L. Kubale, J. H. Yiin, M. M. Dahm, T. R. Hales, D. Baris, S. H. Zahm, J. J. Beaumont, K. M. Waters, and L. E. Pinkerton. 2014. "Mortality and Cancer Incidence in a Pooled Cohort of US Firefighters from San Francisco, Chicago and Philadelphia (1950–2009)." *Occupational and Environmental Medicine* 71 (6): 388–397. https://doi.org/10.1136/oemed-2013-101662.

Davies, I. P., R. D. Haugo, J. C. Robertson, and P. S. Levin. 2018. "The Unequal Vulnerability of Communities of Color to Wildfire." *PLoS One* 13 (11): e0205825. https://doi.org/10.1371/journal.pone.0205825.

Davison, G., K. K. Barkjohn, G. S. W. Hagler, A. L. Holder, S. Coefield, C. Noonan, and B. Hassett-Sipple. 2021. "Creating Clean Air Spaces During Wildland Fire Smoke Episodes: Web Summit Summary." *Frontiers in Public Health*. https://doi.org/10.3389/fpubh.2021.508971.

De Vos, A. J., A. Cook, B. Devine, P. J. Thompson, and P. Weinstein. 2009. "Effect of Protective Filters on Fire Fighter Respiratory Health: Field Validation during Prescribed Burns." *American Journal of Industrial Medicine* 52 (1): 76–87. https://doi.org/10.1002/ajim.20651.

DeFlorio-Barker, S., J. Crooks, J. Reyes, and A. G. Rappold. 2019. "Cardiopulmonary Effects of Fine Particulate Matter Exposure among Older Adults, during Wildfire and Non-Wildfire Periods, in the United States 2008–2010." *Environmental Health Perspectives* 127 (3): 37006. https://doi.org/10.1289/EHP3860.

Delfino, R. J., S. Brummel, J. Wu, H. Stern, B. Ostro, M. Lipsett, A. Winer, D. H. Street, L. Zhang, T. Tjoa, and D. L. Gillen. 2008. "The Relationship of Respiratory and Cardiovascular Hospital Admissions to the Southern California Wildfires of 2003." *Occupational and Environmental Medicine* 66 (3): 189–197. https://doi.org/10.1186/s12940-016-0146-8.

Demers, P. A., D. M. DeMarini, K. W. Fent, D. C. Glass, J. Hansen, O. Adetona, M. H.G. Andersen, L. E. B. Freeman, A. J. Caban-Martinez, R. D. Daniels, T. R. Driscoll, J. M. Goodrich, J. M. Graber, T. L. Kirkham, K. Kjaerheim, D. Kriebel, A. S. Long, L. C. Main, M. Oliveira, S. Peters, L. R. Teras, E. R. Watkins, J. L. Burgess, A. A. Stec, P. A. White, N. L. DeBono, L. Benbrahim-Tallaa, A. de Conti, F. El Ghissassi, Y. Grosse, L. T. Stayer, E. Suonio, S. Viegas, R. Wedekind, P. Boucheron, B. Hosseini, J. Kim, H. Zahed, H. Mattock, F. Madia, and M. K. Schubauer-Berigan. 2022. "Carcinogenicity of occupational exposure as a firefighter." *The Lancet Oncology* 23 (8): 985-986. https://doi.org/10.1016/S1470-2045(22)00390-4.

Diggs, D. L., A. C. Huderson, K. L. Harris, J. N. Myers, L. D. Banks, P. V. Rekhadevi, M. S. Niaz, and A. Ramesh. 2011. "Polycyclic Aromatic Hydrocarbons and Digestive Tract Cancers: A Perspective." *Journal of Environmental Science and Health, Part C: Toxicology and Carcinogenesis* 29 (4): 324–357. https://doi.org/10.1080/10590501.2011.629974.

Dills, R., M. Paulsen, J. Ahmad, D. Kalman, F. Elias, and C. Simpson. 2006. "Evaluation of Urinary Methoxyphenols as Biomarkers of Woodsmoke Exposure." *Environmental Science & Technology* 40. https://doi.org/10.1021/es051886f.

Dreessen, J., J. Sullivan, and R. Delgado. 2016. "Observations and Impacts of Transported Canadian Wildfire Smoke on Ozone and Aerosol Air Quality in the Maryland Region on June 9–12, 2015." *Journal of the Air & Waste Management Association* 66 (9): 842–862. https://doi.org/10.1080/10962247.2016.1161674.

Dugue, P. A., J. K. Bassett, E. M. Wong, J. E. Joo, S. Li, C. Yu, D. F. Schmidt, E. Makalic, N. W. Doo, D. D. Buchanan, A. M. Hodge, D. R. English, J. L. Hopper, G. G. Giles, M. C. Southey, and R. L. Milne. 2021. "Biological Aging Measures Based on Blood DNA Methylation and Risk of Cancer: A Prospective Study." *JNCI Cancer Spectrum* 5 (1): pkaa109. https://doi.org/10.1093/jncics/pkaa109.

DuTeaux, S. 2019. *Final Report of Air Quality in the Vicinity of Camp Fire Debris Removal*. Edited by California Environmental Protection Agency, CalRecycle, and California Governor's Office of Emergency Services. Sacramento, CA: California Environmental Protection Agency.

Eberhart, C. 2010. *Measurements of Air Contaminants during the Cerro Grande Fire at Los Alamos National Laboratory*. Los Alamos, NM: Los Alamos National Laboratory.

Eckstein, S., and G. Peri. 2018. "Immigrant Niches and Immigrant Networks in the U.S. Labor Market." *RSF: The Russell Sage Foundation Journal of the Social Sciences* 4 (1): 1–17. https://doi.org/10.7758/rsf.2018.4.1.01.

Edgeley, C., and T. B. Paveglio. 2019. "Exploring Influences on Intended Evacuation Behaviors during Wildfire: What Roles for Pre-fire Actions and Event-Based Cues?" *International Journal of Disaster Risk Reduction* 37. https://doi.org/10.1016/j.ijdrr.2019.101182.

Eisenman, D. P., D. Glik, R. Maranon, L. Gonzales, and S. Asch. 2009. "Developing a Disaster Preparedness Campaign Targeting Low-Income Latino Immigrants: Focus Group Results for Project PREP." *Journal of Health Care for the Poor and Underserved* 20 (2): 330–345. https://doi.org/10.1353/hpu.0.0129.

Emmerton, C. A., C. A. Cooke, S. Hustins, U. Silins, M. B. Emelko, T. Lewis, M. K. Kruk, N. Taube, D. Zhu, B. Jackson, M. Stone, J. G. Kerr, and J. F. Orwin. 2020. "Severe Western Canadian Wildfire Affects Water Quality Even at Large Basin Scales." *Water Research* 183. https://doi.org/10.1016/j.watres.2020.116071.

EPA (US Environmental Protection Agency). 2009. *Integrated Science Assessment (ISA) for Particulate Matter (Final Report, Dec 2009)*. EPA/600/R-08/139F. Washington, DC: EPA.

EPA. 2018. *Residential air cleaners: A technical summary*. EPA 402-F-09-002. Washington, DC: EPA.

EPA. 2019. *Wildfire Smoke: A Guide for Public Health Officials*. EPA-452/R-19-901. Washington, DC: EPA.

EPA. 2020. *Wildfires and indoor air quality (IAQ)*. https://www.epa.gov/indoor-air-quality-iaq/wildfires-and-indoor-air-quality-iaq (accessed February 9, 2021).

EPA. 2021a. *Learn About Asbestos*. https://www.epa.gov/asbestos/learn-about-asbestos#effects (accessed March 22, 2022).

EPA. 2021b. *Learn About Dioxin*. https://www.epa.gov/dioxin/learn-about-dioxin (accessed March 22, 2022).

EPA. 2021c. *Volatile Organic Compounds' Impact on Indoor Air Quality*. https://www.epa.gov/indoor-air-quality-iaq/volatile-organic-compounds-impact-indoor-air-quality (accessed March 22, 2022).

EPA. 2021d. *Health and Environmental Effects of Particulate Matter (PM)*. https://www.epa.gov/pm-pollution/health-and-environmental-effects-particulate-matter-pm (accessed March 22, 2022).

EPA. 2021e. *Why Wildfire Smoke Is a Health Concern*. https://www.epa.gov/wildfire-smoke-course/why-wildfire-smoke-health-concern (accessed February 19, 2022).

EPA. 2021f. *Comparative Assessment of the Impacts of Prescribed Fire versus Wildfire (CAIF): A Case Study in the Western U.S.* https://cfpub.epa.gov/ncea/risk/recordisplay.cfm?deid=352824.

EPA. 2021g. "Which Populations Experience Greater Risks of Adverse Health Effects Resulting from Wildfire Smoke Exposure?" https://www.epa.gov/wildfire-smoke-course/which-populations-experience-greater-risks-adverse-health-effects-resulting (accessed February 19, 2022).

EPA. 2022a. *Risk Management for Phthalates*. https://www.epa.gov/assessing-and-managing-chemicals-under-tsca/risk-management-phthalates (accessed March 23, 2022).

EPA. 2022b. *How to Evaluate Air Sensors for Smoke Monitoring Webinar Archive*. https://www.epa.gov/research-states/how-evaluate-air-sensors-smoke-monitoring-webinar-archive (accessed March 25, 2022).

Fadadu, R. P., B. Grimes, N. P. Jewell, J. Vargo, A. T. Young, K. Abuabara, J. R. Balmes, and M. L. Wei. 2021. "Association of Wildfire Air Pollution and Health Care Use for Atopic Dermatitis and Itch." *JAMA Dermatology* 157 (6): 658–666. https://doi.org/10.1001/jamadermatol.2021.0179.

Fent, K. W., J. Eisenberg, J. Snawder, D. Sammons, J. D. Pleil, M. A. Stiegel, C. Mueller, G. P. Horn, and J. Dalton. 2014. "Systemic Exposure to PAHs and Benzene in Firefighters Suppressing Controlled Structure Fires." *Annals of Occupational Hygiene* 58 (7): 830–845. https://doi.org/10.1093/annhyg/meu036.

Fent, K. W., B. Alexander, J. Roberts, S. Robertson, C. Toennis, D. Sammons, S. Bertke, S. Kerber, D. Smith, and G. Horn. 2017. "Contamination of Firefighter Personal Protective Equipment and Skin and the Effectiveness of Decontamination Procedures." *Journal of Occupational and Environmental Hygiene* 14 (10): 801–814. https://doi.org/10.1080/15459624.2017.1334904.

Fent, K. W., D. E. Evans, K. Babik, C. Striley, S. Bertke, S. Kerber, D. Smith, and G. P. Horn. 2018. "Airborne Contaminants during Controlled Residential Fires." *Journal of Occupational and Environmental Hygiene* 15 (5): 399–412. https://doi.org/10.1080/15459624.2018.1445260.

Fent, K. W., C. Toennis, D. Sammons, S. Robertson, S. Bertke, A. M. Calafat, J. D. Pleil, M. A. Geer Wallace, S. Kerber, D. L. Smith, and G. P. Horn. 2019a. "Firefighters' and Instructors' Absorption of PAHs and Benzene during Training Exercises." *International Journal of Hygiene and Environmental Health* 222 (7): 991–1000. https://doi.org/10.1016/j.ijheh.2019.06.006.

Fent, K. W., C. Toennis, D. Sammons, S. Robertson, S. Bertke, A. M. Calafat, J. D. Pleil, M. A. G. Wallace, S. Kerber, D. Smith, and G. P. Horn. 2019b. "Firefighters' Absorption of PAHs and VOCs during Controlled Residential Fires by Job Assignment and Fire Attack Tactic." *Journal of Exposure Science and Environmental Epidemiology* 30 (2): 338–349. https://doi.org/10.1038/s41370-019-0145-2.

Fent, K. W., M. LaGuardia, D. Luellen, S. McCormick, A. Mayer, I. Chen, S. Kerber, D. Smith, and G. P. Horn. 2020. "Flame Retardants, Dioxins, and Furans in Air and on Firefighters' Protective Ensembles during Controlled Residential Firefighting." *Environment International* 140. https://doi.org/10.1016/j.envint.2020.105756.

Filonchyk, M., M. P. Peterson, and Dongqi Sun. 2022. "Deterioration of Air Quality Associated with the 2020 US Wildfires." *Science of the Total Environment* 826: 154103. https://doi.org/10.1016/j.scitotenv.2022.154103.

Fussell, E., L. Delp, K. Riley, S. Chavez, and A. Valenzuela. 2018. "Implications of Social and Legal Status on Immigrants' Health in Disaster Zones." *American Journal of Public Health* 108 (12). https://doi.org/10.2105/AJPH.2018.304554.

Gallaher, B. M., and R. J. Koch. 2004. *Cerro Grande Fire Impacts to Water Quality and Stream Flow near Los Alamos National Laboratory: Results of Four Years of Monitoring*. Los Alamos, NM: Los Alamos National Laboratory.

Gan, R. W., B. Ford, W. Lassman, G. Pfister, A. Vaidyanathan, E. Fischer, J. Volckens, J. R. Pierce, and S. Magzamen. 2017. "Comparison of Wildfire Smoke Estimation Methods and Associations with Cardiopulmonary-Related Hospital Admissions." *GeoHealth* 1 (3): 122–136. https://doi.org/10.1002/2017GH000073.

Gan, R. W., J. Liu, B. Ford, K. O'Dell, A. Vaidyanathan, A. Wilson, J. Volckens, G. Pfister, E. V. Fischer, J. R. Pierce, and S. Magzamen. 2020. "The association between wildfire smoke exposure and asthma-specific medical care utilization in Oregon during the 2013 wildfire season." *Journal of Exposure Science & Environmental Epidemiology* 30: 618–628. https://doi.org/10.1038/s41370-020-0210-x.

Gaughan, D. M., J. M. Cox-Ganser, P. L. Enright, R. M. Castellan, G. R. Wagner, G. R. Hobbs, T. A. Bledsoe, P. D. Siegel, K. Kreiss, and D. N. Weissman. 2008. "Acute Upper and Lower Respiratory Effects in Wildland Firefighters." *Journal of Occupational and Environmental Medicine* 50 (9): 1019–1028. https://doi.org/10.1097/JOM.0b013e3181754161.

Glass, D. C., S. Pircher, A. Del Monaco, S. Vander Hoorn, and M. R. Sim. 2016. "Mortality and Cancer Incidence in a Cohort of Male Paid Australian Firefighters." *Occupational and Environmental Medicine* 73: 761–771. http://doi.org/10.1136/oemed-2015-103467.

Gold, A., and T. B. Perera. 2021. *EMS Asphyxiation and Other Gas and Fire Hazards*. Treasure Island, FL: StatPearls Publishing.

Goyer, R. 2004. *Issue Paper on the Human Health Effects of Metals*. Washington, DC: EPA.

Haikerwal, A., M. Akram, A. Del Monaco, K. Smith, M. R. Sim, M. Meyer, A. M. Tonkin, M. J. Abramson, and M. Dennekamp. 2015. "Impact of Fine Particulate Matter (PM2.5) Exposure during Wildfires on Cardiovascular Health Outcomes." *Journal of the American Heart Association* 4 (7). https://doi.org/10.1161/JAHA.114.001653.

Haikerwal, A., M. Akram, M. R. Sim, M. Meyer, M. J. Abramson, and M. Dennekamp. 2016. "Fine Particulate Matter (PM$_{2.5}$) Exposure during a Prolonged Wildfire Period and Emergency Department Visits for Asthma." *Respirology* 21 (1): 88–94. https://doi.org/10.1111/resp.12613.

Hardeman, R. R., P. A. Homan, T. Chantarat, B. A. Davis, and T. H. Brown. 2022. "Improving the Measurement of Structural Racism to Achieve Antiracist Health Policy." *Health Affairs* 21 (2). https://doi.org/10.1377/hlthaff.2021.01489.

HHS (US Department of Health and Human Services). 2018. *Low Income Home Energy Data for Fiscal Year 2017.* https://liheappm.acf.hhs.gov/home-energy-notebooks/ (accessed February 26, 2022).

Hodshire, A. L., A. Akherati, M. J. Alvarado, B. Brown-Steiner, S. H. Jathar, J. L. Jimenez, S. M. Kreidenweis, C. Lonsdale, T. B. Onasch, A. M. Ortega, and J. R. Pierce. 2019. "Aging Effects on Biomass Burning Aerosol Mass and Composition: A Critical Review of Field and Laboratory Studies." *Environmental Science & Technology* 53 (17): 10007–10022. https://doi.org/10.1021/acs.est.9b02588.

Hohner, A., K. M. Cawley, J. Oropeza, R. S. Summers, and F. L. Rosario-Ortiz. 2016. "Drinking Water Treatment Response Following a Colorado Wildfire." *Water Research* 105: 187–198. https://doi.org/10.1016/j.watres.2016.08.034.

Holm, S. M., M. D. Miller, and J. R. Balmes. 2021. "Health Effects of Wildfire Smoke in Children and Public Health Tools: A Narrative Review." *Journal of Exposure Science and Environmental Epidemiology* 31 (1): 1–20. https://doi.org/10.1038/s41370-020-00267-4.

Homan, P., T. H. Brown, and B. King. 2021. "Structural Intersectionality as a New Direction for Health Disparities Research." *Journal of Health and Social Behavior* 62 (3): 350–370. https://doi.org/10.1177/00221465211032947.

Hoppe-Jones, C., S. C. Griffin, J. J. Gulotta, D. D. Wallentine, P. K. Moore, S. C. Beitel, L. M. Flahr, J. Zhai, J. J. Zhou, S. R. Littau, D. Dearmon-Moore, A. M. Jung, F. Garavito, S. A. Snyder, and J. L. Burgess. 2021. "Evaluation of Fire-ground Exposures Using Urinary PAH Metabolites." *Journal of Exposure Science and Environmental Epidemiology* 31 (5): 913–922. https://doi.org/10.1038/s41370-021-00311-x.

Hutchinson, J. A., J. Vargo, M. Milet, N. H. F. French, M. Billmire, J. Johnson, and S. Hoshiko. 2018. "The San Diego 2007 Wildfires and Medi-Cal Emergency Department Presentations, Inpatient Hospitalizations, and Outpatient Visits: An Observational Study of Smoke Exposure Periods and a Bidirectional Case-Crossover Analysis." *PLoS Medicine* https://doi.org/10.1371/journal.pmed.1002601.

IARC (International Agency for Research on Cancer). 2010. "Some Non-heterocyclic Polycyclic Aromatic Hydrocarbons and Some Related Exposures." *IARC Monographs on the Identification of Carcinogenic Hazards to Humans* 92.

IDEPSCA (Instituto de Educacion Popular del Sur de California). 2020. "On the Frontlines: The Role of Domestic Workers and Day Laborers in Confronting Recent Wildfires in Southern California." https://www.cadomesticworkers.org/news/on-the-frontlines-the-role-of-domestic-workers-and-day-laborers-in-confronting-recent-wildfires-in-california/ (accessed February 23, 2022).

Igarashi, Y., Y. Onda, Y. Wakiyama, A. Konoplev, M. Zheleznyak, H. Lisovyi, G. Laptev, V. Damiyanovich, D. Samoilov, K. Nanba, and S. Kirieiev. 2020. "Impact of Wildfire on ^{137}Cs and ^{90}Sr Wash-Off in Heavily Contaminated Forests in the Chernobyl Exclusion Zone." *Environmental Pollution* 259: 113764. https://doi.org/10.1016/j.envpol.2019.113764.

Isaacson, K. P., C. R. Proctor, Q. E. Wang, E. Y. Edwards, Y. Noh, A. D. Shah, and A. J. Whelton. 2020. "Drinking Water Contamination from the Thermal Degradation of Plastics: Implications for Wildfire and Structure Fire Response." *Environmental Science: Water Research & Technology* 7 (2): 274–284. https://doi.org/10.1039/d0ew00836b.

Jaenisch, R., and A. Bird. 2003. "Epigenetic Regulation of Gene Expression: How the Genome Integrates Intrinsic and Environmental Signals." *Nature Genetics* 33: 245–254. https://doi.org/10.1038/ng1089.

Jaffe, D. A., S. M. O'Neill, N. K. Larkin, A. L. Holder, D. L. Peterson, J. E. Halofsky, and A. G. Rappold. 2020. "Wildfire and Prescribed Burning Impacts on Air Quality in the United States." *Journal of the Air & Waste Management Association* 70 (6): 583–615. https://doi.org/10.1080/10962247.2020.1749731.

Jeong, K. S., J. Zhou, S. C. Griffin, E. T. Jacobs, D. Dearmon-Moore, J. Zhai, S. R. Littau, J. Gulotta, P. Moore, W. F. Peate, C. M. Richt, and J. L. Burgess. 2018. "MicroRNA Changes in Firefighters." *Journal of Occupational and Environmental Medicine* 60 (5): 469–474. https://doi.org/10.1097/JOM.0000000000001307.

Johnston, F. H., S. B. Henderson, Y. Chen, J. T. Randerson, M. Marlier, R. S. DeFries, P. Kinney, D. M. J. S. Bowman, and M. Brauer. 2012. "Estimated Global Mortality Attributable to Smoke from Landscape Fires." *Environmental Health Perspectives* 120 (5). https://doi.org/10.1289/ehp.1104422.

Jones, L., E. A. Lutz, M. Duncan, and J. L. Burgess. 2015. "Respiratory Protection for Firefighters— Evaluation of CBRN Canisters for Use During Overhaul." *Journal of Occupational and Environmental Hygiene* 12 (5). https://doi.org/10.1080/15459624.2014.989363.

Jones, P. A., and P. W. Laird. 1999. "Cancer Epigentics Comes of Age." *Nature Genetics* 21 (2). https://doi.org/10.1038/5947.

Joseph, G., P. J. Schramm, A. Vaidyanathan, P. Breysse, and B. Goodwin. 2020. *Evidence on the Use of Indoor Air Filtration as an Intervention for Wildfire Smoke Pollutant Exposure: A Summary for Health Departments*. BRACE technical report series. Atlanta, GA: Centers for Disease Control and Prevention. https://stacks.cdc.gov/view/cdc/108289.

Jung, A. M., J. Zhou, S. C. Beitel, S. R. Littau, J. J. Gulotta, D. D. Wallentine, P. K. Moore, and J. L. Burgess. 2021. "Longitudinal Evaluation of Whole Blood miRNA Expression in Firefighters." *Journal of Exposure Science and Environmental Epidemiology* 31 (5): 900–912. https://doi.org/10.1038/s41370-021-00306-8.

Keir, J. L. A., U. S. Akhtar, D. M. J. Matschke, T. L. Kirkham, H. M. Chan, P. Ayotte, P. A. White, and J. M. Blais. 2017. "Elevated Exposures to Polycyclic Aromatic Hydrocarbons and Other Organic Mutagens in Ottawa Firefighters Participating in Emergency, On-Shift Fire Suppression." *Environmental Science & Technology* 51 (21): 12745–12755. https://doi.org/10.1021/acs.est.7b02850.

Kessler, R. 2012. "Followup in Southern California: Decreased Birth Weight Following Prenatal Wildfire Smoke Exposure." *Environmental Health Perspectives* 120 (9). https://doi.org/10.1289/ehp.120-a362b.

Kim, Y. H., H. Tong, M. Daniels, E. Boykin, Q. T. Krantz, J. McGee, M. Hays, K. Kovalcik, J. A. Dye, and M. I. Gilmour. 2014. "Cardiopulmonary Toxicity of Peat Wildfire Particulate Matter and the Predictive Utility of Precision Cut Lung Slices." *Particle and Fibre Toxicology* 11: 29. https://doi.org/10.1186/1743-8977-11-29.

Kirk, K. M., and M. B. Logan. 2015. "Firefighting Instructors' Exposures to Polycyclic Aromatic Hydrocarbons During Live Fire Training Scenarios." *Journal of Occupational and Environmental Hygiene* 12 (4). https://doi.org/10.1080/15459624.2014.955184.

Kleinman, L. I., A. J. Sedlacek III, K. Adachi, P. R. Buseck, S. Collier, M. K. Dubey, A. L. Hodshire, E. Lewis, T. B. Onasch, J. R. Pierce, J. Shilling, S. R. Springston, J. Wang, Q. Zhang, S. Zhou, and R. J. Yokelson. 2020. "Rapid Evolution of Aerosol Particles and Their Optical Properties Downwind of Wildfires in the Western US." *Atmospheric Chemistry and Physics* 20 (21): 13319–13341. https://doi.org/10.5194/acp-20-13319-2020.

Kochanski, A. K., D. V. Mallia, M. G. Fearon, J. Mandel, A. H. Souri, and T. Brown. 2019. "Modeling Wildfire Smoke Feedback Mechanisms Using a Coupled Fire-Atmosphere Model with a Radiatively Active Aerosol Scheme." *Journal of Geophysical Research: Atmospheres* 124 (16): 9099–9116. https://doi.org/10.1029/2019jd030558.

Kodros, J. K., K. O'Dell, J. M. Samet, C. L'Orange, J. R. Pierce, and J. Volckens. 2021. "Quantifying the Health Benefits of Face Masks and Respirators to Mitigate Exposure to Severe Air Pollution." *GeoHealth* 5 (9): e2021GH000482. https://doi.org/10.1029/2021GH000482.

Kohl, L., M. Meng, J. Vera, B. Bergquist, C. A. Cooke, S. Hustins, B. Jackson, C. W. Chow, and A. W. H. Chan. 2019. "Limited Retention of Wildfire-Derived PAHs and Trace Elements in Indoor Environments." *Geophysical Research Letters* 46 (1): 383–391. https://doi.org/10.1029/2018gl080473.

Kolbe, A., and K. L. Gilchrist. 2009. *An Extreme Bushfire Smoke Pollution Event: Health Impacts and Public Health Challenges*. NSW Public Health Bulletin 20. https://doi.org/10.1071/NB08061.

Kolluri, S. K., U. H. Jin, and S. Safe. 2017. "Role of the Aryl Hydrocarbon Receptor in Carcinogenesis and Potential as an Anti-cancer Drug Target." *Archives of Toxicology* 91 (7): 2497–2513. https://doi.org/10.1007/s00204-017-1981-2.

Koman, P. D., M. Billmire, K. R. Baker, R. de Majo, F. J. Anderson, S. Hoshiko, B. J. Thelen, and N. H. F. French. 2019. "Mapping Modeled Exposure of Wildland Fire Smoke for Human Health Studies in California." *Atmosphere (Basel)* 10 (6). https://doi.org/10.3390/atmos10060308.

Kuligowski, E. 2021. "Evacuation Decision-Making and Behavior in Wildfires: Past Research, Current Challenges and a Future Research Agenda." *Fire Safety Journal* 120. https://doi.org/10.1016/j.firesaf.2020.103129.

Kuligowski, E. D., X. Zhao, R. Lovreglio, N. Xu, K. Yang, A. Westbury, D. Nilsson, and N. Brown. 2022. "Modeling Evacuation Decisions in the 2019 Kincade Fire in California." *Safety Science* 146. https://doi.org/10.1016/j.ssci.2021.105541.

Landis, M. S., E. S. Edgerton, E. M. White, G. R. Wentworth, A. P. Sullivan, and A. M. Dillner. 2018. "The Impact of the 2016 Fort McMurray Horse River Wildfire on Ambient Air Pollution Levels in the Athabasca Oil Sands Region, Alberta, Canada." *Science of the Total Environment* 618: 1665–1676. https://doi.org/10.1016/j.scitotenv.2017.10.008.

Laumbach, R. J. 2019. "Clearing the Air on Personal Interventions to Reduce Exposure to Wildfire Smoke." *Annals of the American Thoracic Society* 16 (7): 815–818. https://doi.org/10.1513/AnnalsATS.201812-894PS.

LeMasters, G. K., A. M. Genaidy, P. Succop, J. Deddens, T. Sobeih, H. Barriera-Viruet, K. Dunning, and J. Lockey. 2006. "Cancer Risk Among Firefighters: A Review and Meta-Analysis of 32 Studies." *Journal of Occupational and Environmental Medicine* 48 (11): 1189–1202. https://doi.org/10.1097/01.jom.0000246229.68697.90.

Li, D., T. J. Cova, and P. E. Dennison. 2019. "Setting Wildfire Evacuation Triggers by Coupling Fire and Traffic Simulation Models: A Spatiotemporal GIS Approach." *Fire Technology* 55 (2): 617–642. https://doi.org/10.1007/s10694-018-0771-6.

Liu, D., I. B. Tager, J. R. Balmes, and R. J. Harrison. 1991. "The Effect of Smoke Inhalation on Lung Function and Airway Responsiveness in Wildland Fire Fighters." *American Review of Respiratory Disease* 146 (6). https://doi.org/10.1164/ajrccm/146.6.1469.

Liu, J. C., A. Wilson, L. J. Mickley, K. Ebisu, M. P. Sulprizio, Y. Wang, R. D. Peng, X. Yue, F. Dominici, and M. L. Bell. 2017. "Who Among the Elderly Is Most Vulnerable to Exposure to and Health Risks of Fine Particulate Matter from Wildfire Smoke?" *American Journal of Epidemiology* 186 (6): 730–735. https://doi.org/10.1093/aje/kwx141.

Magzamen, S., R. W. Gan, J. Liu, K. O'Dell, B. Ford, K. Berg, K. Bol, A. Wilson, E. V. Fischer, and J. R. Pierce. 2021. "Differential Cardiopulmonary Health Impacts of Local and Long-Range Transport of Wildfire Smoke." *GeoHealth* 5 (3). https://doi.org/10.1029/2020gh000330.

Main, L. C., A. P. Wolkow, J. L. Tait, P. Della Gatta, J. Raines, R. Snow, and B. Aisbett. 2020. "Firefighter's Acute Inflammatory Response to Wildfire Suppression." *Journal of Occupational and Environmental Medicine* 62 (2): 145–148. https://doi.org/10.1097/JOM.0000000000001775.

Malig, B. J., D. Fairley, D. Pearson, X. Wu, K. Ebisu, and R. Basu. 2021. "Examining Fine Particulate Matter and Cause-Specific Morbidity during the 2017 North San Francisco Bay Wildfires." *Science of the Total Environment* 787: 147507. https://doi.org/10.1016/j.scitotenv.2021.147507.

McCaffrey, S., R. Wilson, and A. Konar. 2018. "Should I Stay or Should I Go Now? Or Should I Wait and See? Influences on Wildfire Evacuation Decisions." *Risk Analysis* 38 (7): 1390–1404. https://doi.org/10.1111/risa.12944.

McLennan, J., B. Ryan, C. Bearman, and K. Toh. 2018. "Should We Leave Now? Behavioral Factors in Evacuation Under Wildfire Threat." *Fire Technology* 55 (2): 487–516. https://doi.org/10.1007/s10694-018-0753-8.

Melendez, B., S. Ghanipoor Machiani, and A. Nara. 2021. "Modelling Traffic during Lilac Wildfire Evacuation Using Cellular Data." *Transportation Research Interdisciplinary Perspectives* 9. https://doi.org/10.1016/j.trip.2021.100335.

Meng, M. 2018. *Composition of Wildfire-Derived Particulate Matter and Impacts on House Dust after a Major Wildfire.* University of Toronto, Toronto, ON. Masters of Applied Science.

Miranda, A. I., V. Martins, P. Cascao, J. H. Amorim, J. Valente, R. Tavares, C. Borrego, O. Tchepel, A. J. Ferreira, C. R. Cordeiro, D. X. Viegas, L. M. Ribeiro, and L. P. Pita. 2010. "Monitoring of Firefighters Exposure to Smoke during Fire Experiments in Portugal." *Environmental International* 36 (7): 736–745. https://doi.org/10.1016/j.envint.2010.05.009.

Mott, J. A., P. Meyer, D. Mannino, S. Redd, E. M. Smith, C. Gotway-Crawford, and E. Chase. 2002. "Wildland Forest Fire Smoke: Health Effects and Intervention Evaluation, Hoopa, California, 1999." *Western Journal of Emergency Medicine* 176 (3). https://doi.org/10.1136/ewjm.176.3.157.

Mozumder, P., N. Raheem, J. Talberth, and J. Berrens. 2008. "Investigating Intended Evacuation from Wildfires in the Wildland–Urban Interface: Application of a Bivariate Probit Model." *Forest Policy and Economics* 10 (6). https://doi.org/10.1016/j.forpol.2008.02.002.

Murphy, S. F., R. B. McCleskey, D. A. Martin, J. M. Holloway, and J. H. Writer. 2020. "Wildfire-Driven Changes in Hydrology Mobilize Arsenic and Metals from Legacy Mine Waste." *Science of the Total Environment* 743: 140635. https://doi.org/10.1016/j.scitotenv.2020.140635.

Nadler, D. L., and I. G. Zurbenko. 2014. "Estimating Cancer Latency Times Using a Weibull Model." *Advances in Epidemiology* 2014: 1–8. https://doi.org/10.1155/2014/746769.

NASA EO (National Aeronautics and Space Administration Earth Observatory). 2021. *Smoke across North America.* https://earthobservatory.nasa.gov/images/148610/smoke-across-north-america.

NASEM (National Academies of Sciences, Engineering, and Medicine). 2017. *Communities in Action: Pathways to Health Equity.* Washington, DC: The National Academies Press. https://doi.org/10.17226/24624.

NASEM. 2022. *Communities, Climate Change, and Health Equity: Proceedings of a Workshop–In Brief.* Washington, DC: The National Academies Press. https://doi.org/10.17226/26435.

Navarro, K. 2020. "Working in Smoke: Wildfire Impacts on the Health of Firefighters and Outdoor Workers and Mitigation Strategies." *Clinics in Chest Medicine* 41 (4): 763–769. https://doi.org/10.1016/j.ccm.2020.08.017.

Navarro, K. M., M. T. Kleinman, C. E. Mackay, T. E. Reinhardt, J. R. Balmes, G. A. Broyles, R. D. Ottmar, L. P. Naher, and J. W. Domitrovich. 2019. "Wildland Firefighter Smoke Exposure and Risk of Lung Cancer and Cardiovascular Disease Mortality." *Environmental Research* 173: 462–468. https://doi.org/10.1016/j.envres.2019.03.060.

NCI (National Cancer Institute). 2019. *Strong Inorganic Acid Mists Containing Sulfuric Acid.* https://www.cancer.gov/about-cancer/causes-prevention/risk/substances/inorganic-acid (accessed March 22, 2022).

NEPHT (National Environmental Public Health Tracking). 2020. *Air Quality.* https://www.cdc.gov/nceh/tracking/topics/AirQuality.htm#ozone (accessed March 22, 2022).

NIEHS (National Institute of Environmental Health Sciences). 2021a. *Flame Retardants.* https://www.niehs.nih.gov/health/topics/agents/flame_retardants/index.cfm.

NIEHS. 2021b. *About the Worker Training Program (WTP).* https://www.niehs.nih.gov/careers/hazmat/about_wetp/index.cfm (accessed March 24, 2022).

NIOSH (National Institute for Occupational Safety and Health). 2011. *Sodium Cyanide: Systemic Agent.* https://www.cdc.gov/niosh/ershdb/emergencyresponsecard_29750036.html (accessed March 22, 2022).

NIOSH. 2014. *Isocyanates*. https://www.cdc.gov/niosh/topics/isocyanates/default.html (accessed March 22, 2022).

NIOSH. 2019a. *Lead & Other Heavy Metals – Reproductive Health*. https://www.cdc.gov/niosh/topics/repro/heavymetals.html (accessed March 22, 2022).

NIOSH. 2019b. *Fighting Wildfires – Hazards During Cleanup Work*. https://www.cdc.gov/niosh/topics/firefighting/cleanup.html (accessed 2022).

NTP (National Toxicology Program). 2016. *NTP Monograph: Immunotoxicity Associated with Exposure to Perfluorooctanoic Acid or Perfluorooctane Sulfonate*. https://ntp.niehs.nih.gov/ntp/ohat/pfoa_pfos/pfoa_pfosmonograph_508.pdf.

O'Dell, K., R. S. Hornbrook, W. Permar, E. J. T. Levin, L. A. Garofalo, E. C. Apel, N. J. Blake, A. Jarnot, M. A. Pothier, D. K. Farmer, L. Hu, T. Campos, B. Ford, J. R. Pierce, and E. V. Fischer. 2020. "Hazardous Air Pollutants in Fresh and Aged Western US Wildfire Smoke and Implications for Long-Term Exposure." *Environmental Science & Technology* 54 (19): 11838–11847. https://doi.org/10.1021/acs.est.0c04497.

O'Donnell, M. H., and A. M. Behie. 2015. "Effects of Wildfire Disaster Exposure on Male Birth Weight in an Australian Population." *Evolution, Medicine, and Public Health*. https://doi.org/10.1093/emph/eov027.

O'Neill, S. M., M. Diao, S. Raffuse, M. Al-Hamdan, M. Barik, Y. Jia, S. Reid, Y. Zou, D. Tong, J. J. West, J. Wilkins, A. Marsha, F. Freedman, J. Vargo, N. K. Larkin, E. Alvarado, and P. Loesche. 2021. "A Multi-analysis Approach for Estimating Regional Health Impacts from the 2017 Northern California Wildfires." *Journal of the Air & Waste Management Association* 71 (7): 791–814. https://doi.org/10.1080/10962247.2021.1891994.

Onwuka, J. U., D. Li, Y. Liu, H. Huang, J. Xu, Y. Liu, Y. Zhang, and Y. Zhao. 2020. "A Panel of DNA Methylation Signature from Peripheral Blood May Predict Colorectal Cancer Susceptibility. *BMC Cancer* 20 (1): 692. https://doi.org/10.1186/s12885-020-07194-5.

Oregon DCBS (Oregon Department of Consumer and Business Services). 2021. *Adoption of Temporary Rules to Address Employee Exposure to Wildfire Smoke*. https://osha.oregon.gov/OSHARules/adopted/2021/ao9-2021-letter-temp-wildfire-smoke.pdf.

Orr, A., C. A. L. Migliaccio, M. Buford, S. Ballou, and C. T. Migliaccio. 2020. "Sustained Effects on Lung Function in Community Members Following Exposure to Hazardous $PM_{2.5}$ Levels from Wildfire Smoke." *Toxics* 8 (3): 53.

Paveglio, T. B., T. Prato, D. Dalenberg, and T. Venn. 2014. "Understanding Evacuation Preferences and Wildfire Mitigations among Northwest Montana Residents." *International Journal of Wildland Fire* 23 (3): 435–444. https://doi.org/10.1071/WF13057.

Perera, F. P. 2008. "Children Are Likely to Suffer Most from Our Fossil Fuel Addiction." *Environmental Health Perspectives* 116 (8). https://doi.org/10.1289/ehp.11173.

Petrowski, K., S. Bührer, B. Strauß, O. Decker, and E. Brähler. 2021. "Examining Air Pollution (PM_{10}), Mental Health and Well-Being in a Representative German Sample." *Scientific Reports* 11: 18436. https://doi.org/10.1038/s41598-021-93773-w.

Pleil, J. D., M. A. Stiegel, and K. W. Fent. 2014. "Exploratory Breath Analyses for Assessing Toxic Dermal Exposures of Firefighters during Suppression of Structural Burns." *Journal of Breath Research* 8 (3). https://doi.org/10.1088/1752-7155/8/3/037107.

Plumlee, G. S., D. A. Martin, T. Hoefen, R. Kokaly, P. Hageman, A. Eckberg, G. P. Meeker, M. Adams, M. Anthony, and P. J. Lamothe. 2007. *Preliminary Analytical Results for Ash and Burned Soils from the October 2007 Southern California Wildfires*. Report 2007-1407. Reston, VA: United States Geological Survey.

Proctor, C. R., J. Lee, D. Yu, A. D. Shah, and A. J. Whelton. 2020. "Wildfire Caused Widespread Drinking Water Distribution Network Contamination. *AWWA Water Science* 2 (4). https://doi.org/10.1002/aws2.1183.

Pukkala, E., J. I. Martinsen, E. Weiderpass, K. Kjarheim, E. Lynge, L. Tryggvadottir, P. Sparen, and P. A. Demers. 2014. "Cancer Incidence Among Firefighters: 45 Years of Follow-Up in Five Nordic Countries." *Occupational and Environmental Medicine* 71 (6). http://doi.org/10.1136/oemed-2013-101803.

Ray, M. R., C. Basu, S. Mukherjee, S. Roychowdhury, and T. Lahiri. 2005. "Micronucleus Frequencies and Nuclear Anomalies in Exfoliated Buccal Epithelial Cells of Firefighters." *International Journal of Human Genetics* 5 (1): 45–48. https://doi.org/10.1080/09723757.2005.11885915.

Reid, C. E., and M. M. Maestas. 2019. "Wildfire Smoke Exposure under Climate Change: Impact on Respiratory Health of Affected Communities." *Current Opinion in Pulmonary Medicine* 25 (2). https://doi.org/10.1097/MCP.0000000000000552.

Reid, C. E., M. Brauer, F. H. Johnston, M. Jerrett, J. R. Balmes, and C. T. Elliott. 2016a. "Critical Review of Health Impacts of Wildfire Smoke Exposure." *Environmental Health Perspectives* 124 (9): 1334–1343. https://doi.org/10.1289/ehp.1409277.

Reid, C. E., M. Jerrett, I. B. Tager, M. L. Petersen, J. K. Mann, and J. R. Balmes. 2016b. "Differential Respiratory Health Effects from the 2008 Northern California Wildfires: A Spatiotemporal Approach." *Environmental Research* 150: 227–235. https://doi.org/10.1016/j.envres.2016.06.012.

Reinhardt, T. E., and R. D. Ottmar. 2004. "Baseline Measurements of Smoke Exposure Among Wildland Firefighters." *Journal of Occupational and Environmental Hygiene* 1 (9). https://doi.org/10.1080/15459620490490101.

Reisen, F., and S. K. Brown. 2009. "Australian Firefighters' Exposure to Air Toxics during Bushfire Burns of Autumn 2005 and 2006." *Environmental International* 35 (2): 342–352. https://doi.org/10.1016/j.envint.2008.08.011.

Reisen, F., J. C. Powell, M. Dennekamp, F. H. Johnston, and A. J. Wheeler. 2019. "Is Remaining Indoors an Effective Way of Reducing Exposure to Fine Particulate Matter During Biomass Burning Events?" *Journal of the Air & Waste Management Association* 69 (5). https://doi.org/10.1080/10962247.2019.1567623.

Resnick, A., B. Woods, H. Krapfl, and B. Toth. 2015. "Health Outcomes Associated with Smoke Exposure in Albuquerque, New Mexico, during the 2011 Wallow Fire." *Journal of Public Health Management and Practice* 21: S55–S61. https://doi.org/10.1097/PHH.0000000000000160.

Reuter, S., S. C. Gupta, M. M. Chaturvedi, and B. B. Aggarwal. 2010. "Oxidative Stress, Inflammation, and Cancer: How Are They Linked?" *Free Radical Biology and Medicine* 49 (11): 1603–1616. https://doi.org/10.1016/j.freeradbiomed.2010.09.006.

Revi, A., D. Satterthwaite, F. Aragón-Durand, et al. 2014. "Towards Transformative Adaptation in Cities: The IPCC's Fifth Assessment." *Environment and Urbanization* 26 (1): 11–28. https://doi.org/10.1177/0956247814523539.

Robinson, M. S., T. R. Anthony, S. R. Littau, P. Herckes, X. Nelson, G. S. Poplin, and J. L. Burgess. 2008. "Occupational PAH Exposures during Prescribed Pile Burns." *Annals of Occupational Hygiene* 52 (6): 497–508. https://doi.org/10.1093/annhyg/men027.

Rooney, B., Y. Wang, J. H. Jiang, B. Zhao, Z.-C. Zeng, and J. H. Seinfeld. 2020. "Air Quality Impact of the Northern California Camp Fire of November 2018." *Atmospheric Chemistry and Physics* 20 (23): 14597–14616. https://doi.org/10.5194/acp-20-14597-2020.

Rota, M., C. Bosetti, S. Boccia, P. Boffetta, and C. La Vecchia. 2014. "Occupational Exposures to Polycyclic Aromatic Hydrocarbons and Respiratory and Urinary Tract Cancers: An Updated Systematic Review and a Meta-Analysis to 2014." *Archives of Toxicology* 88 (8): 1479–1490. https://doi.org/10.1007/s00204-014-1296-5.

Rothman, N., D. P. Ford, M. E. Baser, J. A. Hansen, T. O'Toole, M. S. Tockman, and P. T. Strickland. 1991. "Pulmonary Function and Respiratory Symptoms in Wildland Firefighters." *Journal of Occupational and Environmental Medicine* 33 (11). https://doi.org/10.1097/00043764-199111000-00013.

Rothman, N., M. C. Poirier, R. A. Haas, A. Correa-Villasenor, P. Ford, J. A. Hansen, T. O'Toole, and P. T. Strickland. 1993. "Assocation of PAH-DNA Adducts in Peripheral White Blood Cells with Dietary Exposure to Polyaromatic Hydrocarbons." *Environmental Health Perspectives* 99: 265–267. https://doi.org/10.1289/ehp.9399265.

Rudolph, L., C. Harrison, L. Buckley, and S. North. 2018. *Climate Change, Health, and Equity: A Guide for Local Health Departments*. Oakland, CA and Washington, DC: Public Health Institute and American Public Health Association.

Rust, A. J., T. S. Hogue, S. Saxe, and J. McCray. 2018. "Post-fire Water-Quality Response in the Western United States." *International Journal of Wildland Fire* 27 (3). https://doi.org/10.1071/wf17115.

Sacks, J. D., L. W. Stanek, T. J. Luben, D. O. Johns, B. J. Buckley, J. S. Brown, and M. Ross. 2011. "Particulate Matter–Induced Health Effects: Who Is Susceptible?" *Environmental Health Perspectives* 119 (4). https://doi.org/10.1289/ehp.1002255.

Samet, J. M., S. M. Holm, and S. Jayaraman. 2022. "Respiratory Protection for the Nation: A Report from the National Academies of Sciences, Engineering, and Medicine." *Journal of the American Medical Association* 327 (11): 1023–1024. https://doi.org/10.1001/jama.2022.1318.

Semmens, E. O., J. Domitrovich, K. Conway, and C. W. Noonan. 2016. "A Cross-Sectional Survey of Occupational History as a Wildland Firefighter and Health." *American Journal of Industrial Medicine* 59 (4). https://doi.org/10.1002/ajim.22566.

Shaw, S. D., M. L. Berger, J. H. Harris, S. H. Yun, Q. Wu, C. Liao, A. Blum, A. Stefani, and K. Kannan. 2013. "Persistent Organic Pollutants Including Polychlorinated and Polybrominated Dibenzo-p-Dioxins and Dibenzofurans in Firefighters from Northern California." *Chemosphere* 91 (10): 1386–1394. https://doi.org/10.1016/j.chemosphere.2012.12.070.

Sheldon, T. L., and C. Sankaran. 2017. "The Impact of Indonesian Forest Fires on Singaporean Pollution and Health." *American Economic Review* 107 (5): 526–529. https://doi.org/10.1257/aer.p20171134.

Shi, Y. L. 2021. *Polycyclic Aromatic Hydrocarbons in Settled House Dust*. University of Toronto, Toronto, ON. Masters of Applied Science.

Singer, B. C., W. W. Delp, D. R. Black, and I. S. Walker. 2017. "Measured Performance of Filtration and Ventilation Systems for Fine and Ultrafine Particles and Ozone in an Unoccupied Modern California House." *Indoor Air* 27: 780–790. https://doi.org/10.1111/ina.12359.

Sparks, T. L., and J. Wagner. 2021. "Composition of Particulate Matter during a Wildfire Smoke Episode in an Urban Area." *Aerosol Science and Technology* 55 (6): 734–747. https://doi.org/10.1080/02786826.2021.1895429.

Staack, S. D., S. C. Griffin, V. S. T. Lee, E. A. Lutz, and J. L. Burgess. 2021. "Evaluation of CBRN Respirator Protection in Simulated Fire Overhaul Settings." *Annals of Work Exposures and Health* 65 (7): 843–853. https://doi.org/10.1093/annweh/wxab004.

Stasiewicz, A. M., and T. B. Paveglio. 2021. "Preparing for Wildfire Evacuation and Alternatives: Exploring Influences on Residents' Intended Evacuation Behaviors and Mitigations." *International Journal of Disaster Risk Reduction* 58. https://doi.org/10.1016/j.ijdrr.2021.102177.

State of California. n.d. *California Code of Regulations*, Title 8, Section 5141.1. "Protection from Wildfire Smoke." https://www.dir.ca.gov/title8/5141_1.html.

State of Washington. 2021. *Wildfire Smoke*. https://lni.wa.gov/rulemaking-activity/AO21-26/2126CR103EAdoption.pdf.

Stec, A. A. 2017. "Fire Toxicity – The Elephant in the Room?" *Fire Safety Journal* 91: 79–90.

Stec, A. A., K. E. Dickens, M. Salden, F. E. Hewitt, D. P. Watts, P. E. Houldsworth, and F. L. Martin. 2018. "Occupational Exposure to Polycyclic Aromatic Hydrocarbons and Elevated Cancer Incidence in Firefighters." *Scientific Reports* 8 (1): 2476. https://doi.org/10.1038/s41598-018-20616-6.

Stein, E. D., J. S. Brown, T. S. Hogue, M. P. Burke, and A. Kinoshita. 2012. "Stormwater Contaminant Loading Following Southern California Wildfires." *Environmental Toxicology and Chemistry* 31 (11): 2625–2638. https://doi.org/10.1002/etc.1994.

Sugerman, D. E., J. M. Keir, D. L. Dee, H. Lipman, S. H. Waterman, M. Ginsberg, and D. B. Fishbein. 2012. "Emergency Health Risk Communication During the 2007 San Diego Wildfires: Comprehension, Compliance, and Recall." *Journal of Health Communication* 17 (6): 698–712. https://doi.org/10.1080/10810730.2011.635777.

Sun, X., X. Luo, C. Zhao, B. Zhang, J. Tao, Z. Yang, W. Ma, and T. Liu. 2016. "The Associations between Birth Weight and Exposure to Fine Particulate Matter ($PM_{2.5}$) and Its Chemical Constituents during Pregnancy: A Meta-Analysis." *Environmental Pollution* 211: 38e47. http://doi.org/10.1016/j.envpol.2015.12.022.

Sun, Y., A. Mousavi, S. Masri, and J. Wu. 2022. "Socioeconomic Disparities of Low-Cost Air Quality Sensors in California, 2017–2020." *American Journal of Public Health* 112 (3): 434–442. https://doi.org/10.2105/AJPH.2021.306603.

Tam, N., and C. Adams. 2019. *Characterization of Air Quality During the 2016 Horse River Wildfire Using Permanent and Portable Monitoring*. Edmonton, AB: Government of Alberta, Ministry of Environment and Parks.

Tinling, M. A., J. J. West, W. E. Cascio, V. Kilaru, and A. G. Rappold. 2016. "Repeating Cardiopulmonary Health Effects in Rural North Carolina Population during a Second Large Peat Wildfire." *Environmental Health* 15 (12). https://doi.org/10.1186/s12940-016-0093-4.

To, P., E. Eboreime, and V. I. O. Agyapong. 2021. "The Impact of Wildfires on Mental Health: A Scoping Review." *Behavioral Sciences (Basel)* 11 (9): 126. https://doi.org/10.3390/bs11090126.

Underwriters Laboratories. 2021. *DIY Box Fan Air Cleaner Safety Tips*. https://chemicalinsights.org/wp-content/uploads/Box_Fan_Handout_090921.pdf (accessed February 23, 2022).

USGS (United States Geological Survey). 2012. *Wildfire Effects on Source-Water Quality—Lessons from Fourmile Canyon Fire, Colorado, and Implications for Drinking-Water Treatment*. Reston, VA: United States Geological Survey.

Uzun, H., R. A. Dahlgren, C. Olivares, C. U. Erdem, T. Karanfil, and A. T. Chow. 2020. "Two Years of Post-wildfire Impacts on Dissolved Organic Matter, Nitrogen, and Precursors of Disinfection By-products in California Stream Waters." *Water Research* 181: 115891. https://doi.org/10.1016/j.watres.2020.115891.

Vanos, J. K. 2015. "Children's Health and Vulnerability in Outdoor Microclimates: A Comprehensive Review." *Environment International* 76: 1–15. https://doi.org/10.1016/j.envint.2014.11.016.

Vinck, P., P. N. Pham, L. E. Fletcher, and E. Stover. 2009. "Inequalities and Prospects: Ethnicity and Legal Status in the Construction Labor Force After Hurricane Katrina." *Organization & Environment* 22 (4): 470–478. https://doi.org/10.1177/1086026609347192.

Wang, L., J. A. Aakre, R. Jiang, R. S. Marks, Y. Wu, J. Chen, S. N. Thibodeau, V. S. Pankratz, and P. Yang. 2010. "Methylation Markers for Small Cell Lung Cancer in Peripheral Blood Leukocyte DNA." *Journal of Thoracic Oncology* 5 (6). https://doi.org/10.1097/jto.0b013e3181d6e0b3.

Ward, D. E., and C. C. Hardy. 1991. "Smoke Emissions from Wildland Fires." *Environment International* 17 (2–3): 117–134. https://doi.org/10.1016/0160-4120(91)90095-8.

Wentworth, G. R., Y. A. Aklilu, M. S. Landis, and Y. M. Hsu. 2018. "Impacts of a Large Boreal Wildfire on Ground Level Atmospheric Concentrations of PAHs, VOCs and Ozone." *Atmospheric Environment* 178: 19–30. https://doi.org/10.1016/j.atmosenv.2018.01.013.

Wettstein, Z. S., S. Hoshiko, J. Fahimi, R. J. Harrison, W. E. Cascio, and A. G. Rappold. 2018. "Cardiovascular and Cerebrovascular Emergency Department Visits Associated with Wildfire Smoke Exposure in California in 2015." *Journal of the American Heart Association* 7 (8). https://doi.org/10.1161/JAHA.117.007492.

Wolf, R. E., S. A. Morman, P. L. Hageman, T. M. Hoefen, and G. S. Plumlee. 2011. "Simultaneous Speciation of Arsenic, Selenium, and Chromium: Species Stability, Sample Preservation, and Analysis of Ash and Soil Leachates." *Analytical and Bioanalytical Chemistry* 401 (9): 2733–2745. https://doi.org/10.1007/s00216-011-5275-x.

Wong, S. D., J. C. Broader, and S. A. Shaheen. 2020. *Review of California Wildfire Evacuations from 2017 to 2019*. Berkeley, CA: University of California Institute of Transportation Studies. https://doi.org/10.7922/G29G5K2R.

Woo, S. H. L., J. C. Liu, X. Yue, L. J. Mickley, and M. L. Bell. 2020. "Air Pollution from Wildfires and Human Health Vulnerability in Alaskan Communities under Climate Change." *Environmental Research Letters* 15 (9). https://doi.org/10.1088/1748-9326/ab9270.

Writer, J. H., A. Hohner, J. Oropeza, A. Schmidt, K. M. Cawley, and F. L. Rosario-Ortiz. 2014. "Water Treatment Implications after the High Park Wildfire, Colorado." *Journal – American Water Works Association* 106 (4): E189–E199. https://doi.org/10.5942/jawwa.2014.106.0055.

Wu, Y., A. Arapi, J. Huang, B. Gross, and F. Moshary. 2018. "Intra-continental Wildfire Smoke Transport and Impact on Local Air Quality Observed by Ground-Based and Satellite Remote Sensing in New York City. *Atmospheric Environment* 187: 266–281. https://doi.org/10.1016/j.atmosenv.2018.06.006.

Xiang, J., C. Huang, J. Shirai, Y. Liu, N. Carmona, C. Zuidema, E. Austin, T. Gould, T. Larson, and E. Seto. 2021. "Field Measurements of $PM_{2.5}$ Infiltration Factor and Portable Air Cleaner Effectiveness during Wildfire Episodes in US Residences." *Science of the Total Environment* 773. https://doi.org/10.1016/j.scitotenv.2021.145642.

Xu, R., P. Yu, M. J. Abramson, F. H. Johnston, J. M. Samet, M. L. Bell, A. Haines, K. L. Ebi, S. Li, and Y. Guo. 2020. "Special Report: Wildfires, Global Climate Change, and Human Health." *New England Journal of Medicine* 383 (22). https://doi.org/10.1056/NEJMsr2028985.

Ye, Q., J. E. Krechmer, J. D. Shutter, V. P. Barber, Y. Li, E. Helstrom, . . . and J. H. Kroll. 2021. "Real-Time Laboratory Measurements of VOC Emissions, Removal Rates, and Byproduct Formation from Consumer-Grade Oxidation-Based Air Cleaners." *Environmental Science & Technology Letters* 8 (12): 1020–1025.

Zauscher, M. D., Y. Wang, M. J. K. Moore, C. J. Gaston, and K. A. Prather. 2013. "Air Quality Impact and Physicochemical Aging of Biomass Burning Aerosols during the 2007 San Diego Wildfires." *Environmental Science & Technology* 47 (14): 7633–7643. https://doi.org/10.1021/es4004137.

Zeliger, H. I. 2003. "Toxic Effects of Chemical Mixtures." *Archives of Environmental Health: An International Journal* 58 (1): 23–29.

Zhou, J., T. G. Jenkins, A. M. Jung, K. S. Jeong, J. Zhai, E. T. Jacobs, S. C. Griffin, D. Dearmon-Moore, S. R. Littau, W. F. Peate, N. A. Ellis, P. Lance, Y. Chen, and J. L. Burgess. 2019. "DNA Methylation among Firefighters." *PLoS One*. https://doi.org/10.1371/journal.pone.0214282.

Zhou, X., K. Josey, L. Kamareddine, M. C. Caine, T. Liu, L. J. Mickley, M. Cooper, and F. Dominici. 2021. "Excess of COVID-19 Cases and Deaths Due to Fine Particulate Matter Exposure during the 2020 Wildfires in the United States." *Scientific Advances* 7 (33): eabi8789. https://doi.org/10.1126/sciadv.abi8789.

7

Measurement of WUI Fires: Emissions and Exposures

This chapter summarizes the measurements and data needed to improve the understanding of the chemistry of wildland-urban interface (WUI) fires and their emissions. Measurement systems and data are needed on fire detection, characterizing fuels and emissions, and exposure assessment, and to improve model prediction. While existing measurement techniques can provide much of the needed data, improved sampling and chemical analysis systems are needed to capture some of the features of fires at the WUI. The chapter concludes by emphasizing the need for coordination across disciplines to collect robust data that present an integrated picture of the chemistry of WUI fires.

AN OVERVIEW OF DATA AND MEASUREMENT SYSTEMS FOR WUI FIRES

Previous chapters have outlined the diverse data needs associated with understanding the chemistry of fires at the WUI, which range from collecting data on material characteristics of heterogeneous structures, to measuring the concentrations, exposures, and health impacts of large numbers of toxicants, many of which will be present at trace levels. Collecting much of these data is challenging, since the unique chemical signatures of WUI fires are distributed in complex ways in the fire plume.

Data and measurement needs are also interconnected, as shown in Figure 7-1. Data on fuel compositions will determine combustion pathways. The chemical species formed during combustion will determine which atmospheric reaction pathways will be most important. The atmospheric chemistry and transport will determine the toxicants to which communities are exposed and the manner of the exposure. Exposures will determine health impacts. Because of this interconnection, coordinated collection of data and measurements, from fuel and emission characterization to exposures, will have the most value. Collection of data on chemical species over their entire cycle from emission to exposure will enhance the value of all of the data that are collected.

Data and measurements are also needed on various timescales. Retrospective studies of materials, fires, emissions, exposures, and health impacts, both in the field and in laboratories, and via modeling, will be important to improve understanding of the chemistry of WUI fires. These types of studies will benefit from coordinated databases and information repositories that allow efficient sharing of information and methods. Some of the information repositories will also need to be broadly and rapidly accessible. For example, decision-makers charged with mitigating the impacts of WUI fires could utilize information on the types of combustible materials at the WUI to provide first responders and the public with information to mitigate exposure to toxicants.

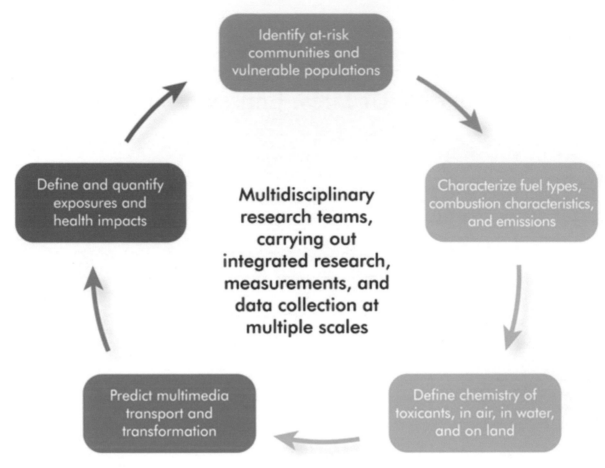

FIGURE 7-1 The interdependence of multidisciplinary WUI fire research activities.

This chapter summarizes the data and measurement needs in each of the steps in the cycle shown in Figure 7-1, from emissions to exposures and impacts. The focus will be on the information needed, the limits of current measurements and data, and prospects for more extensive data collection.

DETECTING AND DETERMINING THE AREAL EXTENT OF WUI FIRES

Several federal agencies are involved in gathering wildfire statistics in the United States, including the National Interagency Fire Center (NIFC). Federal, state, local, and tribal land management agencies report information to the NIFC through established channels using Incident Management Situation Reports. The NIFC's National Interagency Coordination Center compiles the statistics annually, and they are posted online at https://www.nifc.gov/fire-information/statistics (NIFC, 2022), along with daily news updates and other information. The US Fire Administration, an entity of the US Department of Homeland Security's Federal Emergency Management Agency, also has a National Fire Incident Reporting System at its National Fire Data Center, which compiles data. While this data collection encompasses both wildland fires and WUI fires, mapping tools, described in Chapter 2, can be used to convert comprehensive fire reporting to reporting specific to WUIs, if broadly accepted definitions of WUIs can be developed.

Wildfires can also be estimated from satellite detection of fire energetics through thermal anomalies (commonly referred to as hot spots). These are pixels with temperature and energy above a threshold value, signifying an active fire. Thermal anomaly data can also be used to measure fire radiative power or fire radiative energy (the time-integrated fire radiative power) in combination with emission factors specific to ecosystem or land cover type to generate emissions estimates (Wooster et al., 2005). This top-down approach to emissions estimation can provide data in near real time so that it can be used in emergency response as well as in forecasting smoke impacts. However, delays may occur in burn area estimates depending on the data source of the fire radiative power. Fire radiative power can be derived from multiple instruments onboard multiple satellite platforms including the Moderate Resolution Imaging Spectroradiometer (MODIS) on polar orbiting satellites Terra and Aqua, the Visible Infrared Imager Radiometer Suite (VIIRS) on polar orbiting satellite Suomi National Polar-orbiting Partnership, and the Advanced Baseline Imager on geostationary satellites GOES-16 (covering the eastern United States) and GOES-17 (covering the western United States; GOES = Geostationary Operational Environmental Satellite; see Holloway et al. [2021] for more information on satellite data used for fires and air quality). Geostationary satellites can provide continuous data every 15 minutes, while polar orbiting satellites can provide data only during the satellite overpass (Chow et al., 2021). The trade-off with geostationary-derived data is a lower spatial resolution compared to data from polar orbiting satellites (Jaffe et al., 2020).

Combining satellite-derived data with retrospective analysis of burned areas may provide the highest accuracy for estimates of WUI burned areas. For example, the Incident Command System Form 209 reports include information on the fire size and type from ground reports and helicopter perimeters. This information increases the spatial resolution and includes areas missed by satellites during cloud-obstructed times or due to fires not coinciding with the satellite overpass, resulting in a more complete burned-area data set (Larkin et al., 2020).

MEASURING FUELS AND ESTIMATING EMISSIONS

Gathering Data on Fuels for WUI Fires

The fuel loading and consumption for WUI fires are highly heterogeneous and are currently almost always determined through retrospective analysis (e.g., California Department of Forestry and Fire Protection or National Institute of Standards and Technology post-fire damage assessments). For wildfires burning on wildlands, fuel information is typically derived from satellite imagery and augmented by field surveys. Fuel loading data sets derived from satellite imagery are highly uncertain and may not be able to detect the ground fuels obscured by the tree canopy (Kennedy et al., 2020; Peterson et al., 2022). These fuel loading data sets may identify urban areas as unburnable and thus not include any structural or vehicle materials as a "fuel." Current practice does not estimate WUI fire emissions, and new methodology is needed to estimate emissions from the types of mixed fuels important at the WUI. Satellite data sets on buildings may be one way to develop a bulk structure loading data set comparable to the systems available for mapping wildland fuel loading. However, information on the elemental composition of WUI materials and surveys of the materials in WUI areas are still needed.

Research need: Data sets are needed on the materials loadings and compositions in WUI communities.

Estimating WUI Plume Characteristics and Emissions for Forecasts

Predictive models are important information in providing public health guidance (Chapter 4) and measurements play a critical role in evaluating predictive models; measurement versus model comparisons provide important insights as to what measurements are most critically needed and an assessment as to whether key processes are missing from the models. Ye et al. (2021) recently evaluated the forecasting ability of 12 smoke models for a case study on a single wildland fire. They found that all of the models underpredicted the amount and extent of the plume and that surface concentrations of fine particulate matter ($PM_{2.5}$) were both overpredicted and underpredicted. These biases were attributed to uncertainties in the emissions and the vertical distribution of the plume. This study provides some insight into where measurement efforts, in coordination with modeling, could be focused to integrate WUI fires into smoke forecasting.

The data required to estimate WUI plume characteristics and emissions draw upon many different data sources of fuels and burned areas and are paired with smoke models to estimate the amount of fuel consumed and combustion conditions. These smoke models rely on empirical estimates of various parameters, like fuel consumption by phase or plume rise, while other models employ physics-based algorithms to generate estimates (Chow et al., 2021). Computational fluid dynamics models have only recently been developed to include emission chemistry (Josephson et al., 2020) but remain too computationally intensive to be used for large spatial domains and are limited to retrospective analysis. Approaches have independently been developed to estimate emissions from structure and vehicle fires (not in the WUI fire context) using conventional emission inventory methodologies relying on activity data, fuel loadings, and emission factors. The methods used to estimate emissions described by the California Air Resource Board (CARB; Lozo, 1999) and by the SP Technical Research Institute of Sweden (SP; Blomqvist and McNamee, 2009) involve a range of assumptions.

The CARB 1999 method uses constant values for fuel loading, consumption, and emission factors, and determines emissions based on activity data in the form of the number of structures or vehicle fires. The amount of fuel loading is estimated from the average monetary damage claimed in home fires in comparison to the average value of a California residence in 1991. The method assumes an average home size of 1,649 ft^2 with 11 tons of lumber, and thus a combustible content of 7.9 lb/ft^2. Emission factors for CO, SO_x, total organic gas, and particulate matter are taken from emissions measurements from fires of model wood buildings (Butler and Darley, 1972). The NO_x emission factor is derived from municipal refuse (EPA, 1995).

The CARB approach can be contrasted with the SP approach (Blomqvist and McNamee, 2009), developed for fires in residences, schools, and vehicles. In the SP method, the structure fuel load is constant while the content loading depends upon the square footage. The SP method uses an assumed constant distribution of material type (e.g., wood, polyvinyl chloride, rubber) for the structure and contents and applies fuel-specific emission factors derived from Research Institutes of Sweden (RISE) emission measurement studies. The SP method also assumes that different amounts of the combustibles are consumed depending on the size of the fire, with 5 percent of the internal materials consumed when the fire is limited to one room, 35 percent of the internal materials and structure consumed when the fire is limited to several rooms, and 80 percent of the internal materials and structure consumed when the fire spreads to other buildings.

Neither of these methods makes any attempt to specify combustion conditions, which can strongly impact emissions, as described in Chapter 3. Chapter 3 also documents that emission factor data sets used in each method have limited chemical information and do not reflect realistic mixtures of urban and wildland fuels or WUI fire conditions. Additionally, these methods do not estimate the heat release rate, which is a critical factor determining plume rise and atmospheric dispersion.

In addition to the retrospective approaches described above, satellite measurements of fire radiative power could detect WUI fires and be included in top-down emission estimates. However, these approaches often use constant emission factors that are applied broadly across land types and therefore would not distinguish between emissions from structures and those from vegetation. As described in Chapter 3, unique emissions from structure fires or vehicles are not captured in current top-down estimates.

> **Research need:** There is a need to update the descriptions of fuel classes to include urban components in areas where WUI fires are common, to develop a methodology for top-down estimates to include the chemistry of burning structures and vehicles in the WUI, and to improve bottom-up emission estimation methodologies for WUI fires.

MEASURING EMISSIONS IN THE FIELD

Measurements of the composition of emissions from fires are made at a variety of spatial and temporal scales. Some of the measurement platforms that have been used in the near vicinity of fires are shown in Figure 7-2. Platforms include vehicles, aircraft, unmanned aerial systems, fixed or mobile ground instruments, instruments mounted on towers, and satellites. The instruments deployed on these platforms vary depending on the power, space, and weight of the instrument hardware, as well as the scientific objectives of the measurement campaign. An example of an advanced instrument package for a large aircraft is provided by the National Center for Atmospheric Research (NCAR-UCAR, 2018). There is a major distinction in the available measurement capabilities between long-term monitoring provided by ambient networks and satellite measurements and monitoring from intensive

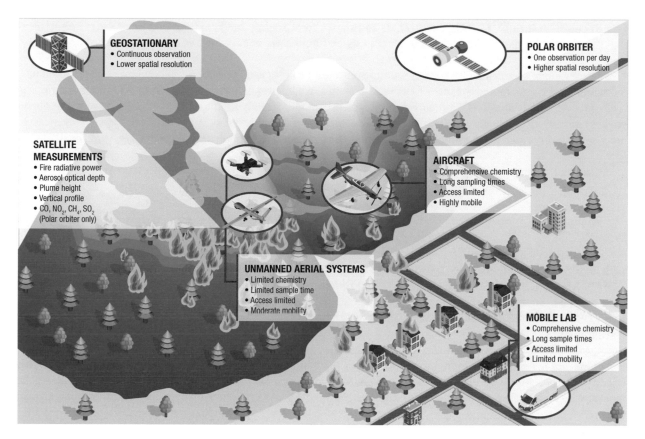

FIGURE 7-2 Measurement platforms for fire detection and emissions characterization. The measurements taken by the two satellite platforms are listed at the left.

research efforts that may be limited to a few weeks, months, or years during a specific study period. Although brief research studies can provide extremely valuable information, the data must generally be extrapolated across larger geographical and temporal scales outside of the study region and duration.

The measurement platform often determines the spatial and temporal resolution and the compounds that can be measured. For example, geostationary satellites measure continuously but have limited spatial resolution and are currently limited to measures such as fire radiative power and aerosol optical depth. Polar orbiting satellites are able to resolve plume heights, vertical profiles, and concentrations of CO, NO_2, CH_4, and SO_2 (Jaffe et al., 2020). Remotely sensed plume heights or pollutants can be measured by polar orbiting satellites only when the satellite overpass coincides with the fire. As a recent example, the Tropospheric Emissions: Monitoring of Pollution (TEMPO) instrument is anticipated to launch in November 2022 and will have spatial resolution as high as 2.0×4.75 km² and temporal resolution of 10 minutes or less. TEMPO will be able to monitor pollutants that may be important for WUI fires like NO_2, SO_2, HCHO, CHOCHO, and BrO and to measure aerosol optical depth, absorbing aerosol index, and single-scattering albedo, all of which may be useful for detecting WUI fires apart from other wildland fires (Naeger et al., 2021). TEMPO will have the capability for "near-real-time emission updates to improve air quality forecasts in modeling systems such as the experimental Rapid Refresh with Chemistry (RAP-Chem) model." Additional future satellite missions like the National Oceanic and Atmospheric Administration's Geostationary Extended Observations (GeoXO) and GOES-T may also increase the ability to track WUI fire smoke plumes.

Finding: Future satellite platforms will provide high spatial and temporal resolution of some chemical constituents that may be useful for tracking emissions from WUI fires.

Recommendation: WUI fire researchers and smoke forecasters should capitalize on advanced satellite measurements and participate as stakeholders to shape future satellite missions and data products.

Other mobile platforms like aircraft and instrumented vehicles can measure a much wider suite of pollutants, including continuous measurements of particulate matter, volatile organic compounds (VOCs), and inorganic gas (HCl, HF, etc.) concentrations. Unmanned aerial systems are increasingly used to measure fire emissions, but due to payload limitations and battery requirements, they can measure only a few compounds continuously, and they measure most emissions through batch samplers, so that emissions may not all be sampled simultaneously (Aurell et al., 2021).

Finding: Even with large aircraft and advanced instrument platforms, the spatial, temporal, and chemical coverage of fire plumes from any single measurement platform is limited.

To provide more spatial, temporal, and chemical coverage, multiple instrument platforms can be deployed, as shown in Figure 7-2. These coordinated deployments require extensive planning and yet must be deployed quickly to fire locations. The expense and difficulty of these campaigns mean that relatively few fires have been characterized with highly coordinated measurements. Recent examples are the Western Wildfire Experiment for Cloud Chemistry, Aerosol Absorption, and Nitrogen (WE-CAN) and Biomass Burn Observation Project (BBOP), which have largely been limited to aircraft measurements in smoke plumes. Some campaigns have focused on prescribed burns to allow for more-detailed advanced planning and a much larger range of data collected, particularly on the fuel composition, consumption, and fire behavior. Examples are the Prescribed Fire Combustion-Atmospheric Dynamics Experiments (Rx-CADRE) and the Fire and Smoke Model and Evaluation Experiment (FASMEE); however, these types of measurements are necessarily restricted to wildland fires. Fire Influence on Regional to Global Environments and Air Quality (FIREX-AQ) is an example of a multi-platform interdisciplinary approach to investigating wildfire smoke that may be a model for future efforts for measuring smoke from WUI fires. However, any future campaigns may require an extended commitment to ensure readiness not just during the traditional fire season (since WUI fires can occur at any time in the year), but also through multiple years to allow for the episodic nature of WUI fires.

Making measurements of the portions of fire plumes that come from the WUI becomes even more challenging. Many WUI fires occur during a very short time frame, often within the span of 24 hours, and the response during that time focuses on preventing loss of life and on defensive actions to protect structures. This rapid evolution of the fire reduces opportunities to capture air emissions from active phases of the fire. However, post-fire assessments are more feasible for measuring emissions from residual smoldering and the composition of deposited ash, residues, soil, and water samples. The National Institute of Standards and Technology has developed methods for collecting data in the field that are integrated into the Incident Command System, including pre-fire training with local partners, which enables rapid (<48 hour) approval of research studies and deployment of study teams into WUI areas (Maranghides et al., 2011).

Finding: WUI fires evolve rapidly, and emergency response activities may limit the ability to deploy and measure emissions from the active phases of the fire.

While field measurement campaigns characterizing wildland fires and WUI fires have limitations, they remain an essential tool for advancing the scientific understanding of wildland and WUI fires. Significant expansion of the data sets on WUI fires will require both continued commitment to existing measurement platforms, and identifying mechanisms for more deployment of instrumentation operating at multiple scales.

Research need: There is a need to develop planned research activities that can be integrated into the Incident Command System before an emergency event to facilitate rapid deployment of the measurement of emissions, residues, and environmental impacts from WUI fires.

Because of the limitations of field measurements of emissions from wildland and WUI fires, and because of the wide variety of fire conditions and fuels that may lead to emissions, laboratory investigations of wildland fire emissions, urban fire emissions, and mixed wildland and urban fire emissions play a critical role in understanding the chemistry of these fires. These research needs are described in Chapter 3.

MONITORING EMISSIONS USING CONTINUOUS/AMBIENT NETWORKS

Over larger spatial and longer temporal scales, fixed monitoring networks, which have been deployed for other purposes, can provide information about the large-scale dispersion of fire plumes. For example, a large number of continuous air quality monitoring stations (CAMSs) are operated throughout the United States by the federal, state, and local governments to assess compliance with National Ambient Air Quality Standards (NAAQS). Because these CAMSs are aimed at assessing NAAQS compliance, they are concentrated in urban areas, and measurements focus on criteria air pollutants such as ozone, particulate matter, and their precursors (e.g., oxides of nitrogen and VOCs), often on an hourly basis (Scheffe et al., 2009). Despite this limited chemical scope, these networks provide important information about the extent of exposure to fire plumes. A limited number of these CAMSs have specific routine measurements aimed at identifying the contributions of biomass burning to measured concentrations. For example, the State of Texas deploys instruments to measure the brown carbon component of particulate matter at multiple CAMS sites to assess contributions of biomass burning (Guagenti et al., 2019; Sheesley et al., 2019).

In addition to CAMS sites, several other ambient monitoring networks measure a larger range of chemical compounds and may be better equipped to detect the potentially unique chemical nature of a WUI fire plume. The national networks have widely available data for a limited number of locations, but many states and local agencies maintain other environmental monitoring networks that measure a variety of chemical species in different environmental matrices (air, water, and soil samples), with varying spatial and temporal coverage. Data from state and local measurements may not be available publicly but can often be obtained by request.

The national networks often operate on a sampling schedule, limiting their ability to capture transient WUI fire plumes. Most air monitoring networks sample for one 24-hour period on a multiday schedule (e.g., 1 day every 3 days, called 1-in-3). The critical review by Hidy et al. (2017) provides a description of air quality monitoring networks, the compounds they measure, and network design and operation in more detail. Table 7-1 lists some of these national air monitoring networks and their details. These types of platforms could be used to expand sampling capabilities for fires, taking advantage of the power, communication, and other infrastructure available at these sites. The opportunities for expanded sampling could also be extended to include national networks for other environmental media, and state and local networks.

Much of the continuous measurement data from national networks is available in near real time and assembled in databases by the US Environmental Protection Agency (EPA). Measurements that require extensive chemical analysis are available several months to a year after the sample was originally taken. These measurements follow

TABLE 7-1 Nationally Supported Ambient Air Monitoring Networks

Monitoring Network	Target Species	Siting Environment	Approximate Number of Sites	Sampling Schedule	Purpose
National Air Toxic Trends Station (EPA, 2021a)	Hazardous air pollutants (VOCs, carbonyls, metals, polycyclic aromatic hydrocarbons [PAHs])	Mostly urban, limited rural	26	1-in-6 or 1-in-12	Regulatory effectiveness
Chemical Speciation Network (EPA, 2021b)	$PM_{2.5}$ mass and chemical composition	Urban	150	1-in-3 or 1-in-6	Long-term air quality trends
Interagency Monitoring of Protected Visual Environments (IMPROVE) (IMPROVE, 2022)	$PM_{2.5}$ and coarse particulate matter (PM_{10}) chemical composition	Rural	110	1-in-3	Long-term visibility trends
National Core Monitoring Network (EPA, 2016)	$PM_{2.5}$ mass and chemical composition, O_3, CO, SO_2, reactive nitrogen	Urban and rural	78	1-in-3	Scientific research
Clean Air Status and Trends Network (EPA, 2021c)	$PM_{2.5}$ mass, ammonia, acidic gases, ions	Rural	100	1-in-7	Long-term acid deposition trends
Atmospheric Science and Chemistry Measurement Network (GATech, 2021)	$PM_{2.5}$ chemical composition and size distribution	Urban and rural	12	Continuous	Scientific research

established EPA methods and quality assurance procedures that ensure the validity of the data. EPA maintains standard measurement methods for a large number of chemical compounds in different environmental matrices (e.g., Compendium of Methods for the Determination of Toxic Organic Compounds in Air) that may need readily applied or adapted WUI fire measurements.

Ambient monitoring networks provide insights into ambient concentrations associated with total fire plumes but are generally not able to distinguish the contributions from fires at the WUI. However, they provide valuable data that can be used to evaluate the performance of air quality models or for studies that examine the sources of emissions and the amount from each source, to constrain estimates of emissions from WUI fires.

Multiple limitations to the current monitoring networks impede the characterization of WUI fire plumes. Most important are the limited spatial and temporal coverage of detailed chemical speciation measurements. Most monitoring sites are in urban areas distant from wildlands that may be less likely to be impacted by a WUI fire plume. Additionally, these networks were designed to monitor long-term trends, and the scheduled sampling time may not coincide with when a plume impacts the site. Samplers may lose power due to the fire or become overloaded due to high concentrations.

> **Finding:** Some ambient monitoring networks measure an extensive range of hazardous air pollutants that may be useful to identify the environmental impacts of WUI fires; however, these networks have limited spatial and temporal resolution.

In addition to the permanent monitoring programs described above, US states and the US Forest Service operate temporary monitoring programs through the Interagency Wildland Fire Air Quality Response Program (USFS, 2021). These deployments target areas where smoke impacts are anticipated and where few or no monitoring stations exist. After the 2018 Camp Fire, the California Air Resources Board identified the need for portable samplers for particulate matter and hazardous air pollutants for specific wildfire events in addition to increasing the sampling frequency and duration in monitoring networks (CARB, 2021).

> **Research need:** Identify possible expansions of the capabilities of national, regional, and local monitoring networks that would enhance information regarding fire emissions and exposures. Evaluate the capabilities and information that could be provided by permanently deployed equipment and equipment that might be portable and deployed on an as-needed basis.

ASSESSING EXPOSURE

As discussed in Chapter 6, exposure is contact with a substance (e.g., WUI fire emissions) via inhalation, ingestion, or touching of the skin. Therefore, emissions data and ambient environment monitoring data to estimate exposures need to be supplemented with the spatial and temporal patterns of human contact to estimate human exposure. Exposure measurement represents individual experience and can capture the intra- and inter-individual variability of impacts within a population. This more accurate representation can be especially important in the case of the dynamic emissions and transport in WUI fires. Thus, exposure measurement is applicable for constructing exposure-response relationships, determining thresholds for exposure controls, and assessing the effectiveness of interventions to reduce exposures.

Exposure is often represented as the concentration of the substance in the relevant environmental matrix/matrices at the location(s) where contact is being made, and its quantification may also take into account the frequency and duration of contact (EPA, 2021d). Measurements of human exposures to environmental contaminants including WUI fire emissions can be conducted through area, personal, and biological monitoring.

Area monitoring for airborne contaminants involves the sampling of air with monitoring equipment at specific indoor or outdoor locations. This is analogous to the collection of samples of water or soil with which persons may interact. Such monitoring discounts the possibility that individuals may interact with multiple microenvironments with varying fire emission concentrations during and after WUI fires.

Personal monitoring eliminates this limitation because of the placement or direct application of the monitoring equipment on the individual. While this approach has been used for firefighters at wildland and structural fires, the unplanned and chaotic nature of fires almost entirely precludes its use for the general population. It can also be impractical due to the cost of implementation for a large population and the intrusiveness that may accompany the wearing of monitoring equipment that may be bulky and/or noisy. However, personal monitoring of first responders such as firefighters is a reasonable option that should be considered.

Rather than the external concentration that is measured with area and personal monitoring, biological monitoring (biomonitoring) involves the measurement of biomarkers, which are substances (e.g., fire emissions or their metabolites) in biological fluid or tissue that are indicative of exposure. While they may more accurately represent internal exposure, biomarkers (e.g., urinary hydroxylated PAHs) that have been used to assess exposures to wildland and structural fires are nonspecific and may be due to exposures from other sources. They also do not discriminate between exposure routes (inhalation, ingestion, or dermal absorption) and are subject to inter-individual variability in their biokinetic behavior.

Regardless of whether area, personal, or biological exposure monitoring is employed, exposure measurement involves sampling the matrix containing the analyte used for quantifying exposure and its instrumental analysis. Methods of sampling for WUI fire emissions are summarized below.

Air Sampling

Generally, sampling for airborne contaminants including WUI fire emissions can be achieved (1) passively by relying on diffusive air flow through a medium that is used to abstract the contaminant from the airstream or (2) actively by pumping air at a specified flow rate through the medium. Published methods for known WUI fire emissions require active sampling, with an exception being the National Institute for Occupational Safety and Health (NIOSH) Method 6700 for NO_2.

The medium used for sampling depends on the physical and chemical characteristics of the contaminant. Particulates including fine particles, asbestos, and metals are sampled for analysis using filters of various membrane materials (e.g., Teflon and polyvinyl chloride). Gas-phase contaminants, including ammonia, inorganic acid and asphyxiant gases, ozone, and VOCs, are sampled by adsorption to (e.g., NIOSH Method 1501 for aromatic hydrocarbons including benzene) or reaction with (e.g., Occupational Safety and Health Administration [OSHA] Method ID-214 for ozone) a solid material, or by reaction in a solvent contained in an impinger (e.g., NIOSH Method 3500 for formaldehyde). Semi-VOCs are sampled using a combination of filters and sorbent material (e.g., NIOSH Method 5515 for PAHs). Whole air samples can also be actively sampled into containers including evacuated canisters and sampling bags (e.g., NIOSH Method 3900 for VOCs).

Although area and personal air monitoring data might represent more accurate exposure information, it is difficult to put resources in place to collect such data for the unplanned, relatively short and episodic air pollution that is caused by WUI fires. Therefore, such data are not typically available.

Soil, Water, and Surface Sampling

Soil, ash, and water for instrumental analysis of WUI fire emissions can be collected from an area/environment that is accessible to individuals by discrete or composite sampling of bulk material into appropriate, precleaned containers. Equipment used to retrieve such samples includes the trowel and auger for soil, brush and spoon for ash, hand-held and weighted bottles for surface water, and bailers and bladder pumps for groundwater. Some automation may be involved in some of the soil and water sampling procedures.

Settled dust on smooth, nonporous surfaces can be sampled for WUI fire emissions using wipes made from various materials including polyester, polypropylene, polyvinyl alcohol, and polyester plus cellulose (OSHA, 2001). The wipes are prewetted, for example, with water for sampling metals (e.g., OSHA Method ID-206) or with organic solvents for sampling organics (e.g., American Society for Testing and Materials [ASTM] Standard Practice D6661-17 [ASTM, 2017]). As an example, Hwang et al. (2019) used Alpha Wipes prewetted with isopropyl alcohol for sampling various surfaces in fire vehicles of structural firefighters for PAHs (Hwang et al., 2019).

Vacuuming can be used to sample settled dust from the floor and porous surfaces (e.g., ASTM Method D5438). Micro-vacuuming settled dust onto a filter membrane cassette is an alternative approach and is especially applied for sampling for non-airborne, settled asbestos (ASTM Method 5755). However, soil, water, and surface monitoring data often indicate potential exposure and may not accurately represent actual exposure. While the lack of baseline measurements can hamper the applicability of soil, water, and surface monitoring for exposure assessment (Murphy et al., 2020), post-fire longitudinal monitoring can inform studies about changes in potential exposure to WUI fire–mediated pollution, as demonstrated in a study of the watersheds impacted by the Horse River Fire (Emmerton et al., 2020).

Sampling for Dermal Absorption

Sampling of clothing and the skin with the same wipe materials and wetting agents that are used for surface sampling can be used for dermal exposure assessment. Corn oil has also been used as a wetting agent (Fent et al., 2014, 2017). Skin and/or cloth wipes have been used to sample for PAHs among structural and wildland firefighters (Banks et al., 2021; Cherry et al., 2021; Fent et al., 2014, 2017) and dioxins, furans, and flame retardants among structural firefighters (Fent et al., 2020). Passive sampling by the placement of an adsorbent, silicone-based wristband or dog tag placed under firefighting clothing (Baum et al., 2020; Caban-Martinez et al., 2020; Poutasse et al., 2020), and tape stripping of the skin, have also been used to assess PAH exposure among structural firefighters (Sjostrom et al., 2019; Strandberg et al., 2018). However, the need to compare with a baseline, to adequately determine exposure, limits the applicability of these dermal sampling approaches to unplanned wildland and WUI fire events. Furthermore, standardized methodology and dermal exposure standards are lacking.

Biological Sampling

Blood, exhaled breath, and urine have been used for the biomonitoring of structural and wildland firefighters in many recent studies (Adetona et al., 2017; Ekpe et al., 2021; Fent et al., 2014, 2019; Mayer et al., 2021; Shaw et al., 2013). Urine and exhaled breath are the preferred biological fluids for short-lived and readily excretable biomarkers that are used for biomonitoring acute exposures such as PAHs and benzene (Adetona et al., 2017; Fent et al., 2014, 2019). Such monitoring often needs a baseline comparison and timely collection of samples. Long-lived and/or lipophilic biomarkers that reflect cumulative chronic exposures such as polychlorinated biphenyls, brominated flame retardants, perfluoroalkyl substances, dioxins, and furans are typically measured in the blood (Ekpe et al., 2021; Park et al., 2015; Shaw et al., 2013). Since these biomarkers are nonspecific, their use for the biomonitoring of exposure to WUI fire smoke may be confounded by other sources of exposure. However, this limitation can partially be addressed by collecting and comparing pre- and post-exposure samples. Biomonitoring also can afford the opportunity to be efficient in the evaluation of exposures and toxic effects in the same biological fluid.

Instantaneous Monitoring by Direct Reading Instruments

Direct reading (instantaneous monitoring) instruments that sample, perform instrumental analysis, and read out almost instantaneously are available for various WUI fire emissions. Examples include single- or multiple-sensor monitors that are used for many inorganic gases, and the electrochemical sensor that is the specified technique in NIOSH Method 6604 for CO monitoring (Wu et al., 2021). Other examples include a nondispersive infrared monitor for CO_2 monitoring, portable gas chromatograph/photoionization detector analyzer for VOC measurement, X-ray fluorescence analyzer for in situ measurement of metals in soil, and real-time meters that are used to measure general water quality parameters such as turbidity, total dissolved solids, and dissolved oxygen.

Many of these instruments can be used for both area and personal monitoring since they are portable. Continuous or integrated monitoring can also be conducted with instantaneous monitoring instruments that are equipped with data loggers. CO from exhaled breath, a biomarker, has also been measured among wildland firefighters using direct reading monitors (Dunn et al., 2009; Miranda et al., 2012). However, many of these instruments are susceptible to interferences, may not be accurate, and have relatively low sensitivity and high limits of detection (Spinelle et al., 2017).

Exposure Modeling

Modeling can potentially be used to improve exposure estimates that are obtained from area and personal monitoring. Statistical models such as land use regression, geographically weighted regression, and random forest have been used to estimate $PM_{2.5}$ concentration across space based on measurements from stationary regulatory ambient air monitoring stations, including during a WUI fire (Gan et al., 2017; Mirzaei et al., 2018; Vu et al., 2022). Such modeling allows for the spatiotemporal variability in concentrations to be better captured and has been used in the assessment of exposure-response relationships related to wildfire-impacted air quality (Gan et al., 2017). These models can include meteorology and land use variables, population density, and satellite-based pollution data as independent variables, with ground-level measurements as the dependent variable.

These models can be adapted for use with area monitoring, as has been demonstrated using data from the crowdsourced network of low-cost direct reading $PM_{2.5}$ PurpleAir sensors (Lu et al., 2022). Moreover, model predictions were improved when the ambient regulatory data and the crowdsourced, area-monitoring data were both used in the models (Lu et al., 2022; Vu et al., 2022).

Even so, measurements based on personal and enhanced area monitoring still have limitations and do not incorporate exposure factors that are important causes of variability in exposure. As previously noted, area monitoring assumes a similar frequency and duration of exposure across different microenvironments, and both approaches do not incorporate factors that determine the uptake of contaminants, such as inhalation rate and gas absorption, or particle deposition rates for inhalation exposures. When appropriate, the integration of these factors with monitoring data will result in more accurate estimations of exposure, improved definition of susceptible and vulnerable populations, more refined strategies for exposure control, and better testing of these strategies' effectiveness.

Research need: There is a need to identify new biomarkers that are more specific for WUI fires and can be used for exposure and toxicity assessment.

SAMPLING AND ANALYTICAL METHODS

From fuels and emissions to exposures, measurement approaches and data rely on advanced analytical techniques. Despite recent advances in analytical chemistry, little progress has been made in the determination and analysis of fire emissions. The major limitations are near-field access to fires and the complexity involved in sampling and measuring fire emissions (Stec, 2017).

The choice of sampling method and analytical method to characterize airborne contaminants at a fire incident depends on the physical nature of the airborne samples (i.e., vapor and/or aerosol), the estimated concentrations of contaminants, and any potential interactions with or interferences from other contaminants. Characterization of the fire environment is therefore based on the identification of which of these predefined compounds is considered to be the most significant or major component of the smoke. The choice of specific chemicals to monitor is also based on the availability of methods that reliably collect and analyze air-contaminant samples in the fire environment. Therefore, sampling and analysis focus on relatively simple analytes (or groups of analytes) of concern.

Over the last few years, the use of low-cost screening devices and sensors has increased. The most common are optical particle counters for measuring the size distribution of particles (particulate matter, such as PM_1, $PM_{2.5}$, and PM_{10}), electrochemical sensors (for CO, O_3, NO, HCl, HBr, NO_2, SO_2, H_2S, VOCs, etc.), photoionization detectors, and infrared spectral photometers. While most of these sensors are portable and user friendly, several limitations need to be considered. These include cross-sensitivity, interference from environmental effects (wind, rain, etc.), baseline drift, linearity, and dynamic range. Selectivity of the analytical method (i.e., avoiding matrix effects and/or interference from other fire species), sensitivity, limits of detection, and quantification, to name a few major issues, also need to be carefully considered when selecting from the large number of analytical methodologies.

Research need: There is a need to develop new analytical capabilities for measuring a wide variety of chemical, particle, and biological indicators of toxicants in emission, exposure, and health effects studies.

Research need: There is a need to optimize analytical methods for field deployment and increased accessibility.

COORDINATING MEASUREMENTS OF WUI FIRES

Some types of data discussed in this chapter characterize the fires and the fuels that are combusted. Other types of data are direct field measurements of WUI fire emissions at a variety of spatial scales. These field measurements complement data from controlled laboratory experiments where fuel and fire conditions can be systematically examined and where a greater variety of chemical measurements can be deployed than are available for field studies. Data on the fate, transport, and transformation of emissions need to be coupled with data that characterize human exposure, including personal exposure measurements and biomonitoring. Because each of these data elements informs other elements, data collected in a coordinated way are most valuable.

REFERENCES

Adetona, O., C. D. Simpson, Z. Li, A. Sjodin, A. M. Calafat, and L. P. Naeher. 2017. "Hydroxylated Polycyclic Aromatic Hydrocarbons as Biomarkers of Exposure to Wood Smoke in Wildland Firefighters." *Journal of Exposure Science and Environmental Epidemiology* 27 (1): 78–83. https://doi.org/10.1038/jes.2015.75.

ASTM (American Society for Testing and Materials). 2017. *Standard Practice for Field Collection of Organic Compounds from Surfaces Using Wipe Sampling.* ASTM D6661-17.

Aurell, J., B. Gullett, and A. Holder. 2021. "Wildland Fire Emission Sampling at Fishlake National Forest, Utah Using an Unmanned Aircraft System." *Atmospheric Environment.* https://doi.org/10.1016/j.atmosenv.2021.118193.

Banks, A. P. W., P. Thai, M. Engelsman, X. Wang, A. F. Osorio, and J. F. Mueller. 2021. "Characterising the Exposure of Australian Firefighters to Polycyclic Aromatic Hydrocarbons Generated in Simulated Compartment Fires." *International Journal of Hygiene and Environmental Health* 231. https://doi.org/10.1016/j.ijheh.2020.113637.

Baum, J. L. R., U. Bakali, and C. Killawala. 2020. "Evaluation of Silicone-Based Wristbands as Passive Sampling Systems Using PAHs as an Exposure Proxy for Carcinogen Monitoring in Firefighters: Evidence from the Firefighter Cancer Initiative." *Ecotoxicology and Environmental Safety* 205. https://doi.org/10.1016/j.ecoenv.2020.111100.

Blomqvist, P., and M. McNamee. 2009. *Estimation of CO_2-Emissions from Fires in Dwellings, Schools and Cars in the Nordic Countries.* SP Technical Note 2009:13. Borås, Sweden: SP Swedish National Testing and Research Institute. https://www.diva-portal.org/smash/get/diva2:961412/FULLTEXT01.pdf.

Butler, C. P., and E. F. Darley. 1972. "Fire Dynamics of Model Wood Buildings Fire and Flammability." *Journal of Fire and Flammability* 3: 330–339.

Caban-Martinez, A., P. Louzado-Feliciano, and K. M. Santiago. 2020. "Objective Measurement of Carcinogens among Dominican Republic Firefighters Using Silicone-Based Wristbands." *Journal of Occupational and Environmental Medicine* 62. https://doi.org/10.1097/JOM.0000000000002006.

CARB (California Air Resources Board). 2021. *Camp Fire Air Quality Data Analysis.* Sacramento, CA: California Air Resources Board. https://ww2.arb.ca.gov/resources/documents/camp-fire-air-quality-data-analysis.

Cherry, N., J. M. Galarneau, D. Kinniburgh, B. Quemerais, S. Tiu, and X. Zhang. 2021. "Exposure and Absorption of PAHs in Wildland Firefighters: A Field Study with Pilot Interventions." *Annals of Work Exposures and Health* 65 (2): 148–161. https://doi.org/10.1093/annweh/wxaa064.

Chow, F. K., K. A. Yu, and A. Young. 2021. "High-Resolution Smoke Forecasting for the 2018 Camp Fire in California." *Bulletin of the American Meteorological Society.* https://doi.org/10.1175/BAMS-D-20-0329.1.

Dunn, K. H., I. Devaux, and A. Stock. 2009. "Application of End-Exhaled Breath Monitoring to Assess Carbon Monoxide Exposures of Wildland Firefighters at Prescribed Burns." *Inhalation Toxicology* 21 (1). https://doi.org/10.1080/08958370802207300.

Ekpe, O. D., W. Sim, and S. Choi. 2021. "Assessment of Exposure of Korean Firefighters to Polybrominated Diphenyl Ethers and Polycyclic Aromatic Hydrocarbons via Their Measurement in Serum and Polycyclic Aromatic Hydrocarbon Metabolites in Urine." *Environmental Science & Technology* 55 (20): 14015–14025. https://doi.org/10.1021/acs.est.1c02554.

Emmerton, C. A., C. A. Cooke, S. Hustins, U. Silins, M. B. Emelko, T. Lewis, M. K. Kruk, N. Taube, D. Zhu, B. Jackson, M. Stone, J. G. Kerr, and J. F. Orwin. 2020. "Severe Western Canadian Wildfire Affects Water Quality Even at Large Basin Scales." *Water Research* 183. https://doi.org/10.1016/j.watres.2020.116071.

EPA (US Environmental Protection Agency). 1995. *Compilation of Air Emissions Factors.* https://www.epa.gov/air-emissions-factors-and-quantification/ap-42-compilation-air-emissions-factors.

EPA. 2016. *NCore Multipollutant Monitoring Network.* https://www3.epa.gov/ttn/amtic/ncore.html.

EPA. 2021a. *Air Toxics Ambient Monitoring.* https://www.epa.gov/amtic/air-toxics-ambient-monitoring.

EPA. 2021b. *Chemical Speciation Network (CSN).* https://www.epa.gov/amtic/chemical-speciation-network-csn.

EPA. 2021c. *Clean Air Status and Trends Network (CASTNET).* https://www.epa.gov/castnet.

EPA. 2021d. *Exposure Assessment Tools by Approaches*. https://www.epa.gov/expobox/exposure-assessment-tools-approaches.

Fent, K. W., J. Eisenberg, J. Snawder, D. Sammons, J. D. Pleil, M. A. Stiegel, C. Mueller, G. P. Horn, and J. Dalton. 2014. "Systemic Exposure to PAHs and Benzene in Firefighters Suppressing Controlled Structure Fires." *Annals of Occupational Hygiene* 58 (7): 830–845. https://doi.org/10.1093/annhyg/meu036.

Fent, K. W., B. Alexander, J. Roberts, S. Robertson, C. Toennis, D. Sammons, S. Bertke, S. Kerber, D. Smith, and G. Horn. 2017. "Contamination of Firefighter Personal Protective Equipment and Skin and the Effectiveness of Decontamination Procedures." *Journal of Occupational and Environmental Hygiene* 14 (10): 801–814. https://doi.org/10.1080/15459624.2017.1334904.

Fent, K. W., C. Toennis, D. Sammons, S. Robertson, S. Bertke, A. M. Calafat, J. D. Pleil, M. A. Geer Wallace, S. Kerber, D. L. Smith, and G. P. Horn. 2019. "Firefighters' and Instructors' Absorption of PAHs and Benzene during Training Exercises." *International Journal of Hygiene and Environmental Health* 222 (7): 991–1000. https://doi.org/10.1016/j.ijheh.2019.06.006.

Fent, K. W., M. LaGuardia, D. Luellen, S. McCormick, A. Mayer, I. Chen, S. Kerber, D. Smith, and G. P. Horn. 2020. "Flame Retardants, Dioxins, and Furans in Air and on Firefighters' Protective Ensembles during Controlled Residential Firefighting." *Environment International* 140. https://doi.org/10.1016/j.envint.2020.105756.

Gan, R. W., B. Ford, W. Lassman, G. Pfister, A. Vaidyanathan, E. Fischer, J. Volckens, J. R. Pierce, and S. Magzamen. 2017. "Comparison of Wildfire Smoke Estimation Methods and Associations with Cardiopulmonary-Related Hospital Admissions." *GeoHealth* 1 (3): 122–136. https://doi.org/10.1002/2017GH000073.

Guagenti, M., Shrestha, S., Flynn, J., Alvarez, S. L., Sheesley, R. J., and Usenko, S. December 2019. "Utilizing Intensive Aerosol Optical Properties for The Detection of Biomass Burning in El Paso, Texas-(BC)^2 El Paso Field Campaign." In *American Geophysical Union, Fall Meeting 2019* abstract #A23L–2967.

GATech (Georgia Institute of Technology). 2021. "$12 Million NSF Grant Will Establish Nationwide Atmospheric Measurement Network." https://research.gatech.edu/12-million-nsf-grant-will-establish-nationwide-atmospheric-measurement-network.

Hidy, G. M., P. K. Mueller, S. L. Altshuler, and J. C. Chow. 2017. "Air Quality Measurements—From Rubber Bands to Tapping the Rainbow. *Journal of the Air & Waste Management Association* 67 (6). https://doi.org/10.1080/10962247.2017.1308890.

Holloway, T., D. Miller, S. Anenberg, M. Diao, B. Duncan, A. M. Fiore, D. K. Henze, J. Hess, P. L. Kinney, Y. Liu, J. L. Neu, S. M. O'Neill, M. T. Odman, R. B. Pierce, A. G. Russell, D. Tong, J. J. West, and M. A. Zondlo. 2021. "Satellite Monitoring for Air Quality and Health." *Annual Review of Biomedical Data Science* 4 (1): 417–447. https://doi.org/10.1146/annurev-biodatasci-110920-093120.

Hwang, J., R. Taylor, and C. Cann. 2019. "Evaluation of Accumulated Polycyclic Aromatic Hydrocarbons and Asbestiform Fibers on Firefighter Vehicles: Pilot Study." *Fire Technology* 55. https://doi.org/10.1007/s10694-019-00851-7.

IMPROVE (Interagency Monitoring of Protected Visual Environments). 2022. "Interagency Monitoring of Protected Visual Environments." http://vista.cira.colostate.edu/Improve/.

Jaffe, D. A., S. M. O'Neill, N. K. Larkin, A. L. Holder, D. L. Peterson, J. E. Halofsky, and A. G. Rappold. 2020. "Wildfire and Prescribed Burning Impacts on Air Quality in the United States." *Journal of the Air & Waste Management Association* 70 (6): 583–615. https://doi.org/10.1080/10962247.2020.1749731.

Josephson, A. J., D. Castano, and E. Koo. 2020. "Zonal-Based Emission Source Term Model for Predicting Particulate Emission Factors in Wildfire Simulations." *Fire Technology* 57. https://doi.org/10.1007/s10694-020-01024-7.

Kennedy, M. C., S. J. Prichard, and D. McKenzie. 2020. "Quantifying How Sources of Uncertainty in Combustible Biomass Propagate to Prediction of Wildland Fire Emissions." *International Journal of Wildland Fire* 29. https://doi.org/10.1071/WF19160.

Larkin, N. K., S. M. Raffuse, and S. Huang. 2020. "The Comprehensive Fire Information Reconciled Emissions (CFIRE) Inventory: Wildland Fire Emissions Developed for the 2011 and 2014 U.S. National Emissions Inventory." *Journal of the Air & Waste Management Association* 70 (11). https://doi.org/10.1080/10962247.2020.1802365.

Lozo, C. 1999. *Structure and Automobile Fires*. https://www.arb.ca.gov/ei/areasrc/fullpdf/full7-14.pdf.

Lu, T., M. J. Bechle, and Y. Wan. 2022. "Using Crowd-Sourced Low-Cost Sensors in a Land Use Regression of PM$_{2.5}$ in 6 US Cities." *Air Quality, Atmosphere & Health* 15: 667–678. https://doi.org/10.1007/s11869-022-01162-7.

Maranghides, A., W. Mell, K. Ridenour, and D. McNamara. 2011. *Initial Reconnaissance of the 2011 Wildland-Urban Interface Fires in Amarillo, Texas*. Gaithersburg, MD: National Institute of Standards and Technology.

Mayer, A. C., K. W. Fent, and I. Chen. 2021. "Characterizing Exposures to Flame Retardants, Dioxins, and Furans among Firefighters Responding to Controlled Residential Fires." *International Journal of Hygiene and Environmental Health* 236. https://doi.org/10.1016/j.ijheh.2021.113782.

Miranda, A. I., V. Martins, and P. Cascao. 2012. "Wildland Smoke Exposure Values and Exhaled Breath Indicators in Firefighters." *Journal of Toxicology and Environmental Health, Part A* 75. https://doi.org/10.1080/15287394.2012.690686.

Mirzaei, M., S. Bertazzon, and I. Couloigner. 2018. "Modeling Wildfire Smoke Pollution by Integrating Land Use Regression and Remote Sensing Data: Regional Multi-Temporal Estimates for Public Health and Exposure Models." *Atmosphere* 9 (9). https://doi.org/10.3390/atmos9090335.

Murphy, S. F., R. B. McCleskey, D. A. Martin, J. M. Holloway, and J. H. Writer. 2020. "Wildfire-Driven Changes in Hydrology Mobilize Arsenic and Metals from Legacy Mine Waste." *Science of the Total Environment* 743: 140635. https://doi.org/10.1016/j.scitotenv.2020.140635.

Naeger, A. R., M. J. Newchurch, T. Moore, K. Chance, X. Liu, S. Alexander, K. Murphy, and B. Wang. 2021. "Revolutionary Air-Pollution Applications from Future Tropospheric Emissions: Monitoring of Pollution (TEMPO) Observations." *Bulletin of the American Meteorological Society* 102 (9): E1735–E1741. https://journals.ametsoc.org/view/journals/bams/102/9/BAMS-D-21-0050.1.xml (accessed July 16, 2022).

NCAR-UCAR (National Center for Atmospheric Research–University Corporation for Atmospheric Research). 2018. *WE-CAN Instrumentation Payload*. https://www.eol.ucar.edu/content/we-can-instrumentation-payload.

NIFC (National Interagency Fire Center). 2022. *Fire Information: Statistics*. https://www.nifc.gov/fire-information/statistics.

OSHA (Occupational Safety and Health Administration). 2001. *Evaluation Guidelines for Surface Sampling Methods*. OSHA T-006-01-0104-M.

Park, J., R. W. Voss, and S. McNeel. 2015. "High Exposure of California Firefighters to Polybrominated Diphenyl Ethers." *Environmental Science & Technology* 49. https://doi.org/10.1021/es5055918.

Peterson, D. L., S. McCaffrey, and T. Patel-Weynand, eds. 2022. *Wildfire Smoke in the United States: A Scientific Assessment*. New York, NY: Springer.

Poutasse, C. M., W. S. C. Poston, and S. A. Jahnke. 2020. "Discovery of Firefighter Chemical Exposures Using Military-Style Silicone Dog Tags." *Environment International* 142. https://doi.org/10.1016/j.envint.2020.105818.

Scheffe, R. D., P. A. Solomon, and R. Husar. 2009. "The National Ambient Air Monitoring Strategy: Rethinking the Role of National Networks." *Journal of the Air & Waste Management Association* 59 (5): 579–590. https://doi.org/10.3155/1047-3289.59.5.579.

Shaw, S. D., M. L. Berger, J. H. Harris, S. H. Yun, Q. Wu, C. Liao, A. Blum, A. Stefani, and K. Kannan. 2013. "Persistent Organic Pollutants Including Polychlorinated and Polybrominated Dibenzo-p-dioxins and Dibenzofurans in Firefighters from Northern California." *Chemosphere* 91 (10): 1386–1394. https://doi.org/10.1016/j.chemosphere.2012.12.070.

Sheesley, R., S. Usenko, and J. Flynn. 2019. "Detecting Events and Seasonal Trends in Biomass Burning Plumes using Black and Brown Carbon: (BC)² El Paso." Proposal submitted to the Texas Air Quality Research Program, http://aqrp.ceer.utexas.edu/.

Sjostrom, M., A. Julander, and B. Strandberg. 2019. "Airborne and Dermal Exposure to Polycyclic Aromatic Hydrocarbons, Volatile Organic Compounds, and Particles among Firefighters and Police Investigators." *Annals of Work Exposures and Health* 63 (5). https://doi.org/10.1093/annweh/wxz030.

Spinelle, L., M. Gerboles, and G. Kok. 2017. "Review of Portable and Low-Cost Sensors for the Ambient Air Monitoring of Benzene and Other Volatile Organic Compounds." *Sensors* 17 (7). https://doi.org/10.3390/s17071520.

Stec, A. A. 2017. "Fire Toxicity – The Elephant in the Room?" *Fire Safety Journal* 91: 79–90. https://doi.org/10.1016/j.firesaf.2017.05.003.

Strandberg, B., A. Julander, and M. Sjostrom. 2018. "An Improved Method for Determining Dermal Exposure to Polycyclic Aromatic Hydrocarbons." *Chemosphere* 198. https://doi.org/10.1016/j.chemosphere.2018.01.104.

USFS (US Forest Service). 2021. *Interagency Wildland Fire Air Quality Response Program*. https://www.wildlandfiresmoke.us/ (accessed 2022).

Vu, B., J. Bi, and W. Wang. 2022. "Application of Geostationary Satellite and High-Resolution Meteorology Data in Estimating Hourly PM$_{2.5}$ Levels during the Camp Fire Episode in California." *Remote Sensing of Environment* 271. https://doi.org/10.1016/j.rse.2022.112890.

Wooster, M. J., G. Roberts, and G. L. W. Perry. 2005. "Retrieval of Biomass Combustion Rates and Totals from Fire Radiative Power Observations: FRP Derivation and Calibration Relationships between Biomass Consumption and Fire Radiative Energy Release." *Journal of Geophysical Research: Atmospheres* 110. https://doi.org/10.1029/2005JD006318.

Wu, C., C. Song, and R. Chartier. 2021. "Characterization of Occupational Smoke Exposure among Wildland Firefighters in the Midwestern United States." *Environmental Research* 193. https://doi.org/10.1016/j.envres.2020.110541.

Ye, X., P. Arab, and R. Ahmadov. 2021. "Evaluation and Intercomparison of Wildfire Smoke Forecasts from Multiple Modeling Systems for the 2019 Williams Flats Fire." *Atmospheric Chemistry and Physics* 21. https://doi.org/10.5194/acp-21-14427-2021.

8

The Future of WUI Fire Research

Wildland-urban interface (WUI) communities are growing rapidly, and the threat of severe fires in these communities is also growing. WUI fires vary in their characteristics and include large wildland fires that burn communities in their path, and fires of smaller spatial extent in intermixed wildland and urban areas. Despite their varied characteristics, WUI fires have the common characteristics of fuel mixes that are distinct from both wildland and urban fires, and a combustion chemistry that is spatially and temporally heterogeneous. These unique characteristics of WUI fires can lead to toxicants and exposures that are largely uncharacterized.

Earlier chapters in this report identified research needs for WUI fires that include the following:

- Identifying at-risk communities and vulnerable populations and evaluating interventions
- Characterizing fuel types, combustion characteristics, and emissions
- Defining the chemistry of toxicants from WUI fires in air, in water, and on land, including ash and debris
- Predicting the multimedia transport and transformation of WUI fire toxicants
- Identifying and quantifying toxicant exposures and health impacts resulting from WUI fires

This chapter summarizes and prioritizes those many research needs into a research agenda. To be most effective, the research agenda should be carried out with multidisciplinary teams who integrate work that ranges from hazard identification and emission characterization to measurement of exposures and health outcomes.

Figure 8-1 maps this range of WUI fire–related research activities and their interdependences. As shown in the figure, WUI communities at significant risk of fire need to be systematically identified, along with vulnerable populations. Once communities are identified, the potential fuels in those communities can be identified and characterized, including an analysis of how those characteristics vary among different populations. Laboratory and field measurements are needed to improve understanding of the combustion conditions that these fuels are exposed to. The fuels and combustion conditions can help reveal important multimedia chemical pathways of toxicant formation in WUI fires. The toxicant formation pathways, in turn, influence the transport and transformation of toxicants as they leave the immediate fire plume and their chemical evolution over time. The transport and chemical evolution of the pollutants, and the distribution of vulnerable populations, will determine potential exposures and health outcomes. These potential exposures and health outcomes must be related back to an understanding of the at-risk communities and vulnerable populations.

Implementing an effective research agenda will be challenging. While fundamental data relevant to wildland and urban fires have been collected for decades, understanding the combustion of mixed wildland and urban fuels

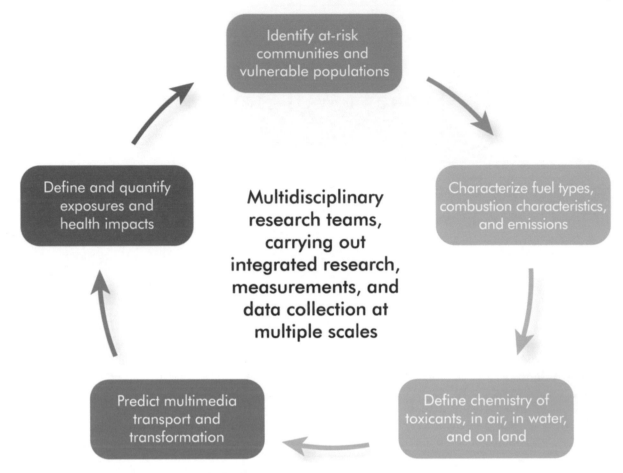

FIGURE 8-1 The interdependence of multidisciplinary WUI fire research activities.

requires new fundamental information that is not currently available. The variability in WUI communities and the types of fire hazards that communities and vulnerable populations are exposed to spans a wide range. Building a comprehensive and fundamental body of knowledge on WUI fires will require a long-term effort.

Field studies of WUI fires, involving both measurements and modeling, are needed to develop a shorter-term understanding and to direct longer-term research efforts. Field studies that characterize fires at the WUI will be challenging, however, because the interface, by definition, is just one part of the fire. Collecting data in targeted WUI areas of rapidly evolving fires is difficult and may require new research approaches. Field studies are essential, however, and need to be implemented despite these challenges.

Finally, the better understanding of WUI fires that will emerge from the research agenda needs to be made relevant to both decision-makers addressing WUI fires and their aftermath, and the public seeking to minimize toxicant exposures and health impacts. Measurement and modeling tools are needed to predict and quickly assess WUI fire chemistry, toxicant formation, and potential exposures. The effectiveness of interventions to minimize potential exposures also needs to be assessed.

Recommendation 1: To understand the chemistry, exposures, and health impacts of toxicants resulting from WUI fires, researchers and agencies that fund research should implement an integrated, multidisciplinary research agenda. Agencies funding, and investigators performing, research on WUI fire emissions should coordinate their research plans, and they should create widely accessible repositories for data and information relevant to WUI fires.

RESEARCH AGENDA

Tables 8-1 and 8-2 summarize the multidisciplinary research agenda recommended by the committee. In Table 8-1, the committee has summarized and prioritized research needs identified in the chapters of this report. The committee has organized research needs into four primary areas (columns) and three complementary approaches (rows). There is also an overarching research need (collecting WUI-specific data) that spans all of the primary areas (columns). This condensed list of research needs was created both by combining multiple, detailed research needs

TABLE 8-1 Research Priorities for Fires at the Wildland-Urban Interface (WUI)

	Fuels and Emissions	Chemistry, Transport, and Transformations	Exposure and Health	Measurement Science and Analytics
Collecting WUI-specific data	Assemble data on fuels, emissions, chemistry, transformations, exposures, and health impacts that are attributable to fires at the WUI, differentiated from wildland fires and urban fires, which will require novel measurements and analyses			
Fundamental measurements and data	- Map WUI communities, and their material loadings and compositions - Identify combustion conditions and emissions typical of WUI fires - Examine interactions between human-made fuels and wildland fire fuels using mechanistic models and experiments at bench, and larger scales	- Identify primary toxicants emerging from WUI fires - Identify secondary species with toxic potential, formed from the atmospheric aging of WUI fire emissions - Gather existing data on air, water, and soil testing associated with WUI fires in an accessible database	- Improve understanding of indoor penetration and composition of WUI fire smoke - Evaluate health implications of smoke constituents, and exposure to constituents in water	- Develop new analytical capabilities for measuring chemical, particle, and biological indicators of WUI fire toxicants in studies on emission, exposure, and health outcome
Field and population studies	- Assess the fuels, consumption, and emissions of WUI fires - Perform coordinated, multi-platform, multimedia studies of WUI fire energetics and emissions	- Identify dominant daytime and nighttime atmospheric oxidants in WUI fire plumes - Identify the key precursors and formation pathways of secondary species with toxic potential, formed by the gaseous, aqueous, multiphase, and catalytic reactions in plumes - Identify key chemical species that can impact water and soils	- Characterize multi-route and multimedia exposures and health impacts - Improve understanding of acute and long-term health effects of WUI fire toxicant exposures - Improve exposure measurements for WUI fire emissions	- Optimize analytical methods for field deployment and increased accessibility - Develop biomarkers specific to WUI fires that can be used for exposure and toxicity assessment - Develop standard procedures for testing water and soil after WUI fires; establish databases of testing studies
Prediction, assessment, and exposure-mitigation capabilities	- Develop risk assessment procedures for WUI fires - Develop predictive models of WUI fire combustion and emissions - Identify strategies at a structure, neighborhood, and community level to mitigate WUI fire risk	- Develop condensed chemical mechanisms and the sub-grid-scale processing needed for regional modeling of WUI fire emissions - Create improved retrospective and prospective models of WUI fire exposures - Evaluate risks to community water systems and response plans	- Measure the effectiveness of interventions for firefighters - Expand identification of vulnerable populations and culturally appropriate interventions - Develop health equity considerations for WUI fire exposures	- Deploy multi-scale sensing capabilities to assess chemical compositions of WUI fire plumes

TABLE 8-2 Factors Influencing the Prioritization of Research Needs in This Report

Factor	Criteria
Decision-makers who are the audience for the research	Decision-makers informed by WUI fire research may range from community planners charged with defining building codes in WUI communities, to public health professionals and regulators, to officials charged with rapidly communicating information to communities impacted by WUI fires. This range of decision-makers will have different portfolios of research needs and will require information on different timescales and at different technical levels.
Timeliness and cost effectiveness	Within a research portfolio, projects will range from those that can be rapidly implemented with immediate impact to those that will require a long-term commitment; they will also vary in cost.
Extent of applicability and magnitude of effect	Research portfolios will have the objectives of being broadly applicable and of addressing the most important questions related to the impacts of WUI fires and the toxicants they emit, including inequitable impacts across populations. Nevertheless, some research will be focused relatively narrowly, such as projects examining particular materials that might be combusted in a WUI fire or particular communities impacted by WUI fire emissions; both narrowly and broadly focused research can vary in the magnitude of the effect that is addressed; for example, the extent of use of materials that might be combusted in WUI fires varies, and the toxicants released in WUI fires vary in their toxicity and health implications and in the nature of their exposure.
State of scientific capabilities	Some research will address focused questions using currently available techniques, while other research might expand the range of questions that could be addressed by improving measurement or other tools used in WUI research.

appearing in individual chapters, and by identifying the research needs that the committee identified as high priority. The factors and criteria used in establishing the high-priority needs are described in more detail in Table 8-2.

Research Priorities Related to Fuels and Emissions

Fundamental Measurements and Data

Research priorities related to fuels and emissions include mapping WUI communities and their material loadings and compositions. As described in Chapters 2 and 3, a need exists for both detailed maps of WUI communities and a characterization of the types of structural materials present in each community. Systematically organized data on the materials present in structures could be coupled with local, regional, and national inventories of structures to develop fuel loading maps in geographic information systems. This type of system is already available for wildland fires and could be developed in an analogous manner for WUI regions. This is identified as a high-priority research area because it is foundational to defining the scope of other research activities (e.g., what types of urban materials should undergo laboratory testing and what types of atmospheric chemical transformations are likely to be important) and will be broadly applicable to many WUI fires.

Research is also needed to improve understanding of the combustion conditions in WUI fires and how meteorology and urban materials' composition and configuration, both within the landscape and within a building, interact to determine the ventilation conditions and the relative contributions of smoldering, glowing, and flaming combustion in different regions of a WUI fire. As described in Chapter 3, the impact of turbulence is difficult to recreate at small scales, and a range of experimental approaches may be needed to fully understand the coupling between the fuel and the environment in a WUI fire. As with the other priority research on fundamental measurements and data, this information is critical for developing predictive models for emissions and plume dispersion.

Urban materials contain varying elemental compositions in different ratios. Laboratory testing and mechanistic modeling, as described in Chapter 3, have demonstrated the importance of fuel composition and ventilation on the resulting emissions. Research can expand on these topics to study the more diverse mixtures relevant to WUI fires using models and testing equipment able to simulate WUI fire conditions. This is identified as a high-priority research area because the information is broadly applicable and because it is essential to understanding the complex fuel mixtures and combustion conditions that may be anticipated in a fire, to provide an initial assessment of what compounds will be emitted into the air and what residues are left behind.

Field Studies

Field measurements provide important insights into WUI chemistries that are occurring and complement the fundamental studies described in the previous section. The research needs described below were identified as high priority; these studies can be used to guide the direction of other research priorities.

Post-fire evaluations of fire exposure and damage assessments have proven a valuable tool for understanding WUI fire ignitions and the features that make a structure most vulnerable to fire. A plausible extension to these assessments is studies to supplement the information with measurements of fuel consumption and residues. An infrastructure for rapid response could allow researchers to sample emissions from smoldering fuels or off-gassing from other materials; these samples would provide important exposure information for workers involved in cleanup activities or people returning to their homes after an evacuation. Opportunistic measurement of WUI fire plumes that impact ambient monitoring locations or long-term atmospheric observatories could also provide valuable information on the composition of these plumes.

Emissions measurements in WUI fire plumes are the only way of capturing the true nature of WUI emissions at scale. Experience with wildland fire research has demonstrated that multiple skill sets are needed for a comprehensive understanding of how the fuels, the meteorology, and the topology couple to influence combustion conditions and the resulting plume emissions. Coordinated multi-platform, multimedia studies of WUI fire energetics and emissions are needed to provide a comprehensive understanding of fire and emissions phenomena. The unpredictable nature of WUI fires and the priority placed on emergency response during a WUI fire will make field studies challenging and require careful advance planning with multiple partners at the federal, state, and local levels.

Prediction, Assessment, and Exposure-Mitigation Capabilities

A comprehensive evaluation of the conditions under which WUI fires have occurred across the United States will provide valuable information on which communities are at greatest risk and under what conditions the threat is largest. WUI fires have shown characteristics distinctive from those of other wildfires, and current fire danger rating systems may not translate to WUI fire danger. A better understanding of the areas and times of highest risk is of immediate importance to develop protective danger ratings and to better target communities with the greatest need for mitigation actions.

Predictive models of wildfire emissions are an essential component of the emergency response used by frontline workers to develop public health guidance. Knowledge of the magnitude and the range of hazardous compounds that can be emitted by a WUI fire is needed in real time to determine exposure reduction actions for first responders and effective public health interventions.

Multiple WUI fire damage mitigation strategies exist, at structure, neighborhood, and community scales, and some, like maintaining defensible space around a structure or using ignition-resistant building materials, are widely adopted in some areas; yet destructive WUI fires persist. Research is needed to identify the most effective mix of mitigation strategies, while acknowledging that resources may be limited. Identifying the environmental conditions, topography, and community characteristics where WUI fires are most likely to occur can support the understanding of which strategies are most effective in differing regions. Additionally, a better understanding of how the choice of materials impacts ignition, combustion conditions, and ultimately emissions is an important step in reducing the risk and hazards associated with WUI fires.

Research Priorities Related to Chemistry, Transport, and Transformations

The research priorites identified for chemistry, transport, and transformations will both be informed by research on fuels and emissions and will inform research related to exposures and health impacts. The rationales for the need for coupled work in fundamental measurements, field studies, and modeling is analogous to the rationales for research on fuels and emissions.

Fundamental Measurements and Data

A combination of approaches is needed to quantify and characterize primary emissions (e.g., reactive species, primary toxicants) emerging from WUI fires, including controlled laboratory combustion of realistic mixtures of WUI materials at realistic combustion conditions, field measurements during plume transects with aircraft and instrumented mobile laboratories, and satellite remote sensing tools.

Identification of the secondary species with toxic potential formed during atmospheric transport and their formation kinetics and mechanisms is needed in order to predict downwind composition, concentrations, and impacts. Controlled laboratory experiments (e.g., flow tube, chamber) are needed to examine the chemistry of individual compounds and realistic WUI fire mixtures for key WUI fire oxidants. Experiments are particularly needed to elucidate the aqueous, multiphase, and catalytic chemistry, as well as secondary aerosol formation from intermediate-volatility organic compounds. Work is needed to identify tracers that assist with determining the relative importance of key chemical pathways in subsequent field campaigns. These laboratory and field data, for air and other media, should be assembled into accessible databases.

Field Studies

Coordinated field measurements in WUI fire plumes at a range of aging times are needed to determine the dominant daytime and nighttime atmospheric oxidants and the concentrations of these oxidants. Measurements should include tracers of oxidant chemistry. This work will help facilitate estimates of atmospheric lifetimes and geographic distributions of primary toxicants.

Making use of the insights from laboratory studies to identify tracers of chemical pathways, coordinated field measurements (near-field through regional or continental scale) are needed to identify the dominant precursors and formation pathways (gaseous, aqueous, multiphase, catalytic) for the major secondary species with toxic potential. These should be prioritized in model development.

Detailed assessment is needed of the potential for water and soil contamination, including standardization of procedures for testing. Efforts should address a wide range of potential chemical contaminants to determine potential areas of concern.

Prediction, Assessment, and Exposure-Mitigation Capabilities

Condensed chemical mechanisms are needed for WUI fire applications in atmospheric chemical transport models. Multimedia partitioning and reactivity should be considered when lumping chemicals to facilitate gaseous, aqueous, multiphase, and catalytic pathways, as appropriate. Sub-grid-scale processing modeling tools are also needed in atmospheric chemical transport models to represent the chemistry in the near field where concentrations are high, since the chemistry is nonlinear.

Development and use of improved retrospective/prospective models to track the atmospheric chemistry of WUI fires are needed. Insights from laboratory and field work concerning the dominant emissions, oxidants, and pathways through which WUI fire emissions are transformed in the atmosphere; condensed chemical mechanisms; and model processing tools should be incorporated.

A need exists for community water systems to evaluate risk and plan for response based on potential and expected levels of contamination after a WUI fire. Plans for evaluating risk should be standardized across local jurisdictions.

Research Priorities Related to Exposure and Health

The research priorites identified for exposure and public health are based on the need for health impact information being a primary driver for the entire research agenda. While these research needs will be informed by research on fuels and emissions, and chemistry, transport, and transformation, research on exposure and health impacts should progress in parallel with research in other areas.

Fundamental Measurements and Data

Assessments of smoke exposure typically rely on outdoor concentrations. A better understanding of the indoor penetration, composition, and persistence of smoke during wildfires in general, and of the chemical emissions profiles of WUI fires in particular, is needed.

Studies that explicitly define the particular type of fire (i.e., wildland fire vs WUI fire) to which the general population is exposed are needed to clearly delineate the types of health effects that could emerge, and to possibly mitigate them, as pollutant emissions (as well as toxicant concentrations) between wildland fires and WUI fires can be quite different.

Additional studies are needed to evaluate the health implications of wildland fire and WUI fire emissions other than particulate matter, such as volatile organic compounds, semi-volatile organic compounds, and polyaromatic compounds, as particulate matter has been the primary metric related to health outcomes in almost all studies up to now. Studies should also examine exposures to contaminants in water.

Field Studies

Research is needed to further characterize the chemicals and health impacts of WUI fire emissions via multimedia routes of exposure, including inhalation, dermal exposure, and ingestion.

A better understanding is needed of the acute and persistent effects of exposures to toxicants related to WUI fires, which may include biomonitoring studies of general populations and vulnerable subpopulations.

A better understanding is needed of the effects of exposures, especially for firefighters and local communities, and should include examining acute, chronic, and delayed health outcomes.

Prediction, Assessment, and Exposure-Mitigation Capabilities

Insufficient research exists on acute exposures and acute and chronic health effects in WUI firefighters and the community. Research is needed on the effectiveness of interventions to mitigate WUI firefighter exposures and the associated adverse health effects.

A need exists to better identify at-risk and vulnerable residents and workers affected by near-field WUI fires and to identify culturally appropriate interventions to reduce adverse health impacts.

A better understanding is needed of the health outcomes for vulnerable populations, resulting from exposure to the complex mixtures of chemical and particle emissions from WUI fires. Health equity considerations for vulnerable populations are needed for WUI fires. Items to consider include disparities related to being a vulnerable population, access to health care and communications, homeownership versus rental or use of a mobile home, occupation, and short- and long-term WUI fire exposures.

Further studies are critical to understanding whether long-term physiological and psychological outcomes associated with regional exposures and the general public are similar for WUI and wildland fires. Moreover, health effects of fire exposures on children are a critical direction for more research.

Research Priorities Related to Measurement Science and Analytics

Fundamental Measurements and Data

Understanding of emissions exposures and health effects associated with WUI fires is often limited by measurement capabilities. As discussed in more detail in Chapter 7, new analytical methods are needed for measuring a wide variety of chemical, particle, and biological indicators of toxicants in studies on emissions, exposures, and health outcomes.

Field Studies

Field measurements of the concentrations of species associated with WUI fires in air, water, and soil involve constraints on the size, power requirements, and other features of the analytical methods that can be deployed. Current and emergent analytical methods should be optimized for field deployment and increased accessibility.

Estimating human exposures requires that air, water, and soil measurements be combined with records of human activity that result in inhalation and ingestion of, and dermal exposure to, toxicants from WUI fires. Biomarkers can reduce the need for some of these data in exposure estimation; however, new biomarkers are needed that are more specific to WUI fires and can be used for exposure and toxicity assessment.

Standardization is needed for the methods used to test water and soil after WUI fires, and databases of information should be established.

Prediction, Assessment, and Exposure-Mitigation Capabilities

Large-scale modeling of WUI fire plumes could be assessed by augmenting ambient monitoring networks or through other sensing systems. Additional sensing capabilities should be deployed to assess chemical compositions of fire plumes.

CROSS-DISCIPLINARY COORDINATION

As shown in Figure 8-1, the information needed to understand the toxicants emerging from wildland fires, structural fires, and WUI fires; their effects; and whom they affect spans a broad range of scientific disciplines. Research findings from each of these disciplines needs to be continuously communicated across disciplinary boundaries because each step in the chain—from hazard identification to quantification of exposures and health effects—depends on information emerging from other steps.

INFORMATION DISSEMINATION

The research agenda outlined in Table 8-1 and described in detail in the chapters of this report has the potential to fill decision-critical gaps in information. This includes information needed by decision-makers charged with mitigating wildfire impacts and information needed by the public to minimize their exposure to WUI fire toxicants. To achieve this potential requires predictive tools, assessment capabilities, and exposure mitigation recommendations. Tools and assessment capabilities need to be applied rapidly, and information emerging from the tools and assessments will need to be synthesized into guidance in near real time.

This need is not unique to WUI fires. Systems have been developed for wildfires to provide information to decision-makers charged with mitigating fire impacts. The research community should build on these existing wildfire tools to provide information frameworks that incorporate the unique attributes of WUI fires.

For example, tools for decision-makers could include plume tracking that incorporates an understanding of the chemistry of WUI fires; it could include summaries of research results relevant to policies such as codes for construction materials used in WUI communities. Public dissemination of research findings could include activities such as expanding the information assembled in resources such as the federal government's Ready campaign website (www.ready.gov/wildfires) to include the unique attributes of WUI fires. This could include public guidance on selecting materials used to build, renovate, or repair WUI homes, or the types of personal protective equipment that are effective against the toxicants generated by WUI fires and that could be included in personal readiness kits.

> **Recommendation 2: Those implementing research programs should design and implement a multidisciplinary WUI research program that includes the development of tools, resources, and messaging designed to inform a wide variety of decision-makers charged with mitigating wildland and WUI fire impacts. They should also create periodic summaries of policy-relevant research findings, and actionable messaging for decision-makers working with at-risk communities and vulnerable populations.**

Policy-relevant research findings and actionable messaging could include recommendations for building materials to be used in WUI communities, public information regarding the effectiveness of measures to mitigate exposures, and community mappings of toxicant precursors, accessible to decision-makers.

More than 40 million homes in the United States are located at the WUI, in communities throughout the United States. Diverse and vulnerable populations will be increasingly exposed to hazards from WUI fires. The development of a multidisciplinary WUI research program, summarized in Table 8-1, will have long-term benefits, but immediate action is also needed. The committee identified a number of areas where rapid action could have immediate benefits. These areas are listed in Box 8-1 and were identified based on their ability to provide foundational information that plays a critical role in defining the scope of the broader research agends. Commitment to both long-term progress and immediate action on improving the understanding of WUI fires would benefit communities throughout the United States and the world.

BOX 8-1
Priorities for Near-Term Research

- Developing data systems to enable communities to predict the chemical composition of materials present in structures at threat from WUI fires; these data systems could include estimates of metal, halogen, and other chemical loadings in structures
- Adding measurements of targeted WUI toxicants to air and water quality monitoring systems; these measurement systems could be rapidly deployed to areas impacted by WUI fires
- Establishing information repositories on toxicant data, best practices for mitigation measures, and best practices for information dissemination; state agencies could lead in the coordination of data collection; data consistency, quality, and access could be addressed at a national level, and at all levels, communication and dissemination strategies for vulnerable and at-risk community populations could be developed

---

Appendix A

Glossary

Acute exposure (Chapter 6): Contact with a substance that occurs once or for only a short time (acute exposure can occur for up to 14 days) compared with intermediate-duration exposure and chronic exposure

Acute health effect (Chapter 6): A health effect that develops immediately or within minutes, hours, or even days after an exposure

Ash (Chapter 3): The solid residue remaining after combustion, generally consisting of minerals and lesser amounts of char; ash may become lofted in the plume to become a component of PM

Atmospheric lifetime (Chapter 4): Average time a molecule of species i remains in the atmosphere

Bioavailability (Chapter 6): The potential for uptake (ability to be absorbed and used by the body) of a substance by a living organism; it is usually expressed as the fraction that can be taken up by the organism in relation to the total amount available

Biomass (Chapter 3): Organic material derived from plants or animals

Black carbon or soot (Chapter 3): A component of PM derived from high-temperature flaming processes; black carbon is composed of aggregates of carbon particles typically 20–40 nm in diameter

Char (Chapter 3): The solid residue remaining after combustion; char generally refers to carbonaceous residues with some minerals

Chronic exposure (Chapter 6): Continuous or repeated contact with a toxic substance over a long period of time (months or years)

Chronic health effect (Chapter 6): An adverse health effect resulting from long-term exposure to a substance; examples could include diabetes, bronchitis, cancer, or any other long-term medical condition

Coarse particulate matter (PM₁₀) (Chapter 6): Inhalable particles with diameters equal to or less than 10 micrometers in diameter

Combustion efficiency (CE) (Chapter 3): The fraction of carbon in the fuel that is emitted as CO_2

Combustion factor (CF) (Chapter 3): The fraction of combustible material exposed to a fire that was actually consumed or volatilized

Combustion zone (Chapter 2): Region in which air temperatures are sufficiently high to drive combustion chemistry

Community water system (CWS) (Chapter 5): A water system serving populations greater than 25 people, year round

Continental scale (Chapters 2, 3, 4, 6): Greater than 1,000 km, where chemistry is driven by processes that occur over days

Dissolved organic matter (DOM) (Chapter 5): A mixture of organic compounds found ubiquitously in surface and groundwaters; derived from terrestrial and aquatic sources

Dry atmospheric deposition (Chapter 4): The removal of particles and gases from the atmosphere to surfaces in the absence of precipitation (i.e., through gravitational settling, impaction, interception, and diffusion)

Emissions (or effluents) (Chapters 3, 5): Species emitted into the air, water, soil, or other media, from a process; these are sometimes called releases

Emission factor (EF) (Chapter 3): The mass of a specific compound (or class of compounds) emitted per kilogram of dry fuel combusted, for a specified dry material or collection of materials

Emission ratio (ER) (Chapter 3): The ratio of the mass of a compound emitted to the mass of a reference compound that is conserved in the plume, often CO or CO_2; it is often reported as an "**enhanced**" ER where the background concentrations of the compound and reference compound have been subtracted

Enclosure fire (Chapter 3): A fire contained within a room or compartment inside a building in which oxygen supply is typically constrained, contrary to open fires; these are sometimes called compartment fires

Energy content (Chapter 3): The amount of energy contained within a mass of fuel; it can be quantified as the higher heating value (or gross calorific value), defined as the amount of heat released from complete combustion of a material when the products are returned to 25°C, or the lower heating value (or net calorific value), defined as the amount of heat released from complete combustion of a dry material initially at 25°C when water as a combustion product remains in the vapor state; other initial and final state temperatures may be found in the literature

Environmental justice (Chapter 6): A social movement developed in response to environmental racism; environmental justice research and practice involves identifying the disproportionate health burdens that populations experience from environmental exposures and social vulnerabilities, and focusing on solutions to alleviate those burdens in partnership with affected communities

Equivalence ratio (φ) (Chapter 3): The ratio of the actual fuel/oxidizer ratio to the stoichiometric fuel/oxidizer ratio; the stoichiometric fuel/oxidizer ratio is the ratio that is theoretically required to fully oxidize the fuel

Fine particulate matter (PM$_{2.5}$) (Chapters 3, 4, 6): Airborne particles with diameters of 2.5 micrometers or less, small enough to enter the lungs and bloodstream, posing risks to human health

Fire plume (Chapters 2, 3, 4): Air mass downwind of combustion zone, containing elevated concentrations of combustion products

Flaming combustion (Chapter 3): Luminous oxidation of gases evolved from the rapid decomposition of a solid biomass fuel

Glowing combustion (Chapter 3): Incandescent heterogeneous oxidation of a solid biomass fuel in which all the volatiles have been driven off

Health equity (Chapter 6): Identifying health disparities between populations that are driven largely by social, economic, and environmental factors, and focusing on solutions to eliminate those disparities

Interface WUI (Chapter 2): Areas characterized by development (housing or other structures) located at the edge of a large area of wildland

Intermediate-volatility organic compounds (IVOCs) (Chapter 4): Compounds with vapor pressures between those of VOCs and semi-VOCs; a suite of compounds that, based on their vapor pressure, tend to evaporate from the particle phase with near-field dilution of plumes

Intermix WUI (Chapter 2): Areas characterized by alternating development (housing or other structures) and wildlands

Local scale (Chapters 2, 3, 4, 6): From 10 to 100 km, where chemistry is driven by processes that occur on a timescale of minutes to hours

Lower heating value or net calorific value (Chapter 3): amount of heat release from complete combustion of material not including the energy required to vaporize products

Modified combustion efficiency (MCE) (Chapter 3): The measured, enhanced emission of CO_2 divided by the sum of the enhanced emission of CO and enhanced emission of CO_2; MCE is typically linearly correlated with CE and used as a proxy for CE since it is easier to measure

Near-field scale (Chapters 2, 3, 4, 6): From 1 to 10 km downwind of the fire, where the plume remains quite concentrated, dilution has a major effect on the gas-particle partitioning, and chemistry is driven by fast processes that occur on a timescale of minutes

Oxidative pyrolysis (Chapter 3): The thermal decomposition of a combustible material in the presence of molecular oxygen in the surrounding atmosphere

Particulate matter (PM) (Chapters 3, 6): A complex mixture of solid particles and liquid droplets found in the air

Partitioning (Chapters 4, 5): The distribution of a species *i* between two media, such as air and particles, or air and water

Plume injection parameters (Chapter 3): The initial characteristics of a fire plume, including its injection altitude and multi-phase chemical composition

Pollutant (Chapters 3, 4, 5, 6): A chemical or biological substance that harms water, air, or land quality

Prescribed burn (Chapter 4): Fire set intentionally for forest or farmland management

Primary organic aerosol (Chapter 4): Organic particulate matter that is emitted from the source (e.g., WUI fire) in particulate form

Primary species with toxic potential (Chapter 4): Toxic substance that is emitted directly from the source (e.g., WUI fire)

Pyrolysis (Chapter 3): The thermal decomposition of a combustible material in the absence of molecular oxygen; the term "pyrolysis" sometimes appears in wildland fire literature to represent oxidative pyrolysis

Regional scale (Chapters 2, 3, 4, 6): From 100 to 1,000 km, where chemistry is driven by processes that occur on a timescale of hours to days

Secondary organic aerosol (Chapters 3, 4): Organic particulate matter that is formed in the atmosphere from precursor gases

Secondary species with toxic potential (Chapter 4): Toxic substances that are formed through atmospheric chemistry

Semi-volatile organic compounds (SVOCs) (Chapters 3, 6): Organic compounds that, based on their vapor pressure, tend to evaporate from the particle phase with near-field dilution of plumes; SVOCs are of concern because of their abundance in the indoor environment and their ability to accumulate and persist in the human body, the infrastructure of buildings, and environmental dust

Smoldering combustion (Chapter 3): Combined processes of thermal decomposition and slow, low-temperature, flameless burning of porous solid biomass fuels; sometimes called glowing combustion

Structural racism (Chapter 6): The totality of ways in which societies foster racial discrimination through mutually reinforcing systems of housing, education, employment, earnings, benefits, credit, media, health care, and criminal justice; these patterns and practices in turn reinforce discriminatory beliefs, values, and the distribution of resources

Toxic (Chapter 6): Related to harmful effects on the body by either inhalation (breathing), ingestion (eating), or dermal absorption of the chemical

Toxic product yield (Chapter 3): The maximum possible mass of a combustion product generated during combustion, per unit mass of test specimen consumed (typically expressed in units of grams per gram or kilograms per kilogram)

Toxicant (Chapters 5, 6): Any chemical that can injure or kill humans, animals, or plants (depending on the magnitude and duration of exposure); a poison

Ultrafine particles (Chapter 6): Particles with a diameter less than or equal to 0.1 micrometers

Urban fire (Chapter 3): Fire that occurs primarily in cities or towns with the potential to rapidly spread to adjoining structures; these fires mostly damage and destroy homes, schools, commercial or industrial buildings, and vehicles

Volatile organic compounds (VOCs) (Chapters 3, 4, 5, 6): Organic compounds with vapor pressures high enough to exist in the atmosphere primarily in the gas phase, typically excluding methane; VOCs can easily become airborne for inhalation exposure

Vulnerable populations (Chapter 6): Individuals or communities at higher risk of adverse health effects from exposures, such as from greater pollutant exposure concentrations, higher health response to a given level of exposure, or reduced capacity to adapt

Wet atmospheric deposition (Chapter 4): The removal of particles and gases to the earth's surface via scavenging by precipitation

Wildfire: Although in common usage, the term wildfire is not used in this report; a wildfire is generally defined as a wildland fire originating from an unplanned ignition, such as lightning, volcanos, unauthorized and accidental human-caused fires, and prescribed fires that are declared wildfires

Wildland fire: Any non-structure fire that occurs in vegetation or natural fuels; includes wildfires and prescribed fires

Wildland-urban interface (WUI): The community that exists where humans and their development meet or intermix with wildland fuel

Appendix B

Committee Biographical Sketches

David T. Allen (Chair), NAE, is the Melvin H. Gertz Regents Professor of Chemical Engineering, and the Director of the Center for Energy and Environmental Resources, at the University of Texas at Austin. His research interests include urban air quality, the engineering of sustainable systems, and the development of materials for environmental and engineering education. Dr. Allen has been a lead investigator for multiple air quality measurement studies, which have had a substantial impact on the direction of air quality policies. Dr. Allen directs the Air Quality Research Program for the State of Texas, and he is the founding Editor-in-Chief of the American Chemical Society's journal *ACS Sustainable Chemistry & Engineering*. He has served on a variety of governmental advisory panels and chaired the US Environmental Protection Agency's Science Advisory Board from 2012 to 2015. Dr. Allen was elected to the National Academy of Engineering in 2017 and has chaired several National Academies of Sciences, Engineering, and Medicine committees. Dr. Allen received his PhD in chemical engineering from the California Institute of Technology.

Olorunfemi Adetona is an assistant professor in the College of Public Health at The Ohio State University. His research focuses on occupational and household exposures to combustion-derived air pollution, including epidemiology- and laboratory-based studies to characterize the exposures of wildland firefighting and the respiratory and systemic responses to such exposures. Specifically, studies explore quantitative occupational (biomarker) exposure-response for cancer and acute cardiovascular effects of wildland firefighting. Dr. Adetona also studies potential agents for mitigating adverse health effects caused by particulate air pollution. Dr. Adetona earned his PhD in toxicology from the University of Georgia.

Michelle Bell, NAM, is the Mary E. Pinchot Professor of Environmental Health at the Yale University School of the Environment, with secondary appointments at the Yale School of Public Health, Environmental Health Sciences Division, and the Yale School of Engineering and Applied Science, Environmental Engineering Program. Her research investigates how human health is affected by atmospheric systems, including air pollution and weather. Other research interests include the health impacts of climate change and environmental justice. Dr. Bell has received numerous awards for her work, including the Prince Albert II de Monaco/Institut Pasteur Award, the Rosenblith New Investigator Award, and the NIH Outstanding New Environmental Scientist (ONES) Award. She was elected to the National Academy of Medicine in 2020. Dr. Bell received her PhD in environmental engineering from Johns Hopkins University.

Marilyn Black is vice president and senior technical advisor for Underwriters Laboratories Inc. (UL), leading its research group Chemical Insights. Her work focuses on improving product design to reduce chemical exposures in home and shared environments. She is the founder and former chairperson for both UL Air Quality Sciences and the GREENGUARD Environmental Institute. She is also the founder of the Khaos Foundation, a nonprofit organization dedicated to protecting the health and well-being of children through education and research. Dr. Black is an active participant in national and international scientific organizational initiatives, research projects, and community outreach programs, and has presented and published over 200 papers on indoor air quality and environmental exposure. She received her PhD in chemistry from the Georgia Institute of Technology.

Jefferey L. Burgess is the Associate Dean for Research for the University of Arizona Mel and Enid Zuckerman College of Public Health. His current translational, occupational, and environmental health research primarily focuses on evaluation and prevention of injurious exposures to firefighters and miners. Dr. Burgess has served as principal investigator for federally funded projects focused on cancer prevention in firefighters, including research grants funded by the Federal Emergency Management Agency, the National Institute of Environmental Health Sciences, and the National Institute for Occupational Safety and Health (NIOSH). He previously served on the National Academies committee to review the NIOSH Mining Safety and Health Research Program. He received his MD from the University of Washington School of Medicine.

Frederick L. Dryer, NAE, is an Educational Foundation Distinguished Research Professor in Mechanical Engineering at the University of South Carolina. He was previously a professor of mechanical and aerospace engineering at Princeton University. Dr. Dryer's research expertise spans a wide range of areas, including thermodynamics, physical chemistry, chemical kinetics, fluid dynamics, heat transfer, abatement of unwanted emissions from energy conversion systems, and understanding and mitigating fire hazards associated with the use of gaseous, liquid, and solid materials on earth and in low-gravity environments. He is currently a Fellow of the International Combustion Institute and is a former associate editor and editorial board member of *Combustion Science and Technology*, as well as a former editorial board member of the *International Journal of Chemical Kinetics* and of *Progress in Energy and Combustion Science*. Dr. Dryer received his PhD in aerospace and mechanical sciences from Princeton University.

Amara Holder is a research mechanical engineer with the US Environmental Protection Agency (EPA) Office of Research and Development and is the EPA lead on the development of a wildland-urban interface emissions inventory. Her research is on discovering the physical, chemical, and optical properties of combustion-generated particles and understanding the processes that determine these characteristics. Dr. Holder has studied numerous combustion systems including wildland fires, woodstoves, cook stoves, motor vehicles, diesel generators, coal-fired power plants, and crude oil burns. She received her PhD in combustion and environmental health from the University of California, Berkeley.

Ana Mascareñas is a consultant with the University of Washington's Center for Anti-Racism and Community Health. She formerly served as the Assistant Director for Environmental Justice, Deputy Director, and Tribal Liaison for the California Environmental Protection Agency's Department of Toxic Substances Control (DTSC). Her work focuses on the intersection of environmental and public health, particularly regarding vulnerable populations and tribal communities in California. Before joining DTSC, she was policy and communications director at Physicians for Social Responsibility Los Angeles. Ms. Mascareñas received her Master of Public Health (MPH) in environmental health sciences from the Fielding School of Public Health at the University of California, Los Angeles.

Fernando Rosario-Ortiz is a professor and director of the Environmental Engineering Program at the University of Colorado Boulder. His current research focuses on environmental photochemistry, the impact of wildfires on water quality and treatment, and characterization of organic matter in different environments. He has won numerous awards, including a National Science Foundation CAREER award for work to identify the impact of effluent organic matter on photochemical processes in surface waters. Dr. Rosario-Ortiz received his doctoral degree in environmental science and engineering from the University of California, Los Angeles.

Anna Agnieszka Stec is a professor in fire chemistry and toxicity at the University of Central Lancashire, United Kingdom. She was the Scientific Coordinator for the "Toxicological and Environmental Aspects" work package of the European Union Cooperation in Science and Technology Action and the UK representative on its Management Committee. Dr. Stec has extensive experience in identifying and understanding toxic hazards (acute and chronic) in and from fires and their effect on humans and the environment. Dr. Stec was appointed by the UK Government's Chief Scientific Advisor to the Scientific Advisory Group to oversee investigation of soil contamination and adverse health effects following the Grenfell Tower fire. She represents the Society of Chemical Industry with the British Standards Hazard to Life from Fire technical committee. Dr. Stec received her PhD in fire chemistry and toxicity from the University of Bolton, United Kingdom.

Barbara J. Turpin is a professor and chair of the Department of Environmental Sciences and Engineering at the University of North Carolina at Chapel Hill. Her research focuses on use of chemical modeling and field research to improve understanding of linkages between air pollution emissions and human exposures. She has a specific focus on particulate matter and aerosols, and how these are affected by human activities and built environments. Dr. Turpin is a fellow of several professional societies, including the American Association for the Advancement of Science and the American Geophysical Union. In 2018, she received the Creative Advances in Environmental Sciences and Technology Award from the American Chemical Society. Dr. Turpin received her PhD in environmental science and engineering from OGI at Oregon Health and Science University.

Judith T. Zelikoff is a professor in the Department of Environmental Medicine at the New York University Grossman School of Medicine. Her primary research interests are in the toxicology of inhaled metals and complex mixtures, including air pollution, tobacco products, and diesel emissions. She has also published on the toxicology of woodburning/woodsmoke and written a number of reviews in the same field. Dr. Zelikoff has published over 150 papers and book chapters in the areas of environmental health and toxicology. She has served as a standing member on two National Institutes of Health study sections and actively participates as an ad hoc reviewer for a variety of institutes. In addition, she has served as a member of the National Toxicology Program Advisory Board. Dr. Zelikoff received her PhD in experimental pathology focusing on viral-induced human diseases at Rutgers Medical School.

Appendix C

Available Data for Example Fires

188

Information Need	Waldo Canyon Fire	Horse River Fire	Gatlinburg Fire	Camp Fire	Marshall Fire
Fire Progression Timeline	- City of Colorado Springs, 2013 - Maranghides et al., 2015	- MNP LLP, 2017	- Culver, 2016 - Guthrie et al., 2017	- Maranghides et al., 2021	- CPR, 2021b - Markus, 2022 - Brown and Paul, 2022
Operational And General Information	- Chin et al., 2016 - Kinoshita et al., 2016	- Government of Alberta, 2018	- Ahillen, 2016 - Flynn, 2017 - Gabbert, 2016 - Guthrie et al., 2017 - Manzello et al., 2018 - Walpole et al., 2020	- CARB, 2021	- Boulder OEM, 2022 - Cobb, 2022 - Vaughan and Jojola, 2022
Fuel Characterization	—	- MNP LLP, 2017	—	- Jin et al., 2019	—
Combustion Characterization	—	—	—	- Simms et al., 2021	—
Emissions	- Colorado Department of Public Health and the Environment staff report that their office is storing unanalyzed Teflon $PM_{2.5}$ filters from sampling in Colorado Springs during this fire	- Adams et al., 2019 - Hsu et al., 2015 - Landis et al., 2018 - Wentworth et al., 2018	—	- CARB, 2021 - EPA emissions inventory - Rooney et al., 2020	—
Atmospheric Chemistry		- Adams et al., 2019 - Wu et al., 2018		- Delp and Singer, 2020 - Ding et al., 2021 - Fadadu et al., 2020 - Li et al., 2020 - Rooney et al., 2020 - Simms et al., 2021 - Vu et al., 2022	—
Water and Soil Chemistry	- CTL Thompson, 2014	- Arciszewski and McMaster, 2021 - Emmerton et al., 2020 - Government of Alberta, 2018, pp. 19, 27		- Proctor et al., 2019 (based on unpublished data from the Paradise Irrigation District) - Proctor et al., 2020 - Solomon et al., 2021 - TetraTech Inc., 2019a	—

Human Exposures/ Health Impacts	—	Physical health: - Cherry et al., 2019 - Cherry et al., 2021a - Kohl et al., 2019 Mental health: - Agyapong et al., 2018 - Agyapong et al., 2019 - Agyapong et al., 2020 - Brown et al., 2019c - Brown et al., 2019b - Brown et al., 2019a - Cherry et al., 2021b - Ritchie et al., 2020 - Verstraeten et al., 2020	—	- CARB, 2021 - DuTeaux, 2019 - Fadadu et al., 2021 - Odimayomi et al., 2021 - Silveira et al., 2021 - Von Behren et al., 2022	- CPR, 2021a - CBS Colorado, 2022
Environmental Impact	- Chin et al., 2016 - The Associated Press, 2017 - Young et al., 2012	- Elmes et al., 2019 - Hebert, 2019 - Zhang et al., 2022	—	- TetraTech Inc., 2019b - Dieckmann et al., 2020	—
Miscellaneous	—	- Omane et al., 2017	—	—	—

SOURCE: Developed as part of the committee's literature survey process and used to inform committee discussions and deliberations.

REFERENCES

Adams, C., C. A. McLinden, M. W. Shephard, N. Dickson, E. Dammers, J. Chen, P. Makar, K. E. Cady-Pereira, N. Tam, S. K. Kharol, L. N. Lamsal, and N. A. Krotkov. 2019. "Satellite-Derived Emissions of Carbon Monoxide, Ammonia, and Nitrogen Dioxide from the 2016 Horse River Wildfire in the Fort McMurray Area." *Atmospheric Chemistry and Physics* 19 (4): 2577–2599. https://doi.org/10.5194/acp-19-2577-2019.

Agyapong, V. I. O., M. Hrabok, M. Juhas, J. Omeje, E. Denga, B. Nwaka, I. Akinjise, S. E. Corbett, S. Moosavi, M. Brown, P. Chue, A. J. Greenshaw, and X. M. Li. 2018. "Prevalence Rates and Predictors of Generalized Anxiety Disorder Symptoms in Residents of Fort McMurray Six Months After a Wildfire." *Frontiers in Psychiatry* 9. https://doi.org/10.3389/fpsyt.2018.00345.

Agyapong, V. I. O., M. Juhás, M. R. G. Brown, J. Omege, E. Denga, B. Nwaka, I. Akinjise, S. E. Corbett, M. Hrabok, X. M. Li, A. Greenshaw, and P. Chue. 2019. "Prevalence Rates and Correlates of Probable Major Depressive Disorder in Residents of Fort McMurray 6 Months After a Wildfire." *International Journal of Mental Health and Addiction* 17 (1): 120–136. https://doi.org/10.1007/s11469-018-0004-8.

Agyapong, V. I. O., A. Ritchie, M. R. G. Brown, S. Noble, M. Mankowsi, E. Denga, B. Nwaka, I. Akinjise, S. E. Corbett, S. Moosavi, P. Chue, X. M. Li, P. H. Silverstone, and A. J. Greenshaw. 2020. "Long-Term Mental Health Effects of a Devastating Wildfire Are Amplified by Socio-Demographic and Clinical Antecedents in Elementary and High School Staff." *Frontiers in Psychiatry* 11. https://doi.org/10.3389/fpsyt.2020.00448.

Ahillen, S. December 28, 2016. "By the Numbers: Gatlinburg Fire." *Knoxville News Sentinel*. https://www.knoxnews.com/story/news/local/tennessee/2016/12/28/numbers-gatlinburg-fire/95847766/.

Arciszewski, T. J., and M. E. McMaster. 2021. "Potential Influence of Sewage Phosphorus and Wet and Dry Deposition Detected in Fish Collected in the Athabasca River North of Fort McMurray." *Environments* 8 (2): 1–21. https://doi.org/10.3390/environments8020014.

Boulder OEM (Boulder Office of Emergency Management). January 6, 2022. "Boulder County Releases Updated List of Structures Damaged and Destroyed in the Marshall Fire." https://www.bouldercounty.org/news/boulder-county-releases-updated-list-of-structures-damaged-and-destroyed-in-the-marshall-fire/ (accessed January 31, 2022).

Brown, J., and J. Paul. 2022. "The Minute-by-Minute Story of the Marshall Fire's Wind-Fueled Tear Through Boulder County." *The Colorado Sun*. https://coloradosun.com/2022/01/06/marshall-fire-boulder-county-timeline/.

Brown, M. R. G., V. Agyapong, A. J. Greenshaw, I. Cribben, P. Brett-MacLean, J. Drolet, C. M. Harker, J. Omeje, M. Mankowsi, S. Noble, D. T. Kitching, and P. H. Silverstone. 2019a. "Significant PTSD and Other Mental Health Effects Present 18 Months after the Fort McMurray Wildfire: Findings from 3,070 Grades 7–12 Students." *Frontiers in Psychiatry* 10: 623. https://doi.org/10.3389/fpsyt.2019.00623.

Brown, M. R. G., V. Agyapong, A. J. Greenshaw, I. Cribben, P. Brett-MacLean, J. Drolet, C. McDonald-Harker, J. Omeje, M. Mankowsi, S. Noble, D. Kitching, and P. H. Silverstone. 2019b. "Correction to: After the Fort McMurray Wildfire There Are Significant Increases in Mental Health Symptoms in Grade 7–12 Students Compared to Controls." *BMC Psychiatry* 19 (1): 97. https://doi.org/10.1186/s12888-019-2074-y.

Brown, M. R. G., V. Agyapong, A. J. Greenshaw, I. Cribben, P. Brett-Maclean, J. Drolet, C. McDonald-Harker, J. Omeje, M. Mankowsi, S. Noble, D. Kitching, and P. H. Silverstone. 2019c. "After the Fort McMurray Wildfire There Are Significant Increases in Mental Health Symptoms in Grade 7–12 Students Compared to Controls." *BMC Psychiatry* 19 (1): 18. https://doi.org/10.1186/s12888-018-2007-1.

CARB (California Air Resources Board). 2021. *Camp Fire Air Quality Data Analysis*. Sacramento, CA: California Air Resources Board. https://ww2.arb.ca.gov/resources/documents/camp-fire-air-quality-data-analysis.

CBS Colorado. January 21, 2022. "Marshall Fire: Investigators Consider Underground Mine Fire in Origin." https://www.cbsnews.com/colorado/news/marshall-fire-boulder-county-underground-mine-fire/.

Cherry, N., Y. A. Aklilu, J. Beach, P. Britz-Mckibbin, R. Elbourne, J. M. Galarneau, B. Gill, D. Kinniburgh, and X. Zhang. 2019. "Urinary 1-Hydroxypyrene and Skin Contamination in Firefighters Deployed to the Fort McMurray Fire." *Annals of Work Exposures and Health* 63 (4): 448–458. https://doi.org/10.1093/annweh/wxz006.

Cherry, N., J. Beach, and J. M. Galarneau. 2021a. "Are Inflammatory Markers an Indicator of Exposure or Effect in Firefighters Fighting a Devastating Wildfire? Follow-up of a Cohort in Alberta, Canada." *Annals of Work Exposures and Health* 65 (6): 635–648. https://doi.org/10.1093/annweh/wxaa142.

Cherry, N., J. M. Galarneau, W. Haynes, and B. Sluggett. 2021b. "The Role of Organizational Supports in Mitigating Mental Ill Health in Firefighters: A Cohort Study in Alberta, Canada." *American Journal of Industrial Medicine* 64 (7): 593–601. https://doi.org/10.1002/ajim.23249.

Chin, A., L. An, J. L. Florsheim, L. R. Laurencio, R. A. Marston, A. P. Solverson, G. L. Simon, E. Stinson, and E. Wohl. 2016. "Investigating Feedbacks in Human-Landscape Systems: Lessons Following a Wildfire in Colorado, USA." *Geomorphology* 252: 40–50. https://doi.org/10.1016/j.geomorph.2015.07.030.

City of Colorado Springs. 2013. *Waldo Canyon Fire Final After Action Report.* Colorado Springs, CO: City of Colorado Springs. https://cdpsdocs.state.co.us/coe/Website/Data_Repository/Waldo%20Canyon%20Fire%20Final%20After%20Action%20Report_City%20of%20Colorado%20Springs.pdf.

Cobb, E. 2022. "'A Hurricane of Fire': How the Marshall Fire Tore through Boulder County." *Daily Camera.* https://www.dailycamera.com/2022/01/30/hurricane-of-fire/.

CPR (Colorado Public Radio). 2021a. "Boulder County Fires: Three Missing from Marshall, Superior Are Suspected Dead; Thousands without Gas and Electricity in Freezing Temps." *CPR News.* https://www.cpr.org/2022/01/01/boulder-county-fires-water-shut-off-in-superior-thousands-without-gas-and-electricity-in-freezing-temps/.

CPR. 2021b. "Boulder County Fires: More than 500 Houses Burn, Tens of Thousands Evacuate as Fires Continue to Spread." *CPR News.* https://www.cpr.org/2021/12/30/boulder-county-grass-fires/.

CTL Thompson. 2014. "Appendix A: Laboratory Test Results." In *Soils Report.* https://coloradosprings.gov/sites/default/files/inline-images/exhibit_8-_soils_report.pdf.

Culver, A. 2016. "Timeline: Gatlinburg Wildfires." *WATE.com.* http://wate.com/2016/12/13/timeline-gatlinburg-wildfires/.

Delp, W. W., and B. C. Singer. 2020. "Wildfire Smoke Adjustment Factors for Low-Cost and Professional PM$_{2.5}$ Monitors with Optical Sensors." *Sensors (Switzerland)* 20 (13): 1–21. https://doi.org/10.3390/s20133683.

Dieckmann, H. G., L. R. R. Costa, B. Martínez-López, and J. E. Madigan. 2020. "Implementation of an Animal Health Database in Response to the 2018 California Camp Fire." *Journal of the American Veterinary Medical Association* 256 (9): 1005–1010. https://doi.org/10.2460/javma.256.9.1005.

Ding, Y., I. Cruz, F. Freedman, and A. Venkatram. 2021. "Improving Spatial Resolution of PM$_{2.5}$ Measurements during Wildfires." *Atmospheric Pollution Research* 12 (5). https://doi.org/10.1016/j.apr.2021.03.010.

DuTeaux, S. 2019. *Final Report of Air Quality in the Vicinity of Camp Fire Debris Removal.* Chico, CA: Butte County Air Quality Management District. https://bcaqmd.org/wp-content/uploads/BCAQMD-Governing-Board-DuTeaux-DROC-12-12-2019.pdf.

Elmes, M. C., D. K. Thompson, and J. S. Price. 2019. "Changes to the Hydrophysical Properties of Upland and Riparian Soils in a Burned Fen Watershed in the Athabasca Oil Sands Region, Northern Alberta, Canada." *Catena* 181. https://doi.org/10.1016/j.catena.2019.104077.

Emmerton, C. A., C. A. Cooke, S. Hustins, U. Silins, M. B. Emelko, T. Lewis, M. K. Kruk, N. Taube, D. Zhu, and B. Jackson. 2020. "Severe Western Canadian Wildfire Affects Water Quality Even at Large Basin Scales." *Water Research* 183: 116071.

Fadadu, R. P., J. R. Balmes, and S. M. Holm. 2020. "Differences in the Estimation of Wildfire-Associated Air Pollution by Satellite Mapping of Smoke Plumes and Ground-Level Monitoring." *International Journal of Environmental Research and Public Health* 17 (21): 8164. https://doi.org/10.3390/ijerph17218164.

Fadadu, R. P., B. Grimes, N. P. Jewell, J. Vargo, A. T. Young, K. Abuabara, J. R. Balmes, and M. L. Wei. 2021. "Association of Wildfire Air Pollution and Health Care Use for Atopic Dermatitis and Itch." *JAMA Dermatology* 157 (6): 658–666. https://doi.org/10.1001/jamadermatol.2021.0179.

Flynn, SE. 2017. *Wildfire: A Changing Landscape.* Boston, MA: Northeastern University, Global Resilience Institute.

Gabbert, B. 2016. "Analyzing the Fire that Burned into Gatlinburg." *Wildfire Today.* https://wildfiretoday.com/2016/12/05/analyzing-the-fire-that-burned-into-gatlinburg/.

Government of Alberta. 2018. *Horse River Wildfire Response: Environmental Monitoring, Public Health Assessment and Ecological Screening Technical Summary Document.* Edmonton, AB: Environmental Public Health Science Unit, H. P. B., Public Health and Compliance Division, Alberta Health.

Guthrie, V., M. Finucane, P. Keith, and D. Stinnett. 2017. *After Action Review of the November 28, 2016, Firestorm.* Spring, TX: ABSG Consulting.

Hebert, C. E. 2019. "The River Runs Through It: The Athabasca River Delivers Mercury to Aquatic Birds Breeding Far Downstream." *PLoS One* 14 (4). https://doi.org/10.1371/journal.pone.0206192.

Hsu, Y. M., T. Harner, H. Li, and P. Fellin. 2015. "PAH Measurements in Air in the Athabasca Oil Sands Region." *Environmental Science & Technology* 49 (9): 5584–5592. https://doi.org/10.1021/acs.est.5b00178.

Jin, Y., E. Scaduto, and B. Chen. 2019. "Fire Behavior Monitoring and Assessment in California with Multi-sensor Satellite Observations." Presented at the International Geoscience and Remote Sensing Symposium (IGARSS).

Kinoshita, A. M., A. Chin, G. L. Simon, C. Briles, T. S. Hogue, A. P. O'Dowd, A. K. Gerlak, and A. U. Albornoz. 2016. "Wildfire, Water, and Society: Toward Integrative Research in the 'Anthropocene.'" *Anthropocene* 16: 16–27.

Kohl, L., M. Meng, J. de Vera, B. Bergquist, C. A. Cooke, S. Hustins, B. Jackson, C. W. Chow, and A. W. H. Chan. 2019. "Limited Retention of Wildfire-Derived PAHs and Trace Elements in Indoor Environments." *Geophysical Research Letters* 46 (1): 383–391. https://doi.org/10.1029/2018GL080473.

Landis, M. S., E. S. Edgerton, E. M. White, G. R. Wentworth, A. P. Sullivan, and A. M. Dillner. 2018. "The Impact of the 2016 Fort McMurray Horse River Wildfire on Ambient Air Pollution Levels in the Athabasca Oil Sands Region, Alberta, Canada." *Science of the Total Environment* 618: 1665–1676. https://doi.org/10.1016/j.scitotenv.2017.10.008.

Li, Y., D. Q. Tong, F. Ngan, M. D. Cohen, A. F. Stein, S. Kondragunta, X. Zhang, C. Ichoku, E. J. Hyer, and R. A. Kahn. 2020. "Ensemble $PM_{2.5}$ Forecasting During the 2018 Camp Fire Event Using the HYSPLIT Transport and Dispersion Model." *Journal of Geophysical Research: Atmospheres* 125 (15). https://doi.org/10.1029/2020JD032768.

Manzello, S. L., R. Blanchi, M. J. Gollner, D. Gorham, S. McAllister, E. Pastor, E. Planas, P. Reszka, and S. Suzuki. 2018. "Summary of Workshop Large Outdoor Fires and the Built Environment." *Fire Safety Journal* 100: 76–92. https://doi.org/10.1016/j.firesaf.2018.07.002.

Maranghides, A., D. McNamara, R. Vihnanek, J. Restaino, and C. Leland. 2015. *A Case Study of a Community Affected by the Waldo Fire – Event Timeline and Defensive Actions*. NIST Technical Note 1910. Gaithersburg, MD: National Institute of Standards and Technology.

Maranghides, A., E. Link, S. Hawks, M. Wilson, W. Brewer, C. Brown, B. Vihnaneck, and W. D. Walton. 2021. "Appendix C. Community WUI Fire Hazard Evaluation Framework." In *A Case Study of the Camp Fire–Fire Progression Timeline*. NIST Technical Note 2135sup

Markus, B. January 5, 2022. "Boulder County Firefighters Lost Crucial Early Minutes Because They Couldn't Find the Start of the Marshall Fire." *CPR News*. https://www.cpr.org/2022/01/05/boulder-county-marshall-fire-timeline/.

MNP LLP. 2017. *A Review of the 2016 Horse River Wildfire*. Edmonton, AB: Alberta Agriculture and Forestry Preparedness and Response. https://www.alberta.ca/assets/documents/Wildfire-MNP-Report.pdf.

Odimayomi, T. O., C. R. Proctor, Q. E. Wang, A. Sabbaghi, K. S. Peterson, D. J. Yu, J. Lee, A. D. Shah, C. J. Ley, Y. Noh, C. D. Smith, J. P. Webster, K. Milinkevich, M. W. Lodewyk, J. A. Jenks, J. F. Smith, and A. J. Whelton. 2021. "Water Safety Attitudes, Risk Perception, Experiences, and Education for Households Impacted by the 2018 Camp Fire, California." *Natural Hazards* 108: 947–975. https://doi.org/10.1007/s11069-021-04714-9.

Omane, D., H. Yu, W. V. Liu, and Y. Pourrahimian. 2017. "Investigation on the Use of Chemical Dust Suppressants on Ash Emissions Due to Fort McMurray Wildfire." Presented at SME Annual Conference and Expo 2017: Creating Value in a Cyclical Environment.

Proctor, C. R., J. Lee, A. Shah, and A. Whelton. 2019. *VOC Fate in Water Systems: Discussion to Support the Water Systems Task Force*. https://engineering.purdue.edu/PlumbingSafety/resources/Camp-Fire-VOC-in-Water-Systems-Purdue.pdf.

Proctor, C. R., J. Lee, D. Yu, A. D. Shah, and A. J. Whelton. 2020. "Wildfire Caused Widespread Drinking Water Distribution Network Contamination." *AWWA Water Science* 2 (4): e1183.

Ritchie, A., B. Sautner, J. Omege, E. Denga, B. Nwaka, I. Akinjise, S. E. Corbett, S. Moosavi, A. Greenshaw, P. Chue, X. M. Li, and V. I. O. Agyapong. 2020. "Long-Term Mental Health Effects of a Devastating Wildfire Are Amplified by Sociodemographic and Clinical Antecedents in College Students." *Disaster Medicine and Public Health Preparedness* 15 (6): 707–717. https://doi.org/10.1017/dmp.2020.87.

Rooney, B., Y. Wang, J. H. Jiang, B. Zhao, Z. C. Zeng, and J. H. Seinfeld. 2020. "Air Quality Impact of the Northern California Camp Fire of November 2018." *Atmospheric Chemistry and Physics* 20 (23): 14597–14616. https://doi.org/10.5194/acp-20-14597-2020.

Silveira, S., M. Kornbluh, M. C. Withers, G. Grennan, V. Ramanathan, and J. Mishra. 2021. "Chronic Mental Health Sequelae of Climate Change Extremes: A Case Study of The Deadliest Californian Wildfire." *International Journal of Environmental Research and Public Health* 18 (4): 1487. https://doi.org/10.3390/ijerph18041487.

Simms, L. A., E. Borras, B. S. Chew, B. Matsui, M. M. McCartney, S. K. Robinson, N. Kenyon, and C. E. Davis. 2021. "Environmental Sampling of Volatile Organic Compounds during the 2018 Camp Fire in Northern California." *Journal of Environmental Sciences (China)* 103: 135–147. https://doi.org/10.1016/j.jes.2020.10.003.

Solomon, G. M., S. Hurley, C. Carpenter, T. M. Young, P. English, and P. Reynolds. 2021. "Fire and Water: Assessing Drinking Water Contamination after a Major Wildfire." *ACS ES&T Water* 1 (8): 1878–1886. https://doi.org/10.1021/acsestwater.1c00129.

TetraTech Inc. 2019a. *Final Assessment of Ash Sampling – Camp Fire Incident*.

TetraTech Inc. 2019b. *Air Quality Consolidated Report for the Butte County Air Quality Management District*. https://bcaqmd.org/wp-content/uploads/Consolidated-Report-for-BCAQMD-041919-to-043019-rev1.pdf.

The Associated Press. 2017. "Waldo Canyon Fire: 20 Percent of Soil So Severely Burned It Is Likened to 'Moonscape.'" Denver, CO.

Vaughan, K., and J. Jojola. January 21, 2022. "Underground Coal Fire Being Investigated as Potential Marshall Fire Ignition Source." *9News.com*. https://www.9news.com/article/news/local/wildfire/marshall-fire/investigators-looking-at-underground-coal-fire-as-possible-cause-of-marshall-fire/73-9fbd71ce-3838-40f3-878c-5b2d5e61680e.

Verstraeten, B. S. E., G. Elgbeili, A. Hyde, S. King, and D. M. Olson. 2020. "Maternal Mental Health after a Wildfire: Effects of Social Support in the Fort McMurray Wood Buffalo Study." *Canadian Journal of Psychiatry* 66 (8): 710–718. https://doi.org/10.1177/0706743720970859.

Von Behren, J., M. Wong, D. Morales, P. Reynolds, P. B. English, and G. Solomon. 2022. "Returning Individual Tap Water Testing Results to Research Study Participants after a Wildfire Disaster." *International Journal of Environmental Research and Public Health* 19 (2): 907.

Vu, B. N., J. Bi, W. Wang, A. Huff, S. Kondragunta, and Y. Liu. 2022. "Application of Geostationary Satellite and High-Resolution Meteorology Data in Estimating Hourly $PM_{2.5}$ Levels during the Camp Fire Episode in California." *Remote Sensing of Environment* 271: 112890. https://doi.org/10.1016/j.rse.2022.112890.

Walpole, E. H., E. D. Kuligowski, L. Cain, A. Fitzpatrick, and C. Salley. 2020. *Evacuation Decision-Making in the 2016 Chimney Tops 2 Fire: Results of a Household Survey.* NIST Technical Note 2103. Gaithersburg, MD: National Institute of Standards and Technology. https://doi.org/10.6028/NIST.TN.2103.

Wentworth, G. R., Y. A. Aklilu, M. S. Landis, and Y. M. Hsu. 2018. "Impacts of a Large Boreal Wildfire on Ground Level Atmospheric Concentrations of PAHs, VOCs and Ozone." *Atmospheric Environment* 178: 19–30. https://doi.org/10.1016/j.atmosenv.2018.01.013.

Wu, Y., A. Arapi, J. Huang, B. Gross, and F. Moshary. 2018. "Intra-continental Wildfire Smoke Transport and Impact on Local Air Quality Observed by Ground-Based and Satellite Remote Sensing in New York City." *Atmospheric Environment* 187: 266–281. https://doi.org/10.1016/j.atmosenv.2018.06.006.

Young, D., B. Rust, and W. C. B. Team. 2012. *Waldo Canyon Fire—Burned Area Emergency Response Soil Resource Assessment.* Redding, CA: US Forest Service, Region 5.

Zhang, Y., R. Pelletier, T. Noernberg, M. W. Donner, I. Grant-Weaver, J. W. Martin, and W. Shotyk. 2022. "Impact of the 2016 Fort McMurray Wildfires on Atmospheric Deposition of Polycyclic Aromatic Hydrocarbons and Trace Elements to Surrounding Ombrotrophic Bogs." *Environment International* 158: 106910. https://doi.org/10.1016/j.envint.2021.106910.

Appendix D

Public Workshop Agenda

10:00 Opening Remarks and Goals of the Workshop
David Allen, Committee Chair

I. COMPOSITION OF URBAN MATERIALS AND THEIR COMBUSTION PRODUCTS

Session Chairs: Marilyn Black and Anna Stec

10:10 The Fuel of Our Homes - from Building Materials to Content
Birgitte Messerschmidt, National Fire Protection Association

10:35 Combustion Product Yields: Basic Principles and Examples from Large Fire Tests
Per Blomqvist, Research Institutes of Sweden

11:00 Behavior of Flame Retardants and Other Chemicals of Concern in Fires and Their Degradation Processes
Richard Hull, University of Central Lancashire, United Kingdom

11:25 Flame Retardants in Building Materials and Consumer Products: Concerns for Exposure
Heather Stapleton, Duke University

11:50 Break

II. EMISSION SOURCES AND POTENTIAL EXPOSURES

Session Chairs: Jeff Burgess and Fernando Rosario-Ortiz

12:30 Exploring the Complexity of Gas and Particle Phase Organic Chemistry and Indoor Infiltration Rates When Wildfire Smoke Arrives in Highly Populated Regions of California
Allen Goldstein, University of California Berkeley

12:55   Residential Indoor Exposure Downwind of Fires
        Shelly Miller, University of Colorado Boulder

1:20   Soil and Combustion Debris as Specific Emission Source Vectors for Water
        Bruce Macler, US Environmental Protection Agency (retired)

1:45   Exposure Routes as a Function of Chemical Types and Target Population: Who Are the Sensitive Populations and What Are Their Major Risks?
        John Balmes, University of California San Francisco & University of California Berkeley

2:10   Break

III. CHEMICAL PROCESSES
Session Chairs: Barbara Turpin and Fred Dryer

2:20   Wildland-Urban Interface (WUI) Fires: Perhaps the Greatest Challenge for Fire Safety Science?
        Samuel Manzello, National Institute of Standards and Technology

2:45   How Do Fire Conditions and Synthetic Materials Affect Near Field Chemistry in Urban Wildfires?
        Eric Guillaume, Efectis

3:10   How Does Regional Chemistry in Urban Wildfire Plumes Differ from Wildland Fires?
        Halogens and Plastics
        Steven Brown, National Oceanic and Atmospheric Association

3:35   How Does Regional Chemistry in Urban Wildfire Plumes Differ from Wildland Fires? Insights from Chemical Transport Modeling
        Christine Wiedinmyer, University of Colorado Boulder

IV. DATA GAPS AND RESEARCH NEEDS
Session Chairs: David Allen and Amara Holder

4:00   A Panel Discussion on Research Needs and Data Gaps
        Per Blomqvist, Research Institutes of Sweden
        Samuel Manzello, National Institute of Standards and Technology
        Steven Brown, National Oceanic and Atmospheric Association
        Kathleen Navarro, National Institute for Occupational Safety and Health
        Peter Lahm, United States Forest Service

Potential Discussion Questions:
  • How do the structures and materials in the built environment impact the chemistry of urban wildfire emissions?
  • How does the fire behavior in an urban wildfire affect the quantity and composition of emissions from burning structures and their contents?
  • How might the composition/chemistry of urban wildfire emissions change once released into the environment?
  • What unique factors of urban wildfires may impact acute and chronic health effects of exposed populations? How do the exposures change for urban wildfires compared to vegetative wildfires? What populations may be most vulnerable to these exposures?
  • How does the incident response of an urban wildfire differ from a vegetative wildfire? What information would be needed, on what time scales, to enable public health communication/interventions for urban wildfires? How might the communication/interventions change for occupational exposures?

5:00   Adjourn

Appendix E

Engineering Calculations for Table 3-2

Table 3-2 presents data describing two characteristics of structural materials—combustible mass and energy content—that relate to how the homes built from those materials would burn and their potential emissions if they were destroyed by a WUI fire. Although the committee searched for this type of data in the existing literature, no data of this type are currently available. The values in the table were therefore calculated by the committee to demonstrate how combustible mass and energy content can be derived from published information on the quantities of different typical materials found in residential construction. The table provides values for two example single-family homes for which construction material quantities were publicly available (EPA, 2016; Messerschmidt, 2021).

The committee wishes to emphasize that multiple assumptions are involved in completing the engineering calculations that resulted in Table 3-2. The data in that table are intended only as an illustrative example, and an uncertainty analysis was therefore not performed. For the reader who may be interested in replicating, improving upon, or extending these calculations to expand the scientific community's understanding of fire loading in WUI fires, the committee's assumptions and calculations are detailed in this appendix. The first section describes combustible mass calculations, and the second section describes energy content calculations.

CALCULATING COMBUSTIBLE MASS

Combustible mass must be obtained from the source or calculated before energy content can be determined. Messerschmidt provided material quantities directly on the basis of combustible mass (Messerschmidt, 2021), and no engineering calculations were needed to provide those values in Table 3-2.

The US Environmental Protection Agency (EPA) provided quantities of each material in terms of weight, area, or length, which can be found in Table 1-1 of the reference (EPA, 2016). The committee applied conversion factors to convert the quantities from source units into mass equivalents (Table E-1). To convert area and length values to a mass basis, the committee made informed assumptions about the specific material composition, when not specified, and sought out sources for the relevant density information.

CALCULATING ENERGY CONTENT

The committee identified the best available sources for net calorific value. When possible, energy content data were derived from the Society of Fire Protection Engineers (SFPE) Handbook of Fire Protection Engineering (SFPE, 2016) or from Xie et al. (2019) to conform with standard fire loading calculation methodology. In all

TABLE E-1 Conversion Factors Used in Calculating Mass-Basis Quantities for EPA Example Home

Material[a]	Source Amount	Source Units	Conversion Factor	Conversion Factor Units	Total Mass (kg)	Combustible Mass (kg)
Lumber (assumed white pine)[b]	13,837	board ft	0.95	kg/board ft	13,159	13,159
Sheathing (assumed 1/2" OSB)[c]	13,118	ft[d]	0.77	kg/ft[b]	10,115	10,115
Exterior wood siding[e]	3,206	ft[d]	1.09	kg/ft[b]	3,485	3,485
Roofing material (assumed ½ inch OSB)[c]	3,103	ft[d]	0.77	kg/ft[b]	2,393	2,393
Roofing asphalt shingles[e]	3,103	ft[d]	1.23	kg/ft[b]	3,805	746
Glass fiber insulation	3,061	ft[d]	_[d]	_[d]	1,830	183[d]
Wall material (assumed 3/8 inch gypsum board)[f]	6,050	ft[d]	0.57	kg/ft[b]	3,429[g]	_[g]
Ceiling material (assumed 3/8 inch gypsum board)[f]	2,335	ft[d]	0.57	kg/ft[b]	1,323[g]	_[g]
Ducting (assumed galvanized steel)	266	linear ft	2.27	kg/ft[b]	603	_[g]
Windows (assumed vinyl)[h]	19	units	20.1	kg/ft[b]	382	192
Concrete	19	tons	907.185	kg/ton	17,237[g]	_[g]
Total					62,168	34,680

NOTE: OSB = oriented strand board.

[a]If the source did not provide a specific material composition for a given construction material, the composition assumed is given in parentheses in the first column.

[b]The linear density of lumber was assumed to be 0.95 kg/board ft, based on the density of white pine (eastern and western averaged) adjusted to a moisture content of 12 percent. Source: https://www.fpl.fs.fed.us/documnts/fplgtr/fpl_gtr190.pdf.

[c]A ½ inch assumption was made for all OSB to match Messerschmidt (2021). Area density (kg/ft²) for ½ inch OSB was obtained from https://roofonline.com/weights-measures/weight-of-plywood-and-osb/.

[d]The conversion factor approach used for most materials was not appropriate for glass fiber insulation because this material is composed of only a fraction of combustible adhesives in addition to the noncombustible glass fibers. For this material, the equivalent combustible mass was obtained by multiplying the replacement mass provided in the source (EPA, 2016) by 10 percent, the combustible fraction according to Messerschmidt (2021) and https://pharosproject.net/common-products/2080060#contents-panel.

[e]EPA provided replacement mass values for wood shingle siding and roofing asphalt shingles (Appendix A1 in EPA, 2016), but the area densities (kg/ft²) that resulted were too high to be reasonable. Instead, the committee obtained area densities for each of these materials from https://www.engineeringtoolbox.com/roofing-materials-weight-d_1498.html. The percentage of combustible material of a shingle was obtained from https://greet.es.anl.gov/greet_building, Building Life-Cycle Analysis with the GREET Building Module: Methodology, Data, and Case Studies, as 20 percent asphalt by mass.

[f]Area density (kg/ft²) of gypsum board was obtained from https://www.certainteed.com/drywall/products/regular-drywall/.

[g]Not a combustible material.

[h]The window materials were obtained from Building Industry Reporting and Design for Sustainability (BIRDS) New Residential Database Technical Manual (NIST Technical Note 1878, https://nvlpubs.nist.gov/nistpubs/TechnicalNotes/NIST.TN.1878.pdf) for a 1 m² vinyl casement window composed of 40 percent vinyl and 10 percent other materials.

SOURCE: EPA, 2016.

TABLE E-2 Net Calorific Values Used to Calculate Energy Content

Material	Mean Net Calorific Value (MJ/kg)	Net Calorific Source
Wood	18.50	Xie et al., 2019; Bain et al., 2003
OSB / particle board	18.54	Phyllis, n.d.
Polyvinyl chloride	10.07	SFPE, 2016, Table A.31
Polyurethane foam	25.60[a]	SFPE, 2016, Table A.31
Fiberglass phenol formaldehyde resin	28.55[a]	SFPE, 2016, Table A.31
Asphalt (from shingles)	40.20	UNSD, n.d.

[a]For these materials, the reference provided ranges, from which mean values were calculated (as the center of the range).

other cases, net calorific values were sourced from measurement data (Phyllis, n.d.) or from widely adopted reference data (United Nations standard net calorific values [UNSD, n.d.]). The applicability of standardized data for representing in-use construction materials requires validation.

The combustible mass values taken from Messerschmidt (2021) and calculated from EPA (2016) were then multiplied by the mean net calorific values in Table E-2 to obtain the energy content data shown in Table 3-2.

REFERENCES

Bain, R. L., W. A. Amos, M. Downing, and R. L. Perlack. 2003. *Biopower Technical Assessment: State of the Industry and Technology*. Golden, CO: National Renewable Energy Laboratory. https://www.nrel.gov/docs/fy03osti/33123.pdf.

EPA (US Environmental Protection Agency). 2016. *Analysis of the Lifecycle Impacts and Potential for Avoided Impacts Associated with Single Family Homes*. EPA Report 530-R13-004. Washington, DC: EPA. https://www.epa.gov/smm/analysis-lifecycle-impacts-and-potential-avoided-impacts-associated-single-family-homes (accessed February 16, 2022).

Messerschmidt, B. 2021. "The Fuel of Our Homes – From Building Materials to Content." Presented at The Chemistry of Urban Wildfires: An Information-Gathering Workshop on June 8, 2021, National Academies of Sciences, Engineering, and Medicine, Washington, DC.

Phyllis. n.d. *Phyllis2 Database for the Physico-chemical Composition of (Treated) Lignocellulosic Biomass, Micro- and Macroalgae, Various Feedstocks for Biogas Production and Biochar*. TNO Biobased and Circular Technologies. https://phyllis.nl/.

SFPE (Society of Fire Protection Engineers). 2016. *SPFE Handbook of Fire Protection Engineering* 5th edition. Edited by M. Hurley. Gaithersburg, MD: SFPE.

UNSD (United Nations Statistics Division). n.d. *Standard Net Calorific Values*. New York, NY: United Nations Statistics Division. https://unstats.un.org/unsd/energy/balance/2014/05.pdf.

Xie, Q., et al. 2019. "Probabilistic Analysis of Building Fire Severity Based on Fire Load Density Models." *Fire Technology* 55: 1349–1375. https://doi.org/10.1007/s10694-018-0716-0 (accessed April 1, 2022).